Metals: Properties and Applications

Metals: Properties and Applications

Editor: Ricky Peyret

NY RESEARCH
P R E S S

New York

Published by NY Research Press
118-35 Queens Blvd., Suite 400,
Forest Hills, NY 11375, USA
www.nyresearchpress.com

Metals: Properties and Applications
Edited by Ricky Peyret

International Standard Book Number: 978-1-63238-636-6 (Hardback)

Cataloging-in-Publication Data

Metals : properties and applications / edited by Ricky Peyret.
p. cm.
Includes bibliographical references and index.
ISBN 978-1-63238-636-6
1. Metals. 2. Materials. I. Peyret, Ricky.
TA459 .M48 2019
620.16--dc23

Contents

Preface

This book has been an outcome of determined endeavour from a group of educationists in the field. The primary objective was to involve a broad spectrum of professionals from diverse cultural background involved in the field for developing new researches. The book not only targets students but also scholars pursuing higher research for further enhancement of the theoretical and practical applications of the subject.

Any material in a solid state which is hard and exhibits characteristics like electrical conductivity, shine and capacity is referred to as a metal. The atoms of metals are arranged in either body-centered cubic, hexagonal close-packed or face-centered cubic structure. Metals are fundamental to different fields of science. They play a central role in the rapid industrial progress and technological advancement. This book aims to highlight certain key concepts related to the study of metals and explain their properties and applications in an elaborate manner. From theories to research to practical applications, studies related to all contemporary topics of relevance to this field have been included in this book. It elucidates new techniques and their applications in a multidisciplinary manner. It is designed to serve as a resource guide for students and experts alike and contribute to the growth of this discipline.

It was an honour to edit such a profound book and also a challenging task to compile and examine all the relevant data for accuracy and originality. I wish to acknowledge the efforts of the contributors for submitting such brilliant and diverse chapters in the field and for endlessly working for the completion of the book. Last, but not the least; I thank my family for being a constant source of support in all my research endeavours.

Editor

Multi-Track Friction Stir Lap Welding of 2024 Aluminum Alloy: Processing, Microstructure and Mechanical Properties

Shengke Zou [1], Shuyuan Ma [1], Changmeng Liu [1,*], Cheng Chen [1], Limin Ma [2], Jiping Lu [1] and Jing Guo [1]

[1]   School of Mechanical Engineering, Beijing Institute of Technology, Beijing 100081, China;
      zsk210@bit.edu.cn (S.Z.); bitmc@bit.edu.cn (S.M.); 2220150085@bit.edu.cn (C.C.); jipinglu@bit.edu.cn (J.L.);
      guojingcn@hotmail.com (J.G.)
[2]   Beijing Aeronautical Science & Technology Research Institute of COMAC, Beijing 102211, China;
      liminmacn@outlook.com
*   Correspondence: liuchangmeng@bit.edu.cn

Academic Editor: Manoj Gupta

Abstract: Friction stir lap welding (FSLW) raises the possibility of fabricating high-performance aluminum components at low cost and high efficiency. In this study, we mainly applied FSLW to fabricate multi-track 2024 aluminum alloy without using tool tilt angle, which is important for obtaining defect-free joint but significantly increases equipment cost. Firstly, systematic single-track FSLW experiments were conducted to attain appropriate processing parameters, and we found that defect-free single-track could also be obtained by the application of two-pass processing at a rotation speed of 1000 rpm and a traverse speed of 300 mm/min. Then, multi-track FSLW experiments were conducted and full density multi-track samples were fabricated at an overlapping rate of 20%. Finally, the microstructure and mechanical properties of the full density multi-track samples were investigated. The results indicated that ultrafine equiaxed grains with the grain diameter about 9.4 μm could be obtained in FSLW samples due to the dynamic recrystallization during FSLW, which leads to a yield strength of 117.2 MPa (17.55% higher than the rolled 2024-O alloy substrate) and an elongation rate of 31.05% (113.84% higher than the substrate).

Keywords: friction stir lap welding; 2024 aluminum alloy; tool tilt angle; two-pass processing; microstructure; mechanical properties

1. Introduction

With the increasing requirements of lightweight and economization in many manufacturing industries, aluminum alloys have gradually replaced iron alloys as the most widely used metal material, because of their excellent characteristics such as high crustal content, high specific strength, good machining property, and so on [1]. As a typical high strength aluminum alloy, 2024 aluminum alloy has been gaining commercial importance in aerospace, automotive, and other industries [2]. Increasing interest has been placed on high strength aluminum alloys, apparently due to their combined advantage of strength and lightweight. It is well known that aluminum alloys are difficult to join by fusion welding techniques because of their hot cracking sensitivity, especially for some aluminum alloys with precipitation hardening (such as 2XXX and 7XXX series aluminum alloys). Therefore, for aluminum components fabricated by fusion based welding process, the major concerns include porosity, coarse grains and other solidification related defects [2].

Friction stir lap welding (FSLW) is a solid-state processing technique based on the principle of friction stir welding (FSW). FSLW has a huge potential to fabricate aluminum components with high mechanical performances [3]. The FSLW process is schematically shown in Figure 1. A non-consumable rotating tool with custom-designed pin and shoulder is inserted into the overlapping surfaces of plates, and subsequently traversed along the configured joint line. This makes the plates join together. Compared with other metal welding technologies, FSLW does not have an independent heat source, because the necessary heat to weld aluminum substrates is provided by the frictional heat between the rotating tool and work piece. The frictional heat is controlled by changing the process parameters. During the whole FSLW process, the welding temperature is controlled to be below the melting point of the material. Therefore, the metal substrates are plasticized instead of melting in the FSLW process. It can efficiently avoid all solidification defects. The plasticized materials are extruded from the advancing side to the retreating side under the action of the rotation and the traverse movement of the pin. The weld nugget is formed under the axial pressure provided by the shoulder.

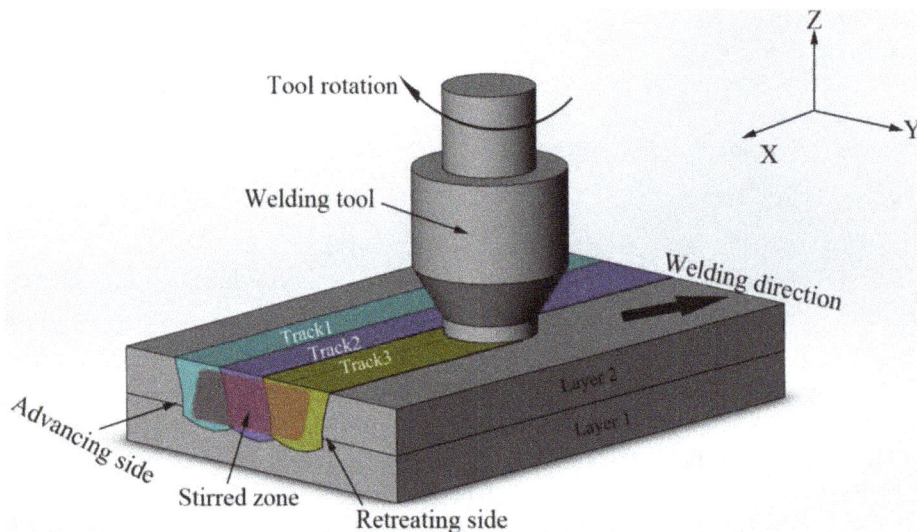

Figure 1. Schematic of the multi-track friction stir lap welding (FSLW) process.

Friction stir welding has been studied intensively in recent years due to its importance in industrial applications [4–6]. The majority of these studies have been based on butt joint configuration, and friction stir lap welding has received considerably less attention [7]. Limited articles related to FSLW were provided by few researchers. For example, Zhang et al. [8] have studied the effect of welding speed on mechanical performance of friction stir lap welded 7B04 aluminum. Ghosh et al. [9] have studied the microstructure and mechanical properties of high strength steels FSLW under different heat input and cooling rates. However, almost all these FSLW studies focus on single-track processing. Multi-track FSLW has potential to fabricate large-scale aluminum monolithic components. It is also the key technology for the friction stir additive manufacturing (FSAM), which is an important direction for the development of friction stir technology [10–12]. On the other hand, the tilt angle is normally adopted in these reports about FSLW, because tilt angle is widely believed to be an important processing parameter to obtain defect-free joint for friction stir technologies [13–15]. Considering that the tilt angle must maintain the same direction as the processing direction, using tilting angle introduces significant complications in processing path and equipment cost. Therefore, it is meaningful to explore the multi-track FSLW processing without tilt angle.

In this study, the formation characters, microstructure and mechanical properties of the multi-track 2024 aluminum alloy specimen are investigated based on the improved FSLW technique without tilt angle.

2. Experimental Procedures

2.1. Experimental Setup and Manufacturing Process

The substrates used in this study were rolled 2024 aluminum alloy plates with 5-mm thickness in annealed condition (2024-O). According to the experimental requirements, the plates with a size of $200 \times 200 \times 5$ mm are used for multi-track FSLW on a modified machining center (Maker: Makino, Tokyo, Japan; Model: FNC-86). Strict clamping and temperature sensitivity are two challenging points of friction stir technology. Therefore, a fixture with circulating liquid cooling was designed in this study for firm clamping and cooling. The schematic diagram of clamping was shown in Figure 2a. The welding tool is the most important part in FSLW, which provides the required heat and pressure for forming. As shown in Figure 2b, the tool used in this study was made of H13 tool steel with a left cylindrical threaded pin. The end surface of shoulder consists of concentric rings with decreasing height from outside to inside. The surface presents concave-shape in general. The shoulder diameter is 20 mm, and the pin diameter and pin height are 8 and 6 mm, respectively. In order to increase the flowability of plastic material, the pin was provided with three slots.

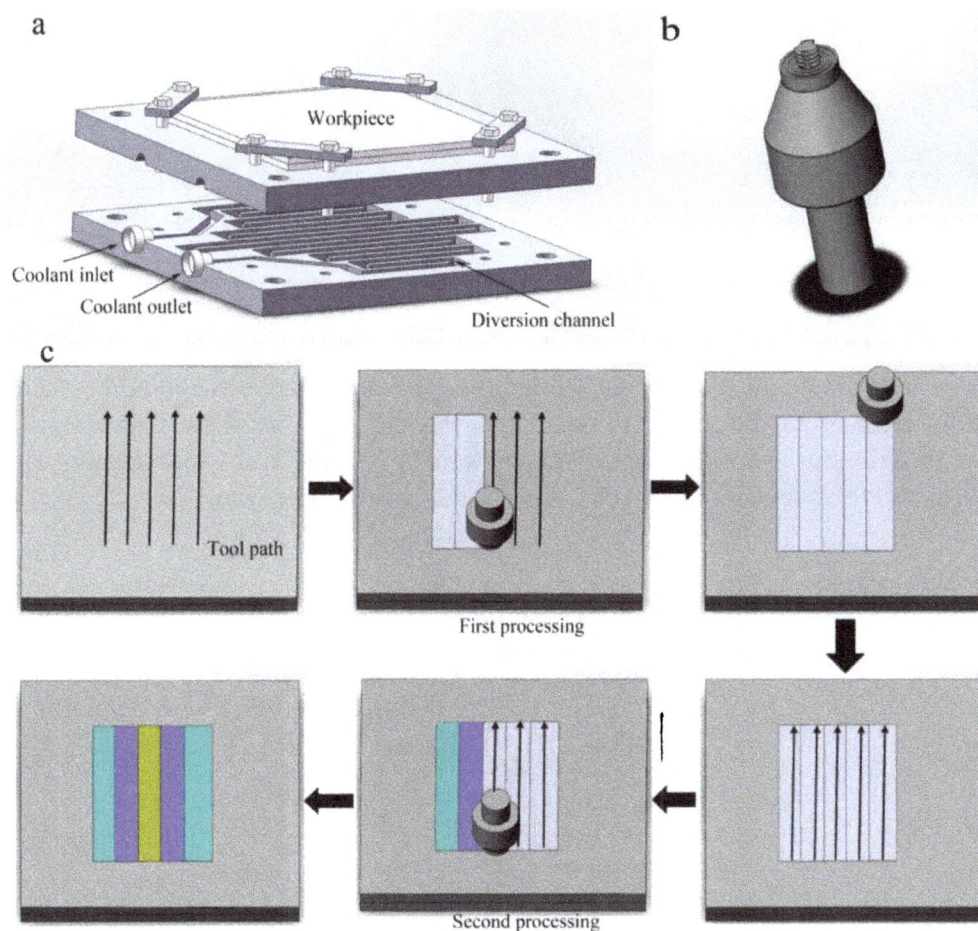

Figure 2. Schematic diagram of (**a**) clamping; (**b**) tool and (**c**) experimental procedure.

It is well known that friction stir welding and related technologies are sensitive to the variation of parameters. Therefore, systematic experimental studies were conducted to get appropriate parameters, as shown in Table 1. Based on these test experiments, a rotation speed of 1000 rpm, a traverse speed of 300 mm/min and a plunge depth of 0.2 mm were adopted. More specific analysis was presented in Section 3.1.1 of this article.

Table 1. Experimental parameters used to get appropriate parameters.

Parameters	No.	Welding Speed, v (mm/min)	Rotation Speed, ω (rpm)	Plunge Depth, h (mm)
	a1	300	500	0.2
	a2	300	750	0.2
	a3	300	1000	0.2
Constant v	a4	300	1250	0.2
	a5	300	1500	0.2
	a6	300	1750	0.2
	a7	300	2000	0.2
	b1	100	1000	0.2
	b2	150	1000	0.2
	b3	200	1000	0.2
	b4	250	1000	0.2
	b5	300	1000	0.3
Constant ω	b6	350	1000	0.2
	b7	400	1000	0.2
	b8	450	1000	0.2
	b9	500	1000	0.2
	b10	550	1000	0.2
	c1	150	500	0.2
	c2	225	750	0.2
	c3	300	1000	0.4
Constant ω/v	c4	375	1250	0.2
	c5	450	1500	0.2
	c6	525	1750	0.2
	c7	600	2000	0.2

In particular, the tilt angle was not used in order to reduce the cost of equipment and avoid the difficulty in controlling tilt angle direction during the forming process. The detailed reason why controlling tilt angle can be avoided is discussed in detail in Section 3.1.2 of this article. According to the results above, the main FSLW parameters used in this study are shown in Table 2.

Table 2. FSLW parameters used in this study.

FSLW Parameters	Values
Rotation speed	1000 rpm
Traverse speed	300 mm/min
Plunge depth	0.2 mm
Tilt angle	0
Overlapping rate	20%
Tool path pattern	Same direction
Special process	Two-pass processing

Finally, a multi-tracked build is obtained with an effective width of about 38 mm (see Figure 3a), and the experimental simulation diagram of experimental procedure is shown in Figure 2c.

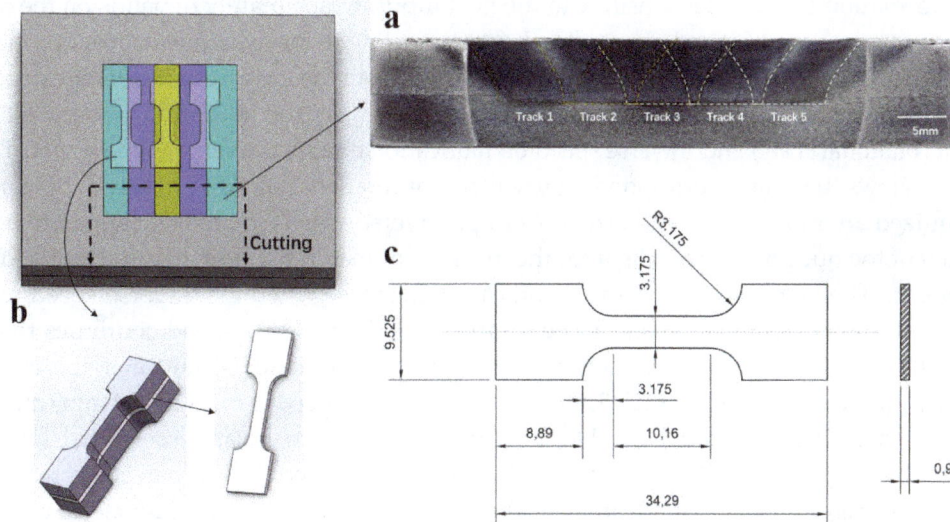

Figure 3. Sample preparation: (**a**) cross sections for microstructure observations; (**b**) manufacturing procedure of tensile specimens and (**c**) dimensions of tensile specimens.

2.2. Microstructural Characterization and Mechanical Testing

With respect to the microstructural characterization and secondary phase particles analysis, the specimens were cross-sectioned perpendicular to the welding direction and suffered grinding following a standard metallographic process as shown in Figure 3a. Because the microstructure in the near surface region (such as the grain morphology of 2024 after FSLW) is difficult to observe, both mechanical polishing and electrochemical polishing were taken to obtain a higher surface quality. The electrochemical polishing was performed at a temperature of $-15\,^{\circ}\text{C}$ and a voltage of 20 V for 16 s with the chemical solution composed by 10% perchloric acid and 90% alcohol. Then, the specimens were etched in a mixture of 10 mL nitric acid, 6 mL hydrochloric acid, 4 mL hydrofluoric acid, and 180 mL water for duration of 30 s. The microstructures of specimens were characterized by optical microscopy (Maker: Leica, Wetzlar, Germany; Model: Leica DM4000M). The secondary phase particles and the fracture surfaces of the tensile samples were analyzed by scanning electron microscope (SEM, JEOL, Tokyo, Japan; Model: JSM-6610LV).

Figure 3b presents the manufacturing procedure of the tensile specimens. These specimens were cut from the middle parts of the plates along the welding direction and then machined to the required dimensions by electric discharge wire cutting. The dimensions of tensile specimens are shown in Figure 3c. The tensile properties were evaluated by an electronic universal material testing machine (Maker: Instron, Darmstadt, Germany; Model: 5966) at a displacement rate of 0.01 mm/s at room temperature. The average value of three specimens was used.

3. Results and Discussion

3.1. Single-Track Friction Stir Lap Welding

3.1.1. Single-Track FSLW with One-Pass Processing

Figure 4 shows the macrostructures of the stirred zone cross sections of one-pass FSLW components under diverse process parameters. It is clear from the macrographs that obvious defects are observed in the samples at most process parameters. Actually, FSLW was developed based on the principle of FSW, and FSW has been widely found to have the characteristic of narrow parameter range [16–20]. For the current case, it is more difficult to produce high-quality welds by FLSW compared with the widely studied friction stir butt welding. It makes the FSLW more highly dependent on the welding parameters [21].

The formation of defects is sensitive to the heat input, which mainly depends on the rotational speed (ω), traverse speed (v) and plunge depth (h). According to the experimental results, the effect of the plunge depth on the defect formation is relatively slight, but the increase of the plunge depth largely raises the load of machine tool. Therefore, the plunge depth is fixed at 0.2 mm, and the effect of the change of rotational speed and traverse speed on fabrication quality is mainly investigated. The Line 1 in Figure 4 shows the results corresponding to a constant rotational speed of 1000 rpm (this parameter was optimized after lots of trial and error). At high traverse speeds, there were apparent holes near the bottom of the nugget. This results from the insufficient material fluidity, caused by the deficiency of heat input. The hole size is found to significantly decrease when the traverse speed reduces to 300 mm/min, owing to the augment of heat input. As the traverse speed continues to decrease, apparent defects appear again, due to the abnormal stirring. Sometimes, the material is even melted induced by the excess heat input [22]. Therefore, the traverse speed is fixed to 300 mm/min, and the rotation speed is changed, as indicated by the Line 2 of Figure 4. It can be seen that some serious defects, such as groove-type defect, appear at low rotational speed (lower than 500 rpm) because of the serious shortage of heat input. Meanwhile, the hole size decreases as the rotational speed grows up to 1000 rpm. At the rotational speed between 1000 rpm and 1500 rpm, the hole is not discernible but could be observed under the microscope. After the rotational speed increases to larger than 1500 rpm, the apparent hole appears again. Therefore, the processing parameters are chosen as v 300 mm/min and ω 1000 rpm.

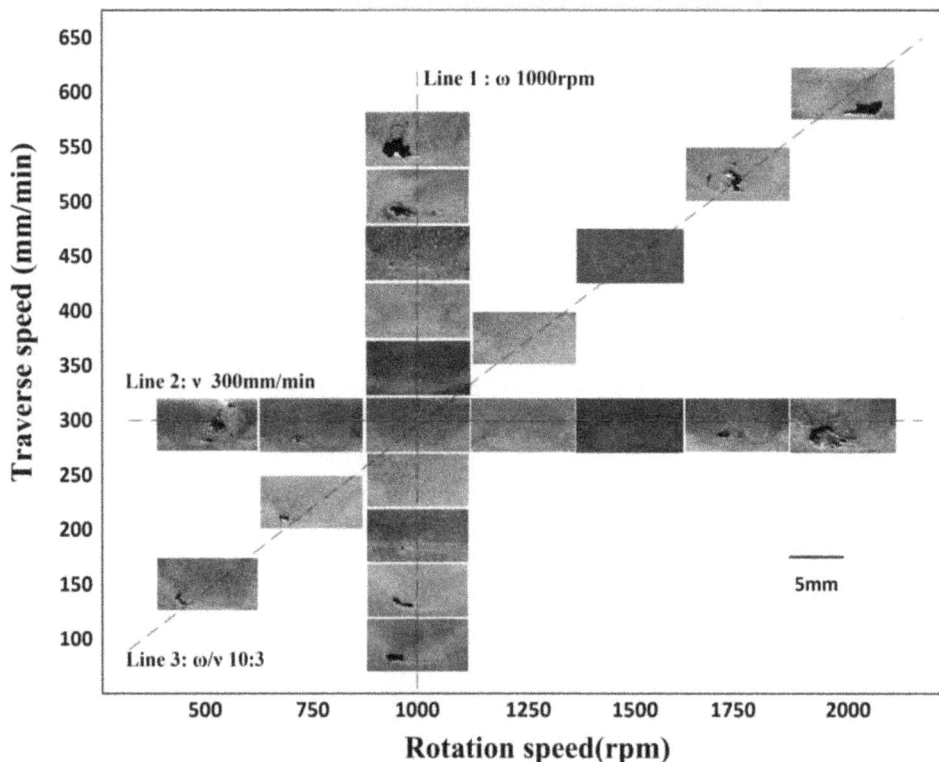

Figure 4. The stirred zone cross sections of one-pass FSLW components under diverse process parameters.

As maintained by related FSW studies [23,24], ω/v is an important parameter to characterize heat input for 2024 aluminum alloy fabricated by FSW. According to the experimental results above, a relatively suitable heat input is obtained at a ω/v ratio of 10:3. Therefore, good molding quality is expected, if rotational speeds and traverse speeds are varied but their ratio ω/v keeps 10:3. However, as indicated along the Line 3 in Figure 4, apparent holes are still found for many cases. The best possible result was still obtained under v 300 mm/min and ω 1000 rpm. The forming process of FSLW technology is complicated, which involves complex material flow and thermal cycling. The above results suggest that it is hard to find a specific parameter (such as ω/v) to characterize the molding quality in FSLW.

3.1.2. Single-Track FSLW with Two-Pass Processing

Figure 5 shows the macrostructures of the cross-sections, as well as the higher magnification image, for FSLW fabricated samples under diverse process parameters. Figure 5a clearly shows that some hole defects are obviously observed at the inappropriate parameter of v 300 mm/min and ω 1250 rpm. With the appropriate parameter of v 300 mm/min and ω 1000 rpm, the hole defects disappear, but kiss bonding defects are still present, as shown in Figure 5b. The kiss bonding defect is very common in FSW, and it mainly results from the complicated material flow and insufficient pressure during FSLW. Furthermore, the interface of the plates is vertical to the tool, which makes it more difficult to sufficiently pulverize the oxide layers at the interfaces. In considerable repeated experiments, it is common to observe different degrees of kiss bonding defects with the optical microscope. Some special attempts were also taken in order to increase the material flow and pressure, such as slotted stirring pin, increasing tooth pitch and depth, using special shoulder surface. However, it is still difficult to realize non-defect forming. The main cause of kiss bonding defects in this study is abandoning the use of tilt angle, which is an important parameter for controlling defects [25–27].

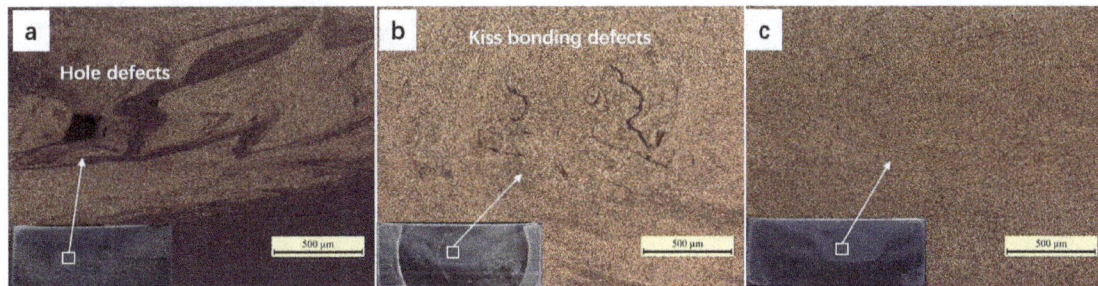

Figure 5. Local magnifications of nugget at diverse process parameters: (**a**) v 300 mm/min and ω 1250 rpm with one-pass processing; (**b**) v 300 mm/min and ω 1000 rpm with one-pass processing; (**c**) v 300 mm/min and ω 1000 rpm with two-pass processing.

The functional principle of the tilt angle is shown in Figure 6. During FSLW, the formation of the stir zone is attributed to the following processes. Firstly, the stirred materials become plastic under the action of compression and frictional heat. Secondly, the plastic regions of the lower plates move upwards and combine with the upper plate. Then, the combined materials move downwards along the left-hand thread pin. Finally, the combined materials are released from the pin owing to the motion of the tool, and the stir zone forms under the pressure provided by the shoulder [10]. On one hand, the existence of tilt angle increases the degree and scope of material flow, especially for the axial flow [28]. On the other hand, the sloping shoulder can provide not only the wrapped force F_1 owing to the concave surface of the tool, but also an extra thrust force F_2 due to the extrusion between the incline and the plastic material as shown in Figure 6a. Therefore, the material flow and the force in the FSLW process without tilt angle are less than that with tilt angle to a certain extent as shown in Figure 6b. The axial material flow in the process without tilt angle is provided only by the left-hand thread pin. It is not enough to crush the oxide layers of the plates and completely fill up the original gap between the plates. Therefore, the coarse alumina particles and kiss bonding defects are left.

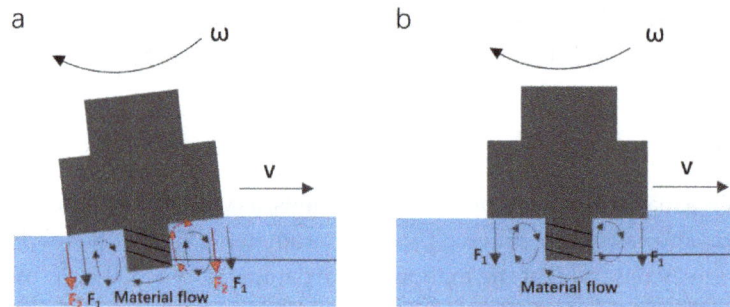

Figure 6. Schematic of metal flow and molding pressure: (**a**) FSLW process with tilt angle and (**b**) FSLW process without tilt angle.

The coarse alumina particles and kiss bonding defects will seriously affect the performance of the components, which must be avoided by using some special methods. In this study, the method of two-pass processing was taken in this study to eliminate the kiss bonding defects and refine the oxide particles, which is also used in friction stir processing by a few researchers [29,30]. The schematic illustration of two-pass processing during FSLW is shown in Figure 7. Some coarse alumina particles and various degrees of kiss bonding defects are left after the first pass processing due to the above-mentioned factors as shown in Figure 7a. Then the second pass processing with the same technological parameters was conducted. The aim of this processing was not deleting the whole oxide layers and original gap between the plates, but destructing the coarse alumina particles and kiss bonding defects. Accordingly, the material flow and the pressure provided by the second pass processing are sufficient to eliminate the kiss bonding defects and refine the oxide particles as shown in Figure 7b. This idea is also verified by our experiments. According to the microstructure shown in Figure 5c, defect-free component is obtained after two-pass processing.

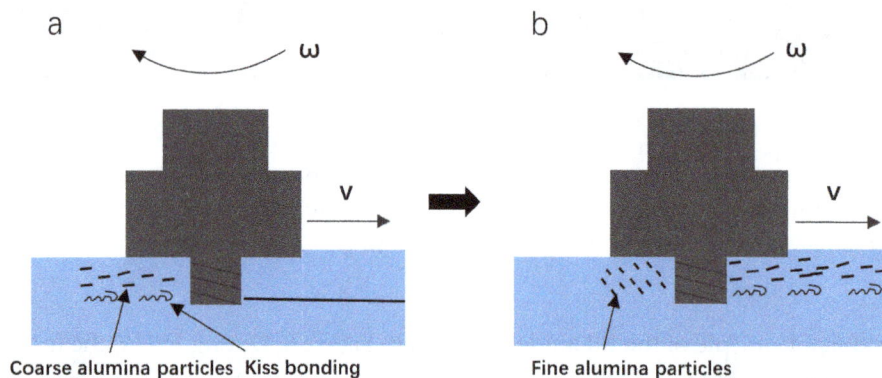

Figure 7. Schematic of two-pass processing for defect elimination: (**a**) first processing and (**b**) second processing.

3.2. Multi-Track Friction Stir Lap Welding

The existing investigations about the microstructure and mechanical properties of FSLW fabricated samples mainly focus on the single-track multilayer processing; studies on multi-track processing are relatively scarce. In this study, multi-track FSLW processing of 2024-O aluminum alloy is investigated. The theoretical width of the stirred zone after single-track FSLW should be eight millimeters, because the pin diameter used in the processing is eight millimeters, but a distinct unbonded defect is found between the adjacent tracks in multi-track FSLW under a hatch distance of eight millimeters (offset between two neighboring tracks), as shown in Figure 8a. The hook is a kind of inherent feature in friction stir lap welding process [31], which is characterized by a crack-like unbonded defect on both the advancing side and retreating side of the deformed faying surfaces. Accordingly, the actual effective width of the stirred zone after single-track FSLW is smaller than the pin diameter owing to the hook. The unbonded defects between the adjacent tracks are significantly diminished under a 7.2 mm hatch distance (10% overlapping rate), as shown in Figure 8b. When it reached 6.4 mm (20% overlapping rate), the unbonded defects disappeared and a component with dense microstructure was formed, as shown in Figure 8c.

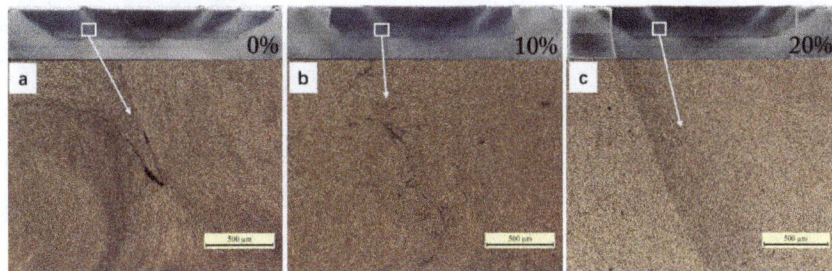

Figure 8. Local magnifications of overlap regions at different overlapping rates in multi-track samples: (**a**) 0%; (**b**) 10%; and (**c**) 20%.

3.3. Microstructure and Mechanical Properties

These components with full density were analyzed for the characterization of microstructure, mechanical properties and fracture surface. To show the influence of FSLW processing on the microstructure, the components of the FSLW fabricated sample and rolled substrate are cross-sectioned to observe the microstructure by optical microscope, as shown in Figure 9a,b, respectively. The original microstructure of the rolled substrate is characterized by coarse banded grains (see Figure 9b), and the FSLW fabricated sample is characterized by a typical feature with ultrafine equiaxed grains (see Figure 9a). The micrographs show that the average grain size reduced from about 150 μm to 9.4 μm after processing owing to dynamic recrystallization. Moreover, the microstructure evolution is a very complex process described as follows: (1) Firstly, the parent grains split into coarse band structures; (2) Secondly, some new elongated fibrous grains are formed owing to the increasing of the strain, and then further grain subdivision occurs continuously; (3) Then, fine nugget scale grains are formed with the increasing of temperature. The bands of fine grains are forced together and increase in volume fraction with strain; (4) Finally, the unstable fibrous grain fragments form a full nugget-like microstructure consisting of low aspect ratio ultrafine grains [10].

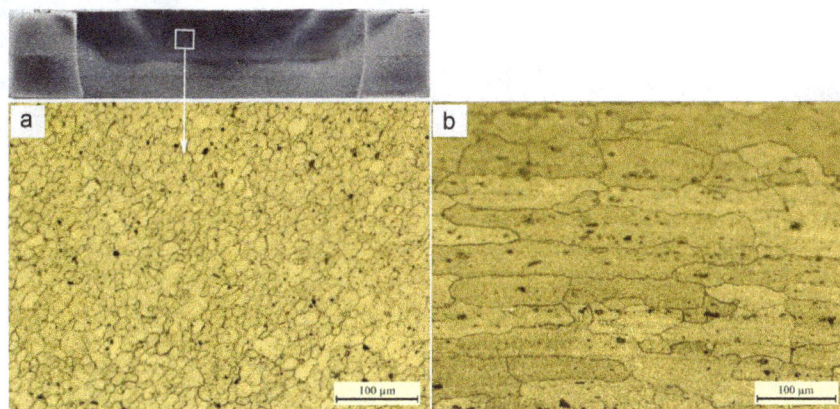

Figure 9. Optical microscope images showing the microstructures of (**a**) FSLW 2024 sample; (**b**) rolled 2024-O substrate.

Then, the mechanical properties of the FSLW sample and rolled substrate are measured by tensile test. Figure 10 gives the results of the yield strength, ultimate tensile strength, and the elongation rate of the tensile tests. Compared with the rolled substrate, all three mechanical properties of FSLW samples increase observably. The FSLW tensile specimens shows an average yield stress of about 117.2 MPa and a tensile stress of about 227.8 MPa, which are about 17.55% and 6.45% higher than those of rolled substrate, respectively. In particular, the elongation rate increases from 14.52% of the rolled substrate to about 31.05% of FSLW samples, which is about two times that of rolled substrate.

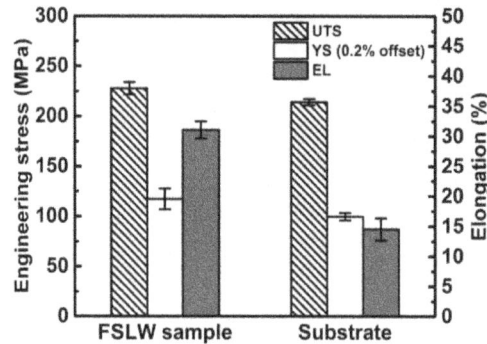

Figure 10. Tensile properties of FSLW samples and rolled 2024-O substrate.

The simultaneous increase in strength and ductility after FSLW can be attributed to the radical changes in the microstructure. Many researchers have studied the relationship between mechanical properties and microstructure. The mechanical property of the alloy is directly affected by the average grain size, according to the Hall-Petch equation [32]. Therefore, the formation of finer grains of the samples contributes to higher tensile strength and coarser grains easily lead to poor tensile properties. During the FSLW process, the severe plastic deformation and dynamic recrystallization lead to the ultrafine equiaxed grains that result in the significant improvement of mechanical properties. Furthermore, the fracture surface morphologies of the tensile specimens are characterized by SEM. In comparison with the fracture surface of the substrate sample (see Figure 11b), the FSLW fabricated samples have deeper and finer dimples (see Figure 11a), consistent with their higher ductility. In addition, it should be noted that the strength of all the samples is not very high, because the 2024-O alloy substrate has a lack of precipitation strengthening [33].

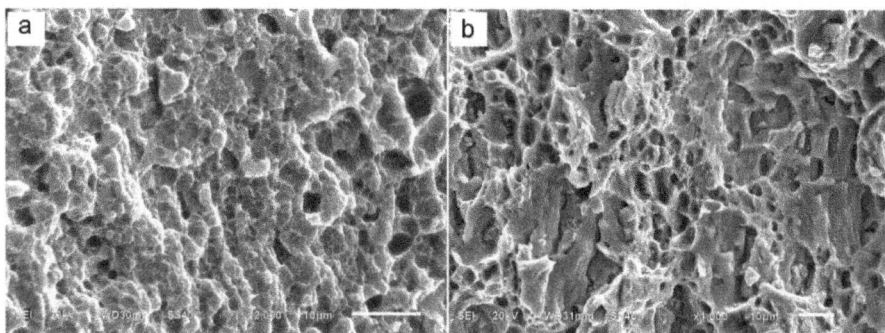

Figure 11. Scanning electron microscopy (SEM) images showing the tensile fracture surface of (**a**) FSLW samples and (**b**) rolled substrate.

4. Conclusions

In this study, a full density multi-track 2024-O aluminum alloy component was successfully manufactured by FSLW without the application of tool tilt angle. The corresponding processing, defects, microstructures and tensile properties were investigated. According to the experimental results and theoretical analysis, the main conclusions are summarized as follows:

(1) Defects generated during FSLW are very sensitive to the variation of parameters. Systematic studies were carried out to determine the suitable processing parameters. The rotational speed and traverse speed are found to play an important role in hole generation.

(2) Tilt angle has a certain positive effect on defect elimination, but significantly increases equipment cost. Alternatively, two-pass processing is found to be able to replace the tilt angle and ensure defect-free forming. Under two-pass processing with welding speed 300 mm/min, rotation rate 1000 rpm and plunge depth 0.2 mm, defect-free samples were successfully fabricated.

(3) Hook-shaped defects are an inherent feature of FSLW, which combine to become crack-like unbonded defects in the overlapping zone during multi-track processing. The overlapping rate of 20% is found to be enough to eliminate these unbonded defects, and obtain full density multi-track samples.

(4) Ultrafine equiaxed grains with the diameter about 9.4 μm are obtained in FSLW samples, because of the dynamic recrystallization during FSLW. Accordingly, the strength of the multi-track FSLW samples is much higher than that of the rolled 2024-O substrate. Meanwhile, it exhibits a high ductility of the elongation about 31.05%, which is about two times that of the rolled 2024-O substrate.

Acknowledgments: The work was financially supported by the National Natural Science Foundation of China (51505033), Beijing Natural Science Foundation (3162027), and Excellent Young Scholars Research Fund of Beijing Institute of Technology (2015YG0302).

Author Contributions: Shengke Zou performed most of experiments and wrote this manuscript. Changmeng Liu designed the research, helped analyze the experimental data and gave some constructive suggestions. Limin Ma and Jing Guo helped do some experiments. Shuyuan Ma, Cheng Chen and Jiping Lu participated in the discussion on the results and guided the writing of the article.

Conflicts of Interest: The authors declare no conflict of interest.

References

1. Hu, Z.; Yuan, S.; Wang, X.; Liu, G.; Huang, Y. Effect of post-weld heat treatment on the microstructure and plastic deformation behavior of friction stir welded 2024. *Mater. Des.* **2011**, *32*, 5055–5060. [CrossRef]

2. Brandl, E.; Heckenberger, U.; Holzinger, V.; Buchbinder, D. Additive manufactured AlSi10Mg samples using selective laser melting (SLM): Microstructure, high cycle fatigue, and fracture behavior. *Mater. Des.* **2012**, *34*, 159–169. [CrossRef]

3. Kwon, J.W.; Kang, M.S.; Yoon, S.O.; Kwon, Y.J.; Hong, S.T.; Kim, D.I.; Lee, K.H.; Seo, J.D.; Moon, J.S.; Han, K.S. Influence of tool plunge depth and welding distance on friction stir lap welding of AA5454-O aluminum alloy plates with different thicknesses. *Trans. Nonferrous Met. Soc. Chin.* **2012**, *22*, 624–628. [CrossRef]

4. Mishra, R.S.; Ma, Z.Y. Friction stir welding and processing. *Mater. Sci. Eng. R Rep.* **2005**, *50*, 13–58. [CrossRef]

5. Celik, S.; Cakir, R. Effect of friction stir welding parameters on the mechanical and microstructure properties of the Al-Cu butt joint. *Metals* **2016**, *6*, 133. [CrossRef]

6. Moreira, P.M.G.P.; Figueiredo, M.A.V.D.; Castro, P.M.S.T.D. Fatigue behaviour of fsw and mig weldments for two aluminium alloys. *Theor. Appl. Fract. Mech.* **2007**, *48*, 169–177. [CrossRef]

7. Chen, Z.W.; Yazdanian, S. Friction stir lap welding: Material flow, joint structure and strength. *J. Achieve Mater. Manuf. Eng.* **2012**, *55*, 629–637.

8. Zhang, H.; Wang, M.; Zhang, X.; Zhu, Z.; Yu, T.; Yang, G. Effect of welding speed on defect features and mechanical performance of friction stir lap welded 7b04 aluminum alloy. *Metals* **2016**, *6*, 87. [CrossRef]

9. Ghosh, M.; Kumar, K.; Mishra, R.S. Friction stir lap welded advanced high strength steels: Microstructure and mechanical properties. *Mater. Sci. Eng. A* **2011**, *528*, 8111–8119. [CrossRef]

10. Mao, Y.; Ke, L.; Huang, C.; Liu, F.; Liu, Q. Formation characteristic, microstructure, and mechanical performances of aluminum-based components by friction stir additive manufacturing. *Int. J. Adv. Manuf. Technol.* **2015**, *83*, 1637–1647.

11. Palanivel, S.; Sidhar, H.; Mishra, R.S. Friction stir additive manufacturing: Route to high structural performance. *JOM* **2015**, *67*, 616–621. [CrossRef]

12. Palanivel, S.; Nelaturu, P.; Glass, B.; Mishra, R.S. Friction stir additive manufacturing for high structural performance through microstructural control in an mg based WE43 alloy. *Mater. Des.* **2015**, *65*, 934–952. [CrossRef]

13. Hamid, H.A.D.; Roslee, A.A. Study the role of friction stir welding tilt angle on microstructure and hardness. *Appl. Mech. Mater.* **2015**, *799–800*, 51–60. [CrossRef]

14. Badheka, V.J. Effects of tilt angle on properties of dissimilar friction stir welding copper to aluminum. *Adv. Mater. Manuf. Process.* **2016**, *31*, 255–263.

15. Latif, A.; Fadhil, M. *Friction Stir Welding (FSW): The Effect of Tilting Angle*; Universiti Teknologi Petronas: Seri lskandar, Malaysia, 2013.

16. Luo, C.; Li, X.; Song, D.; Zhou, N.; Li, Y.; Qi, W. Microstructure evolution and mechanical properties of friction stir welded dissimilar joints of Mg–Zn–Gd and Mg–Al–Zn alloys. *Mater. Sci. Eng. A* **2016**, *664*, 103–113. [CrossRef]

17. Salari, E.; Jahazi, M.; Khodabandeh, A.; Ghasemi-Nanesa, H. Influence of tool geometry and rotational speed on mechanical properties and defect formation in friction stir lap welded 5456 aluminum alloy sheets. *Mater. Des.* **2014**, *58*, 381–389. [CrossRef]

18. Song, Y.; Yang, X.; Cui, L.; Hou, X.; Shen, Z.; Xu, Y. Defect features and mechanical properties of friction stir lap welded dissimilar AA2024–AA7075 aluminum alloy sheets. *Mater. Des.* **2014**, *55*, 9–18. [CrossRef]

19. Arbegast, W.J. A flow-partitioned deformation zone model for defect formation during friction stir welding. *Scr. Mater.* **2008**, *58*, 372–376. [CrossRef]

20. Colligan, K.J.; Mishra, R.S. A conceptual model for the process variables related to heat generation in friction stir welding of aluminum. *Scr. Mater.* **2008**, *58*, 327–331. [CrossRef]

21. Soundararajan, V.; Yarrapareddy, E.; Kovacevic, R. Investigation of the friction stir lap welding of aluminum alloys AA 5182 and AA 6022. *J. Mater. Eng. Perform.* **2007**, *37*, 74–76. [CrossRef]

22. Kim, Y.G.; Fujii, H.; Tsumura, T.; Komazaki, T.; Nakata, K. Three defect types in friction stir welding of aluminum die casting alloy. *Mater. Sci. Eng. A* **2006**, *415*, 250–254. [CrossRef]

23. Ren, S.R.; Ma, Z.Y.; Chen, L.Q. Effect of welding parameters on tensile properties and fracture behavior of friction stir welded Al–Mg–Si alloy. *Scr. Mater.* **2007**, *56*, 69–72. [CrossRef]

24. Radisavljevic, I.; Zivkovic, A.; Radovic, N.; Grabulov, V. Influence of FSW parameters on formation quality and mechanical properties of Al 2024-T351 butt welded joints. *Trans. Nonferrous Met. Soc. Chin.* **2013**, *23*, 3525–3539. [CrossRef]

25. Reddy, P.J.; Kailas, S.V.; Srivatsan, T.S. Effect of tool angle on friction stir welding of aluminum alloy 5052: Role of sheet thickness. *Adv. Mater. Res.* **2011**, *410*, 196–205. [CrossRef]

26. Bilgin, M.B.; Meran, C.; Canyurt, O.E. Effect of tool angle on friction stir weldability of AISI 430. *Weld. J.* **2013**, *92*, 42–46.

27. Behmand, S.A.; Mirsalehi, S.E.; Omidvar, H.; Safarkhanian, M.A. Single- and double-pass fsw lap joining of AA5456 sheets with different thicknesses. *Mater. Sci. Technol.* **2016**. [CrossRef]

28. Mishra, R.S.; Mahoney, M.W. *Friction Stir Welding and Processing II*; Springer: Berlin, Germany, 2014; pp. 13–58.

29. El-Rayes, M.M.; El-Danaf, E.A. The influence of multi-pass friction stir processing on the microstructural and mechanical properties of aluminum alloy 6082. *J. Mater. Process. Technol.* **2012**, *212*, 1157–1168. [CrossRef]

30. Aktarer, S.M.; Sekban, D.M.; Saray, O.; Kucukomeroglu, T.; Ma, Z.Y.; Purcek, G. Effect of two-pass friction stir processing on the microstructure and mechanical properties of as-cast binary Al–12Si alloy. *Mater. Sci. Eng. A* **2015**, *636*, 311–319. [CrossRef]

31. Aldanondo, E.; Arruti, E.; Alvarez, P.; Echeverria, A. *Mechanical and Microstructural Properties of Fsw Lap Joints*; Springer International Publishing: Berlin, Germany, 2013; pp. 35–43.

32. Zhu, Y.Z.; Wang, S.Z.; Li, B.L.; Yin, Z.M.; Wan, Q.; Liu, P. Grain growth and microstructure evolution based mechanical property predicted by a modified hall–petch equation in hot worked Ni76Cr19AlTiCo alloy. *Mater. Des.* **2014**, *55*, 456–462. [CrossRef]

33. Ramesh, K.N.; Pradeep, S.; Pancholi, V. Multipass friction-stir processing and its effect on mechanical properties of aluminum alloy 5086. *Metall. Mater. Trans. A* **2012**, *43*, 4311–4319. [CrossRef]

The 2D Finite Element Microstructure Evaluation of V-Shaped Arc Welding of AISI 1045 Steel

Omer Eyercioglu [1,*], Ahmed Samir Anwar [2], Kursat Gov [3] and Necip Fazil Yilmaz [4]

[1]   Department of Mechanical Engineering, Gaziantep University, Gaziantep 27310, Turkey
[2]   Department of Mechanical Engineering, Salahaddin University, Erbil 44001, Iraq; eyeroglu@gmail.com
[3]   Department of Aeronautics and Astronautics Engineering, Gaziantep University, Gaziantep 27310, Turkey; gov@gantep.edu.tr
[4]   Department of Mechanical Engineering, Gaziantep University, Gaziantep 27310, Turkey; nfyilmaz@gantep.edu.tr
*   Correspondence: eyercioglu@gantep.edu.tr

Academic Editors: Halil Ibrahim Kurt, Adem Kurt and Necip Fazil Yilmaz

Abstract: In the present study, V-shaped arc welding of the AISI 1045 steel is modeled by using 2D Finite Element Model (FEM). The temperature distribution, microstructure, grain growth, and the hardness of the heat-affected zone (HAZ) of the welding are simulated. The experimental work is carried out to validate the FE model. The very close agreement between the simulation and experimental results show that the FE model is very effective for predicting the microstructure, the phase transformation, the grain growth and the hardness. The effect of preheat temperature on the martensite formation is analysed, and it is shown that 225 °C preheating completely eliminates the martensite formations for the 12 mm thick plate.

Keywords: arc welding; AISI 1045 steel; DEFORM™ FE; pre-heating; HAZ; microstructure

1. Introduction

AISI 1045 steel is a medium carbon steel that is used for a wide range of engineering applications such as structural works, machine parts, gears studs, axels, cold extrude, etc. Thus, in most cases, parts are welded together to form permanent joints, but, unfortunately, due to its carbon content, the martensite phase is liable to form in the heat affected zone (HAZ) and may cause brittle fracture of the joint.

A lot of researchers have explored the theoretical and experimental works to conclude the effects of welding parameters (the welding process type that was employed, welding current, voltage, and the arc travelling speed, etc.) on the quality of the weldment. Modelling and simulation is becoming very useful in welding analysis and is preferred for reducing cost and time consumption of experiments [1]. Because of the complexity of the problem, the experimental work is required at a minimum to validate the simulation results. By the progress of the software technology now, it is possible to use numerical techniques like the Finite Element Method (FEM). Sattari-Far and Farahani have used FEM to analyze the thermo mechanical behavior and residual stress in butt welded pipes of 6 and 10 mm thickness of different groove shapes. They stated that welding parameter has a significant effect on the magnitude and distribution of the residual stress in butt weld pipes [2].

Barsoum and Lundbäck studied two- and three-dimensional welding simulations by using MSC-Marc and ANSYS finite element packages for the T-type fillet weld. They concluded that the prediction of the residual stress on the 2D package shows suitable and good agreement for the welding process [3]. Tsai et al. presented numerical modelling for analysis of the heat treatment on the properties of the magnesium alloy in Tungsten inert gas (TIG) welding. They found that increasing the time and temperature of tempering led to high tensile strength and elongation [4]. Unfried, Garzon, and Giraldo represented evaluation of the microstructure after arc welding for the armor steel plate. The approach to the reliability is approved by using the case study methodology. The comparison is also carried out between the experimental work and the modelling for the microstructure evaluation and micro-hardness on the MIL A46100 armor steel plates (low alloy steel) welded by AWS E11018M covered electrode [5]. Zhang et al. modeled the temperature and residual stress due to the welding process on Cr5Mo steel tube by using an ABAQUS package. They noticed that welding residual stress is very large at the beginning stage and it then becomes relaxed in a short time at a high temperature [6].

In this study, V-shaped butt welding of AISI 1045 steel is simulated by the FEM using the DEFORM™ package. The microstructure evaluation of the AISI 1045 steel during two-pass arc welding is modeled by using 2D FEM. The effect of pre-heating on the martensite formation is also analyzed. The experimental work was carried out to validate the results.

2. The FE Modelling

The V-shaped butt welding model is prepared from 12 mm thick AISI 1045 steel plate. The 2D FEM model is used according to the results of Ref [3]. The geometry of the V-groove is selected according to ISO 9692 standard [7] as shown in Figure 1. The welding simulation is carried out in two successive passes and the welding pool is taken as a heat source. The heat transfer through parent metal is analyzed by time for heating (welding) and cooling stages. The deformation, heat transfer, phase transformation, and diffusion are coupled with each other in DEFORMTM simulations. For example, for carbon steel, the material properties are functions of the carbon content. As the concentration of carbon may be changed by diffusion, the mechanical and thermal properties of the steel will change accordingly.

Elasto-Plastic mechanical analysis and the Newton–Raphson iteration solution method have been selected. The material properties of the AISI 1045 steel are taken according to [8,9]. The thermal conductivity has been defined as a function of temperature. The environmental temperature is taken as 20 °C and the welding and cooling is taken place at the environment temperature. The maximum temperature during welding is 1400 °C and it is defined in the FE model. Due to the symmetry of the geometry, one-half of the shape is modeled and meshed to 3000 4-node 2D plane elements.

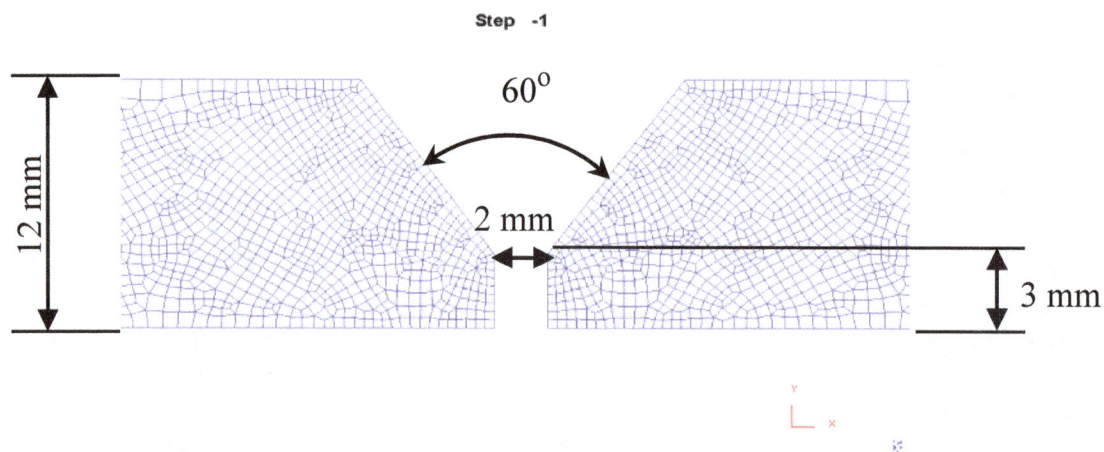

Figure 1. Dimensions of the V-shape welding groove according to ISO 9692 [7].

The flow stress of the material is defined as a function of effective (true) strain and temperature as shown in Figure 2.

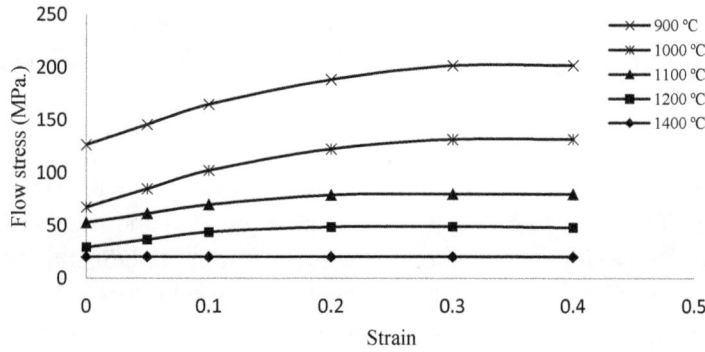

Figure 2. The flow stress and temperature as a function of effective (true) strain.

The thermal conductivity data of the AISI 1045 steel [8] have been defined as shown in Figure 3.

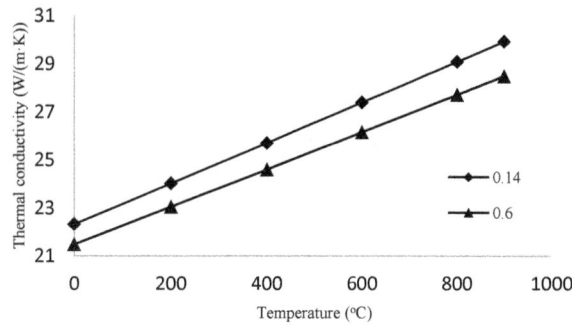

Figure 3. Thermal conductivity of the AISI 1045 steel with respect to the temperature for atom contents (here is the carbon content) of 0.14 and 0.6 percent.

For the phase transformation and the microstructure evaluation in the DEFORM™ 2D, it is necessary to define all of the phases (austenite, bainite and martensite). The phases for the AISI 1045 steel have been defined and the phase transformation of the material was determined by the volumetric weighting of each phase. The transformation of one phase to another is defined as a mother and child relationship [8–10]. The time temperature transformation (TTT) diagram of AISI 1045 steel has been defined as shown in Figure 4.

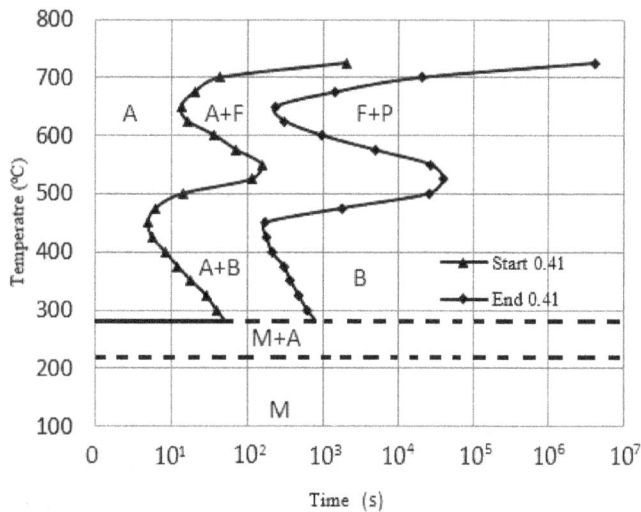

Figure 4. Time Temperature Transformation (TTT) diagram of AISI 1045 steel [9].

The Avrami equation was used for solving volume fraction and transformation which is playing an important role in the transformation process and grain size. This is shown in Figure 5, which has the following form:

$$\xi = 1 - \exp\left(-kt^n\right) \tag{1}$$

where ξ is the volume fraction transformed, t is the time, and k and n are constants (n being the Avrami number). In terms of TTT data, two curves (i.e., transformation start and transformation finish curves) are required in order to solve for k and n. If only one curve is input to DEFORM™ (Scientific Forming Technologies Corporation, Columbus, OH, USA), the Avrami number has to be defined. For the grain modeling, the grain size, the peak strain and the strain rate boundary are defined as a function of the temperature in the DEFORM™ heat treatment lab.

The hardness distribution in the model was estimated based on the Jominy curve by computing the cooling rate for each element [8]. The Jominy curve, hardness as a function of distance, is shown in Figure 6.

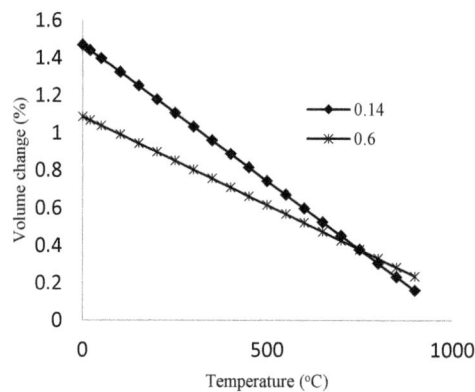

Figure 5. Volume fraction in percent and transformation per temperature for AISI 1045 steel according to 0.14 and 0.6 carbon content [8,10].

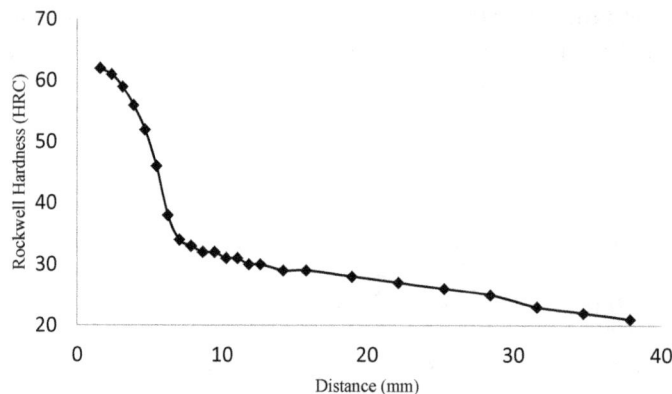

Figure 6. Jominy test by Rockwell hardness, as a function of distance for AISI 1045 steel [8].

3. Materials and Methods

The main purpose of the experimental work is to validate the results of the finite element model. For this reason, a 12 mm thick AISI 1045 steel plate was taken. The chemical composition of the material was analyzed by using a SPECTROMAX 3x metal spectrometer (Spectromax Solutions Ltd, London, UK) and the results are given in Table 1. According to the specification given in the ISO 9692 standard, a V-shaped weld groove was prepared. In the shielded metal arc welding process, depending on the type and the thickness of the material, the specifications such as number of passes, electrode size and type, current and welding speed have to be taken into account. The welding process was carried out for 0.1 cm/s welding speed; 5 mm diameter E7018 electrode and 200 A current. The welded specimen was prepared according to ASTM E3 for metallographic inspection [11]. The macro- and micro-structural features of HAZ of the welding were investigated by using an 8 HD mega-pixel digital cameras and a compound optical metallurgical microscope (equipped with 5 mega-pixel eyepiece digital cameras) that is connected to the PC. Grain size measurements of the HAZ and the base metal for welded joint were done by using the linear intercept method as it is specified in ASTM E112 [12]. After that, the welded parts were sliced and then the pieces were machined to the required size to prepare the sample for micro-hardness testing, which was conducted according to ASTM E92-03 [13]. The micro-hardness measurements were taken as 3 mm, 6 mm and 9 mm above the weld root of the specimen perpendicular to the welding direction using a diamond pyramid indenter.

Table 1. Chemical composition of AISI 1045.

Element	C %	Fe %	Mn %	P %	Si %	Cu %	Cr %	Ti %	S %
Percentage	0.41	98.31	0.849	0.0005	0.161	0.0119	0.0323	0.0005	0.008

4. Experimental Work

4.1. Microstructure and Grain Size

During welding, the molten pool moves through the material, the growth rate and temperature gradient varies considerably across the weld pool. This situation has a significant effect on the phase transformation changes and the average grain size growth in the HAZ area. A transverse macro-section through the HAZ of AISI 1045 steel in the microstructure shows that the average grain sizes are adjacent to the fusion line is large. Meanwhile, the temperature increase in the HAZ triggers the martensite phase formation during cooling. The chemical composition of the medium carbon steel makes significant changes in the curve shape of the Time Temperature Transformation (TTT) diagram and the phase transformation time of the steel. In our case, (AISI 1045 steel), martensite phase transformation has been seen at the different points in the HAZ area during the normal air cooling process. Grain growth and the formation of the martensite phase can cause surface cracking, which is clearly shown in Figure 7. The average grain size distribution found in the finite element model is shown in Figure 8. From the results of the grain size investigation of the AISI 1045 steel, it is found that the grain size of the FE model at the adjacent point to the fusion line for the welding was 1258 μm, whereas the experimental result was 1200 μm. Similar results were obtained for different regions of HAZ of the welding as shown in Figure 9. Therefore, there is good agreement between the experimental and simulation results.

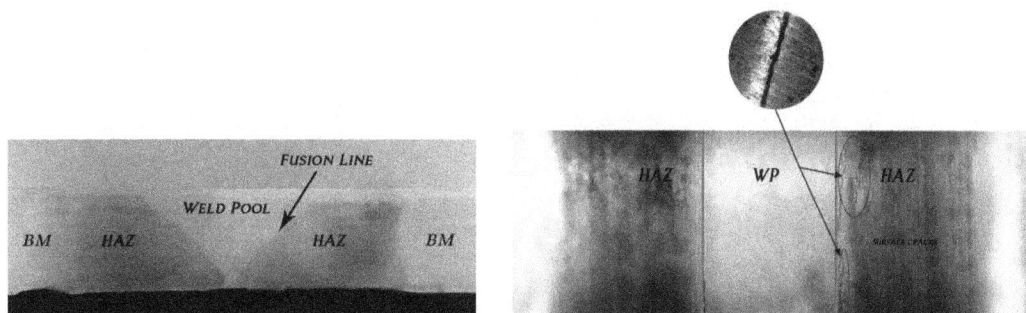

Figure 7. The effect of the shielded metal arc welding on the Heat Affected Zone (HAZ) area and surface crack formation after the welding process for AISI 1045.

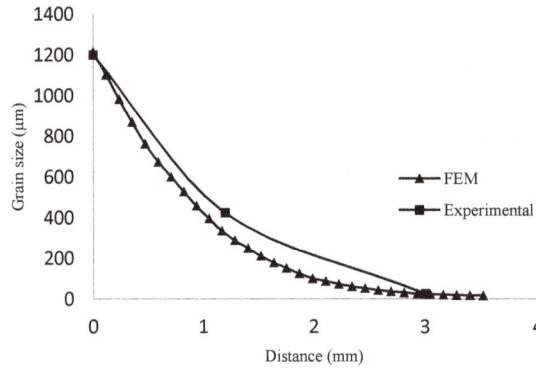

Figure 8. Average grain size taken from adjacent to fusion line through HAZ for AISI 1045 steel.

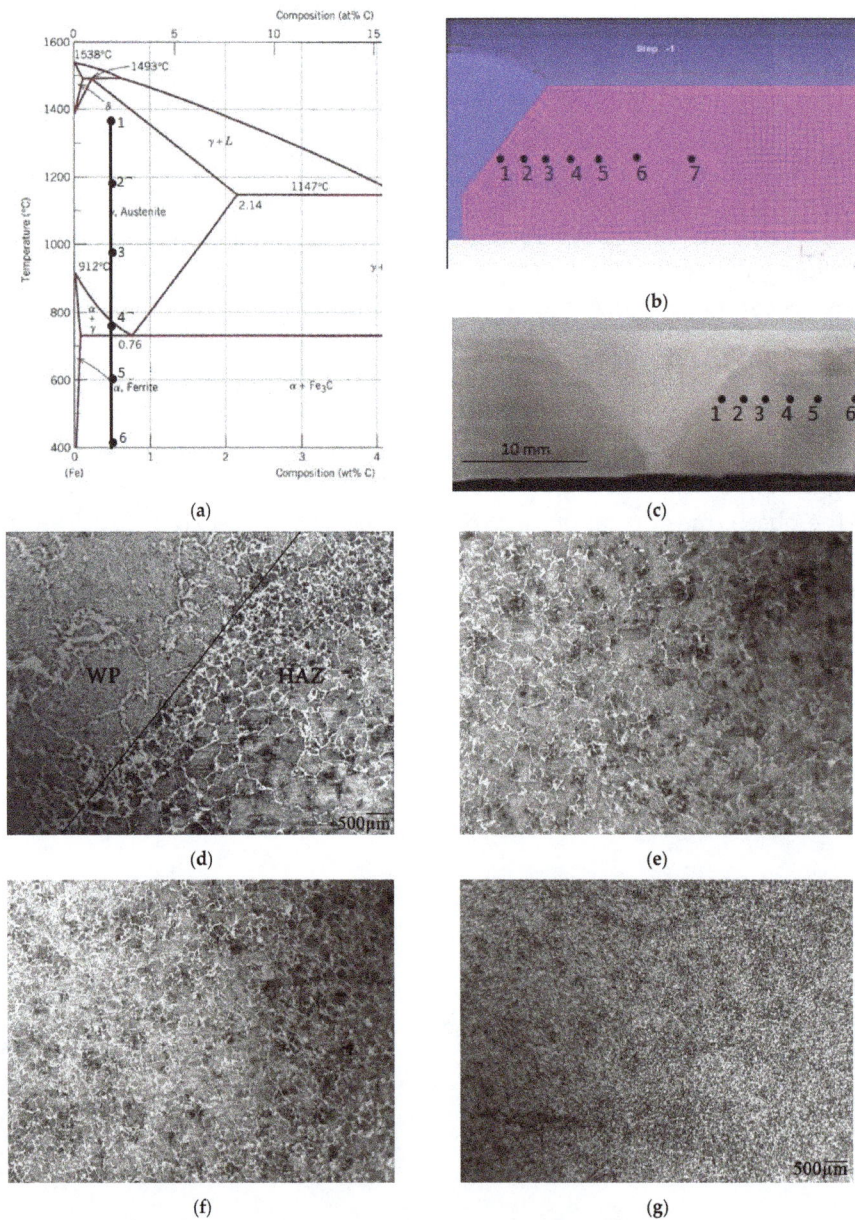

Figure 9. (**a**) Fe-C diagram (**b**) Finite element model (**c**) macrograph, and microstructures around (**d**) point 1 (**e**) point 2 (**f**) point 3 (**g**) point 4 on the transverse section of the welded AISI 1045 steel.

On the other hand, it is possible to investigate the temperature distribution in HAZ of the prepared model during welding and the cooling simulations, which are beneficial for the extraction of the microstructure of the selected points. The recorded temperature distribution of the finite element model is shown in Figure 10.

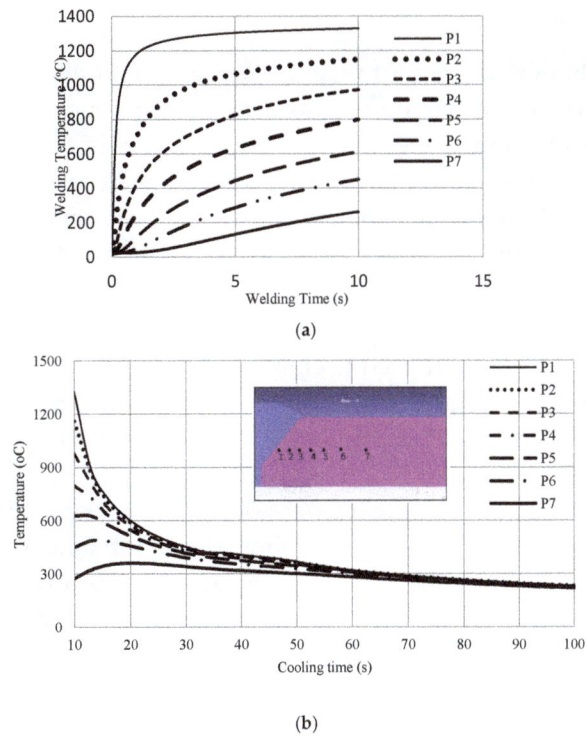

(a)

(b)

Figure 10. Temperature distribution of the HAZ region of the finite element model during (**a**) welding time and (**b**) cooling in still air.

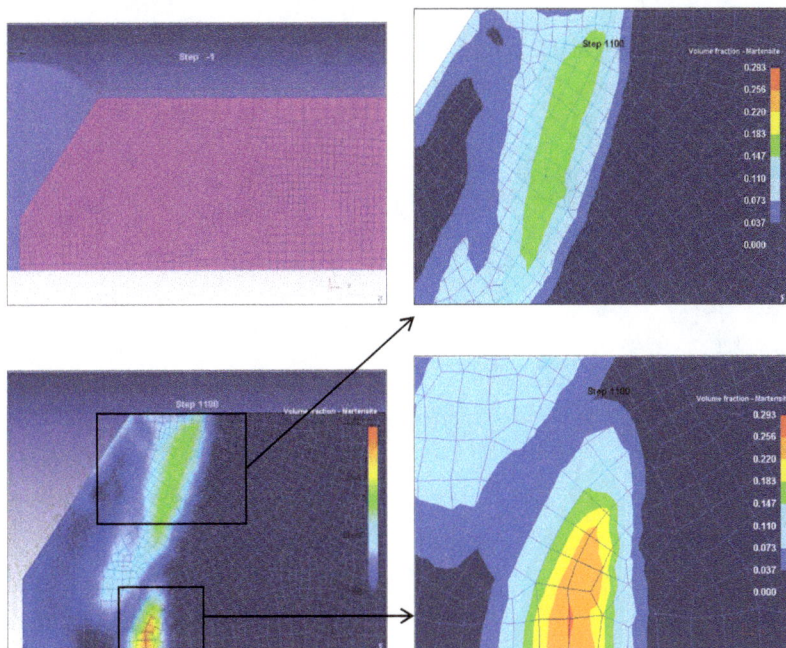

Figure 11. Volume fraction of martensite of HAZ after cooling to room temperature.

The microstructural evolution of the HAZ was determined by the FE model based on the TTT diagram and the cooling curves given in Figure 10. The formation of the martensite phase has been noticed at room temperature as is shown in Figure 11 during cooling stages in normal exposed air cooling processes. The maximum volume fractions of martensite are 0.256 in regions in Figure 11.

4.2. Hardness Test

The hardness distribution of the FE model perpendicular to the welding direction is given in Figure 12. The hardness values were taken as 3 mm, 6 mm and 9 mm above the weld root. The micro-hardness measurements were done on the same area of the specimen and are plotted in Figure 12. The micro-hardness measurements were converted to Rockwell C for comparison. The results show that the FE model and the experimental measurements are in good agreement. The maximum difference between the experimental and FE model is lower than 2 HRC.

4.3. Preheating Process

Preheating processes were investigated to eliminate the martensite formation. The preheating temperatures were increased gradually starting from 100 °C, and, for 225 °C, the formation of martensite was totally eliminated. The resulting simulation of martensite formation is given in Figure 13.

From Figure 13, the maximum volume fractions of martensite were 0.195 and 0.0079 for 100 °C and 200 °C preheating temperatures, respectively. It was found that no martensite was formed for 225 °C preheating temperature. The hardness and the average grain size distribution adjacent to the fusion line, 6 mm above the weld root in the HAZ region, are plotted in Figures 14 and 15 at different preheating temperatures. From Figure 14, the maximum hardness was found as 34.21 HRC for no preheating, and they were 31.95 HRC, 30.2 HRC and 30 HRC for the 100 °C, 200 °C and 225 °C preheating temperatures, respectively. In Figure 15, the peak grain size for the no preheating condition was 1258 μm and gradually reduced with increasing preheating temperature: 900 μm for 100 °C, 738 μm for 200 °C and 447 μm for 225 °C. According to the FE results, it is clear now to say that the 225 °C preheating process for the 12 mm thick, V-shape groove AISI 1045 steel can be recommended to prevent martensite formation.

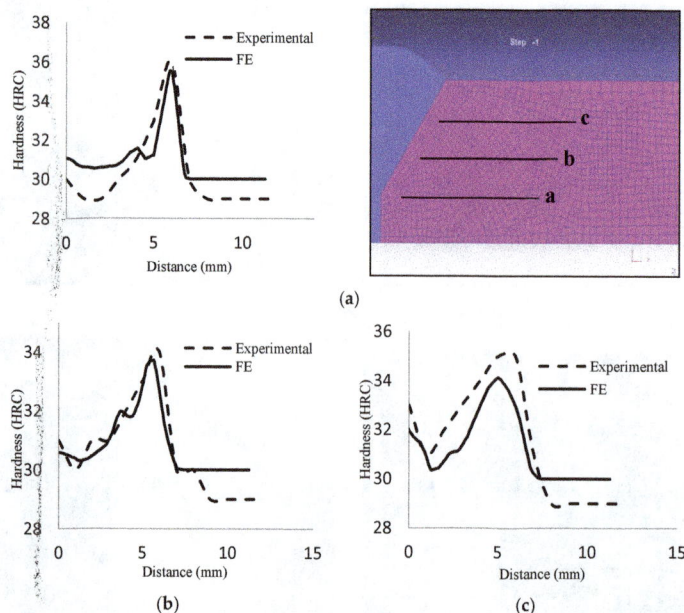

Figure 12. Distribution of hardness for (**a**) 3 mm, (**b**) 6 mm and (**c**) 9 mm above the weld root through the HAZ.

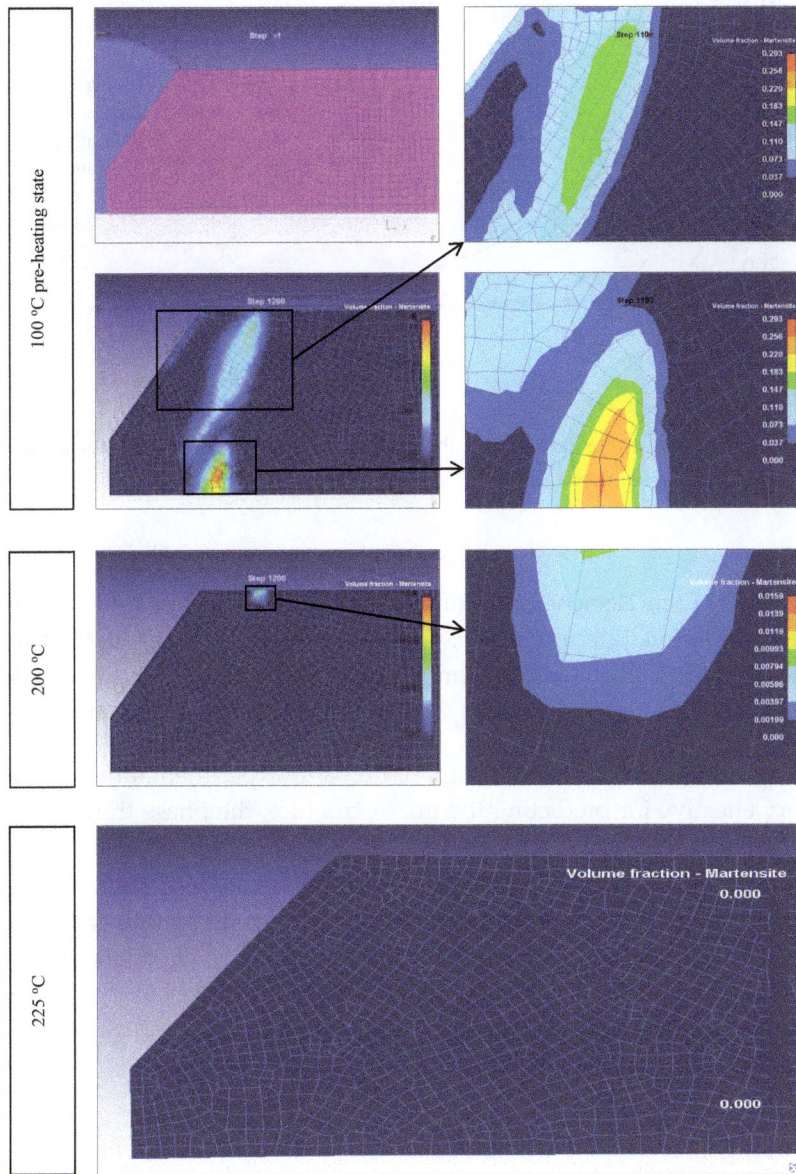

Figure 13. Elimination of martensite formation phase transformation of the AISI 1045 steel for different pre-heating temperatures.

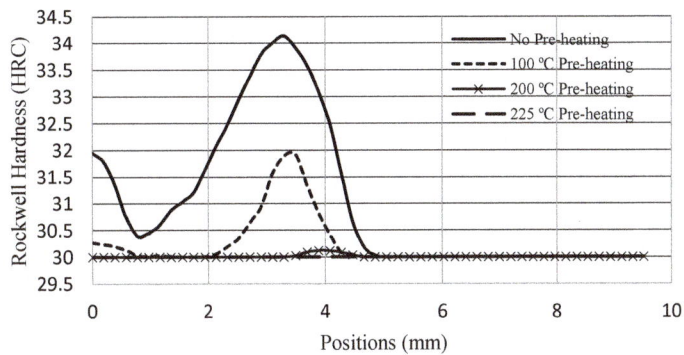

Figure 14. Distribution of hardness at the mid transverse direction of the welding line through the HAZ for different states of preheating.

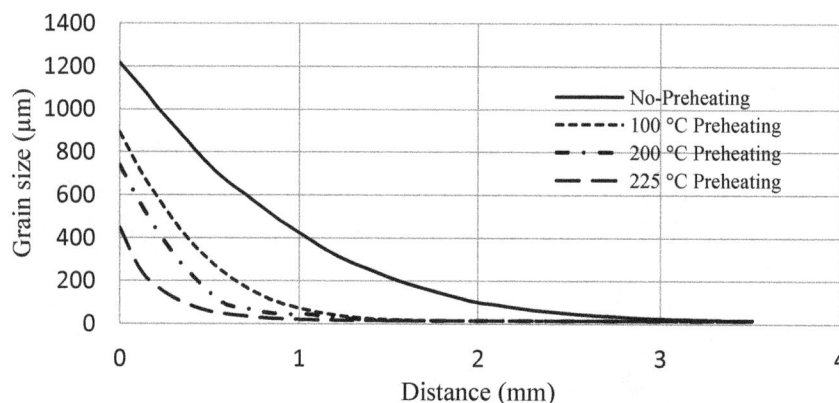

Figure 15. Distribution of average grain size at the mid transverse direction of the welding line through the HAZ for different states of preheating.

5. Conclusions

In this study, the 2D finite element model of V-shaped butt welding of AISI 1045 steel is presented. The temperature distribution, microstructure, grain growth, and the hardness of the heat affected zone (HAZ) were simulated. The results of simulation were compared with the experimental ones. According to the results obtained in the study, the following may be concluded:

1. The very close agreement between the simulation and experimental results show that the FE model is very effective for predicting the microstructure, the phase transformation, the grain growth and the hardness.
2. One of the most practical methods for eliminating martensite formation during cooling of the welding is preheating. The proper selection of the preheat temperature is important in terms of time and cost. It was found that preheating the material to 225 °C eliminates martensite formation completely for 12 mm thick, V-shaped butt welding of AISI 1045 steel.
3. The presented FE model can be used easily for different thicknesses and groove shapes to evaluate the quality of the welding process.

Acknowledgments: We thank the Scientific Research Projects (BAP) unit of Gaziantep University for supporting this study.

Author Contributions: O.E. supervised the whole study and participated the numerical simulations and experimental studies. A.S.A. carried out the finite element analyses and verification of the simulation results. K.G. designed and performed experimental works. N.F.Y. provided material data and analyzed the experimental findings.

Conflicts of Interest: The authors declare no conflict of interest.

References

1. Finite Element Method: An Introduction. Available online: faculty.ksu.edu.sa/rizwanbutt/Documents/FiniteElementsmethods.pdf (accessed on 1 February 2017).
2. Sattari-Far, I.; Farahani, M.R. Effect of the weld groove shape and pass number on residual stresses in butt-welded pipes. *Int. J. Press. Vessel. Pip.* **2009**, *11*, 723–731. [CrossRef]
3. Barsoum, Z.; Lundbäck, A. Simplified FE welding simulation of fillet welds—3D effects on the formation residual stresses. *Eng. Fail. Anal.* **2009**, *7*, 2281–2289. [CrossRef]
4. Tsai, T.C.; Chou, C.C.; Tsai, D.M.; Chiang, K.T. Modeling and analyzing the effects of heat treatment on the characteristics of magnesium alloy joint welded by the tungsten-arc inert gas welding. *Mater. Des.* **2011**, *8–9*, 4187–4194. [CrossRef]
5. Garzon, C.M.; Giraldo, J.E. Numerical and experimental analysis of microstructure evolution during arc welding in armor plate steels. *J. Mater. Process. Technol.* **2009**, *4*, 1688–1700.

6. Zhang, G.; Zhou, C.; Wang, Z.; Xue, F.; Zhao, Y.; Zhang, L. Numerical simulation of creep damage for low alloy steel welded joint Considering as-welding residual stress. *Nucl. Eng. Des.* **2012**. [CrossRef]

7. International Standard Organization, for Welding Groove Shapes. Available online: http://www.iso.org (accessed on 2 February 2017).

8. DEFORM™ User's Manual V. 10.0, Released 2011. Available online: www.deform.com (accessed on 2 February 2017).

9. Vander Voort, G.F. *Atlas of Time-Temperature Diagrams for Irons and Steels*; ASM International: Cleveland, OH, USA, 1991.

10. ASM International Handbook Committee. *ASM Handbook, Volume 04—Heat Treating*; ASM International: Cleveland, OH, USA, 1991; pp. 12–22.

11. American Standard Test Material, ASTM Microstructure Preparation Standard, ASTM E3-11 2012. Available online: www.astm.org/Standards (accessed on 2 February 2017).

12. American Standard Test Material, ASTM Grain Size Measurement Technique, ASTM E112-13 2012. Available online: www.astm.org/Standards (accessed on 2 February 2017).

13. American Standard Test Material, ASTM Hardness Test, ASTM E18-16, 2012. Available online: www.astm.org/Standards (accessed on 2 February 2017).

3

Effect of a Minor Sr Modifier on the Microstructures and Mechanical Properties of 7075 T6 Al Alloys

Shaoming Ma [1,2], **Youhong Sun** [1,2,*], **Huiyuan Wang** [3], **Xiaoshu Lü** [1,4], **Ming Qian** [1], **Yinlong Ma** [1,3], **Chi Zhang** [1,3] and **Baochang Liu** [1,2]

[1] School of Construction Engineering, Jilin University, Changchun 130026, China; masm14@mails.jlu.edu.cn (S.M.); xiaoshu.lu@aalto.fi (X.L.); mqian@jlu.edu.cn (M.Q.); ylma@jlu.edu.cn (Y.M.); zhangchi15@mails.jlu.edu.cn (C.Z); liubc@jlu.edu.cn (B.L.)

[2] Key Laboratory of Drilling and Exploitation Technology in Complex Conditions, Ministry of Land and Resources, China No. 938 Ximinzhu Street, Changchun 130026, China

[3] Key Laboratory of Automobile Materials of Ministry of Education & School of Materials Science and Engineering, Jilin University, No. 5988 Renmin Street, Changchun 130025, China; wanghuiyuan@jlu.edu.cn

[4] Department of Civil and Structural Engineering, School of Engineering, Aalto University, Helsinki 02015, Finland

[*] Correspondence: syh@jlu.edu.cn

Academic Editor: Hugo F. Lopez

Abstract: The influence of a minor strontium (Sr) modifier on the microstructures and mechanical properties of 7075 Al alloys was investigated in this paper. The grain size of cast 7075 Al alloys was refined from 157 μm to 115 μm, 108 μm, and 105 μm after adding 0.05 wt. %, 0.1 wt. %, and 0.2 wt. % Sr, respectively. The extruded 7075 Al alloys was refined with different degrees of Sr modifier. The mechanical properties were optimum when adding 0.1 wt. % Sr. The ultimate tensile strength (σ_b) increased from 573 to 598 MPa and the elongation-to-failure (δ_f) was raised from 19.5% to 24.9%. The microhardness increased from 182 to 195 Hv. The tensile fracture surface via scanning electron microscopy (SEM) revealed a transition from brittle fracture to ductile fracture as Sr increased from 0 wt. % to 0.2 wt. %. The result in this paper proved that the modifier can improve the properties of 7075 Al alloy.

Keywords: 7075 Aluminum alloy; Sr modifier; mechanical properties

1. Introduction

The insufficient mechanical properties of conventional steel drill pipes pose a challenge to the deep and ultra-deep well industry because of the high density of steel. High-strength aluminum alloys, such as 7075 and 2024 Al alloy, are preferred over steel for making drill pipes for deep oil and gas wells due to their better strength to weight ratio, lower stiffness, and higher corrosion resistance [1,2]. At present, aluminum alloy drilling pipes (ADP) have been proved promising for making drilling pipes worldwide, especially in countries such as America, Japan, France, and Russia. ADP has been successfully applied in some world record deep wells, such as SG-3 in Russia, the BD-04A well in Qatar, and the OP-11well on Sakhalin Island.

The 7075 Al alloy, which is one of the 7000 series (Al–Zn–Mg–Cu) ultra-high strength alloys, have been extensively used for structural components in aerospace and automobile industries [3,4]. Generally, the casting methods to produce 7075 Al alloy are simple and economical due to the possibility of utilizing conventional casting equipment without limitation in size and shape of the components [5]. Fine-grain strengthening during casting, including ultrasonic vibrations [6,7], electromagnetic stirring [8], and modification [9], is a good way to simultaneously achieve higher tensile strength and ductility for alloys at present.

Modification during casting is a simple and effective way to control grain size, through which the growth of crystal is inhibited by poisoning its surface with the help of certain modifying elements and, thus, refine the grain size. It is advantageous for increasing the tensile strength and ductility at the same time after extrusion by decreasing the grain size during casting. In 1921, Pacz [9] first applied modified treatment to melted Al–Si alloys with alkali fluoride. For the last decades, modification with other elements has been widely applied for grain refinement, plasticity improvement, phase transformation, and many other fields [10–13]. Sr, which is in the form of conventional Al–10Sr master alloy, exhibits a relatively good and long-lasting modification effect and has, therefore, been extensively studied in both Al and Mg alloys [14–17]. It is well known that the microstructures of Al–10Sr master alloy is composed of α-Al and Al$_4$Sr phase which is a body-centered tetragonal structure (a = 4.46 Å and c = 11.07 Å) [17]. However, Al$_4$Sr phase could not directly influence the refinement unless the free Sr could be obtained by the dissolution of the phase [18]. Unfortunately, to the best of our knowledge, there are only a few reports on the modification effect of Sr on 7075 Al alloy [19,20].

The goal of the present study is to clarify whether Sr modification has effect on the microstructures and mechanical properties of 7075 Al alloy, then reveal the reinforcement mechanism of the Sr modifier preliminarily. It is expected that the results could be helpful in promoting the development of high-strength 7075 Al alloy, thus providing guidance for manufacturing high-strength aluminum alloy drilling pipes for ultra-deep exploration, as well as other industries.

2. Experimental Procedure

Commercial 7075 Al alloy ingots and Al–10Sr master alloy rod were used as starting materials to prepare experimental alloys. First, a 2.5 kg commercial 7075 Al alloy ingot was melted at 720 °C for 10 min in a clay crucible in an electric resistance furnace of 5 kW. Then Al–10Sr preheated at 200 °C in the box-type resistance furnace was added to the melt. The melts were manually stirred for about 2 min using a stainless steel impeller to facilitate incorporation and uniform distribution of Al–10Sr in melts. After that, the melts were held at 720 °C for about 20 min, during which time the melts were stirred every 5 min and deslagged before finally being poured into a cylindrical steel mold which had been preheated at 200 °C to produce 7075–Sr alloy with the primary sample size of 90 mm in diameter and 100 mm in height. The 7075 alloy with different Sr contents were prepared in the same way by adding different amount of Al–10Sr. The designed composition of Sr in melts were 0, 0.05 wt. %, 0.1 wt. %, and 0.2 wt. %.

Cylinder samples with diameter of 90 mm and height of 100 mm were prepared for extrusion process. The samples were homogenized at 460 °C for 6 h, and then extruded at 480 °C to obtain 40 mm × 5 mm plates. After that the samples were solution treated at 470 °C for 1 h and then aged at 120 °C for 24 h. Metallographic samples of cast sample with a size of 10 mm × 10 mm × 10 mm and the ND–TD (normal direction–transverse direction) of extruded T6 heat-treated 7075 Al alloy with a size of 10 mm × 10 mm × 4 mm were prepared in accordance with standard procedures used for metallographic preparation of metal samples. Then the samples were etched with Keller reagent (1.0 mL HF + 1.5 mL HCl + 2.5 mL HNO$_3$ + 95 mL H$_2$O) for about 15 s at room temperature. The microstructures and phase were investigated by optical microscopy (OM) (Carl Zeiss–Axio Imager A$_2$m, Gottingen, Germany). The statistics grain size is obtained by the Nano Measure 2.1 (SJTU, Shanghai, China) and simply fitted with a Gaussian curve with Origin 8.0 software (OriginLab, Hampton, MA, USA). The scanning electron microscopy (SEM) (ZEISS EVO18, Mainz, Germany) fitted an Oxford Inca energy dispersive spectrometer (EDS) (Oxford Instruments, Oxon, London, UK) for further microanalysis. Phase constituents of extruded T6 samples were analyzed by X-ray diffraction (XRD) (D/Max 2500PC, Rigaku, Tokyo, Japan) using Cu Kα radiation in step mode from 20° to 80° with a scanning speed of 4°/min. Thermal analysis was carried out using a SDT-Q600 differential scanning calorimeter (DSC) apparatus (TA Instruments Inc., New Castle, PA, USA) to obtain the freezing temperature of alpha-Al and secondary phases of the extruded samples at a cooling rate of 10 °C/min. Samples of the material (30 mg) were put into an alumina pan and then heated to

700 °C and then cooling at 10 °C/min under air atmosphere. The dimensions and morphologies of the precipitates are only a few tens of nanometer which can only be revealed by the Transmission electron microscopy (TEM) technique (JEM-2100, JEOL, Tokyo, Japan) equipped with an EDS analyzer (Oxford Instruments, London, UK). TEM sample preparation was performed by successive mechanical grinding, with an operated voltage of 200 kV.

The tensile strength and fracture elongation were tested at room temperature by an electronic universal test machine (DDL 100, CIMACH, Changchun, China) at the speed of 0.18 mm/min. The tensile specimens were obtained parallel to the extruding direction, and at least three specimens were tested for each condition. The 7075 with 0.1 wt. % Sr sample was analyzed by SEM (Hitachi S–4800, Tokyo, Japan) and electron backscatter diffracting (EBSD) (NordlysNano, London, UK). The fracture morphology was observed by SEM (EVO18, ZEISS, Mainz, Germany) and the microhardness of extruded 7075 T6 Al alloy were tested by a microhardness tester (1600–5122VD Microment 5104, Buehler Ltd., Chicago, IL, USA) under an applied load of 50 g for 15 s on the Al matrix. At least seven measurements were done for each condition to ensure the accuracy of results.

3. Results and Discussion

As-cast microstructures of 7075 alloys without and with 0.05 wt. %, 0.1 wt. %, and 0.2 wt. % of Sr addition are shown in Figure 1a–d. The grain size distribution is obtained from OM images by Nano Measure 2.1 (SJTU, Shanghai, China) and fitted by Origin 8.0 software (OriginLab, Hampton, MA, USA) with a Gaussian curve (seen in the inset of Figure 1). As can be seen, the grain size decreases by different degree after adding minor Sr modifier. The refined grain can benefit for improving the mechanical properties of extruded 7075 Al alloy subsequently. As the alloys have not been solution or aging heat-treated, no $MgZn_2$ can be found in the OM microstructures.

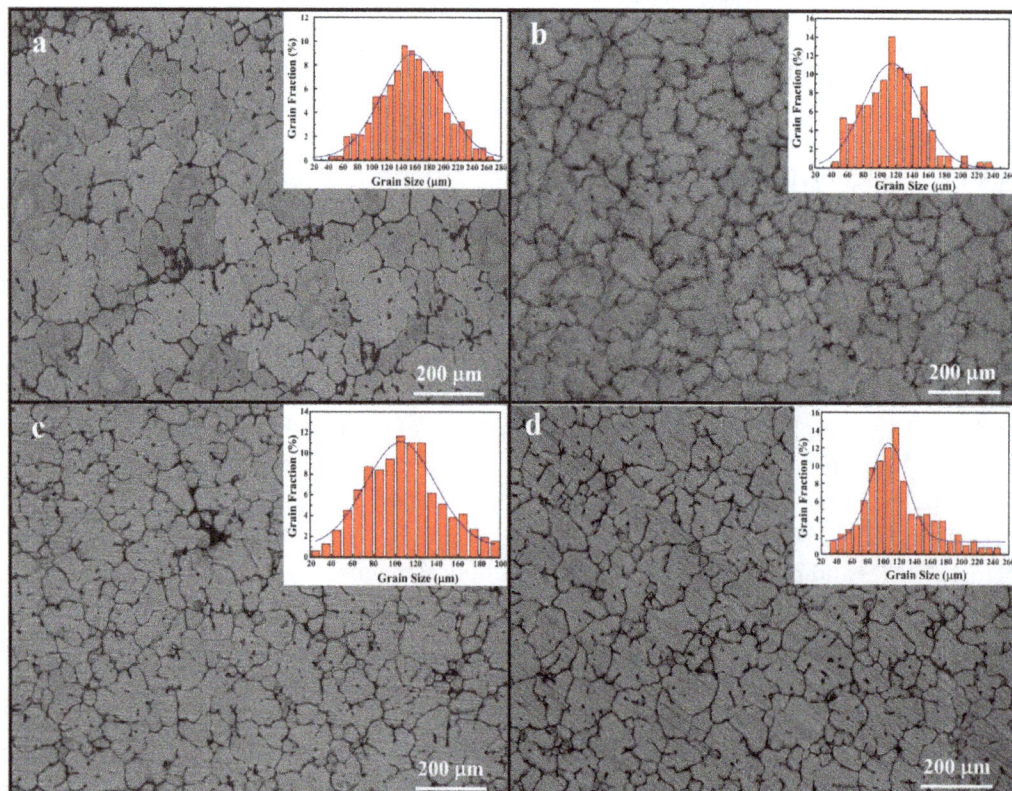

Figure 1. OM microstructures of as-cast 7075 Al alloys without and with various contents of Sr addition: (**a**) 0; (**b**) 0.05 wt. %; (**c**) 0.1 wt. %; and (**d**) 0.2 wt. % Sr (The grain size distribution is obtained from OM images by Nano Measure 2.1 and fitted by Origin 8.0 software with a Gaussion curve).

Figure 2 shows the change of the mean grain size of as-cast 7075 Al without and with different contents of Sr based on the statistical result of Figure 1. By adding 0.05 wt. %, 0.1 wt. %, and 0.2 wt. % Sr, the mean grain size of 7075 Al reduces from 157 μm to 115 μm, 108 μm, and 105 μm respectively. The equation to measure grain size in Nano Measure 2.1 is:

$$F = \frac{\sum_{N=1}^{N} 4\pi A / P^2}{N}$$

where A and P are the area and perimeter of the grains, respectively, and N is the number of grains. For each sample, measurements are taken from the 100 times magnified images.

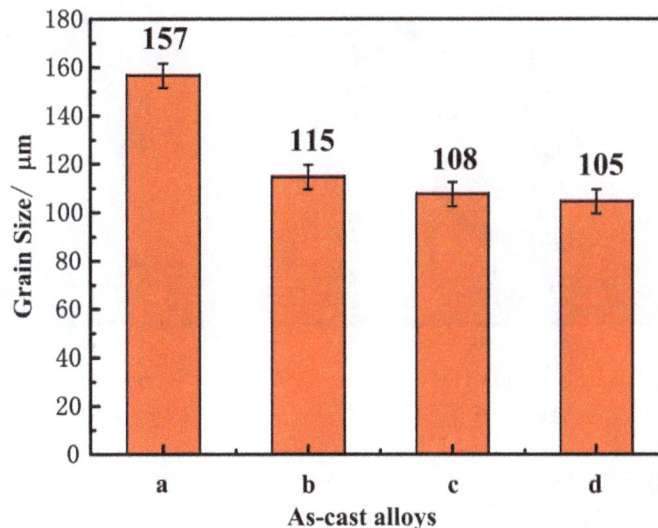

Figure 2. The change of the mean grain size of as-cast 7075 Al alloys without and with various contents of Sr addition: (**a**) 0 wt. %; (**b**) 0.05 wt. %; (**c**) 0.1 wt. %; and (**d**) 0.2 wt. % Sr.

Figure 3a–d shows the microstructures of ND–TD surface of extruded 7075 T6 Al alloys without and with different contents of Sr (0.05 wt. %, 0.1 wt. % and 0.2 wt. %). After extrusion and T6 heat treatment, the globular grains of the alloys are compressed to lamella in the ND–TD direction. The thickness of α-Al lamella and the sizes of strength phase (AlCuMg, $MgZn_2$) decrease and are better distributed (Figure 3b–d) than 7075 Al alloy without modification (Figure 3a).

Figure 4 shows the SEM images of 7075 Al alloys without and with various contents Sr addition. The precipitates are identified as AlCuMg by EDS with a size of ~1–5 μm, which agrees well with the result of OM in Figure 3. It is well known that when the Zn:Mg ratios are between 1:2 and 1:3 in the 7075 aluminum alloys, $MgZn_2$ precipitates are produced at aging temperatures below 200 °C and are the main strengthening factor in 7075 alloys [19], so further experiments are needed to prove the existence of $MgZn_2$.

The TEM micrographs of 7075 Al alloys after T6 treatment with 0.1 wt. % Sr are shown in Figure 5. We found that only the finer dark portion (~30–100 nm) is $MgZn_2$. A great amount of polygon $MgZn_2$ precipitates are found in both samples. It has been concluded that Orowan dislocation bypassing is the operative mechanism, and the increase in strength can be determined [21]. It can be seen that the precipitation plays a key role in strengthening the alloy. Some coarse phases in the grains makes parts of precipitates transform and grow, which is beneficial for the ductility of the specimen [22,23]. However, the relationship between the size of $MgZn_2$ and the ultimate tensile strength is not discussed in this paper.

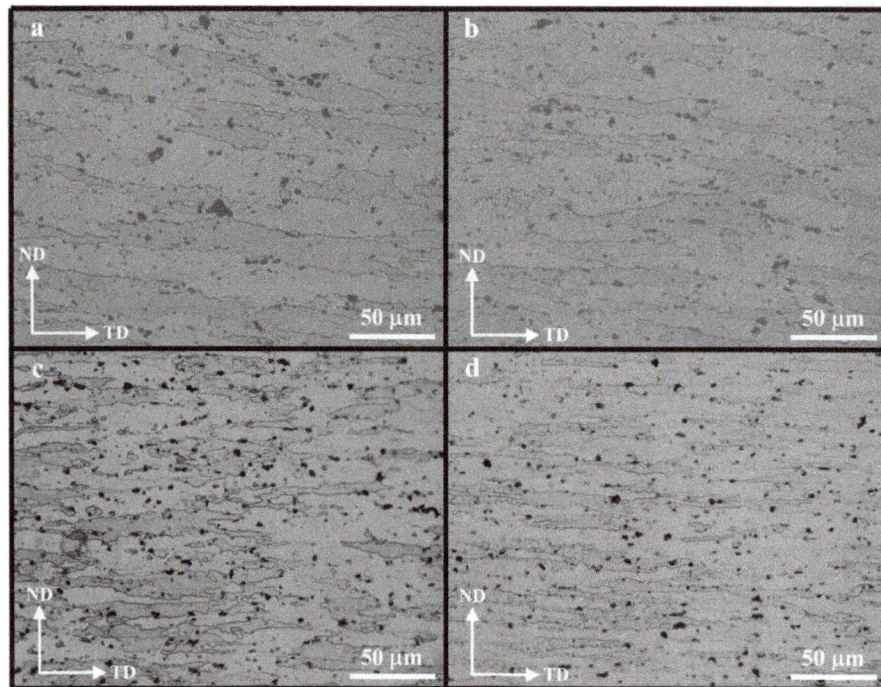

Figure 3. OM microstructures of ND-TD surface for 7075 T6 alloy without and with different contents of Sr addition: (**a**) 0 wt. %; (**b**) 0.05 wt. %; (**c**) 0.1 wt. %; and (**d**) 0.2 wt. % Sr.

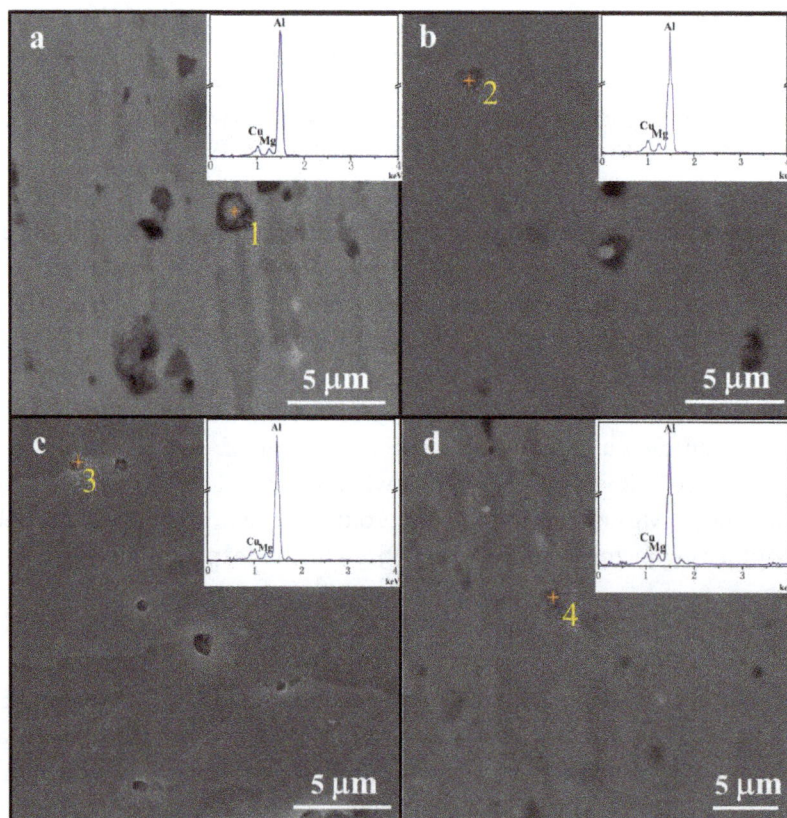

Figure 4. The SEM images of 7075 Al alloys without and with various contents Sr addition: (**a**) 0 wt. %; (**b**) 0.05 wt. %; (**c**) 0.1 wt. %; and (**d**) 0.2 wt. % Sr (the inserts are EDS results for strengthen phases).

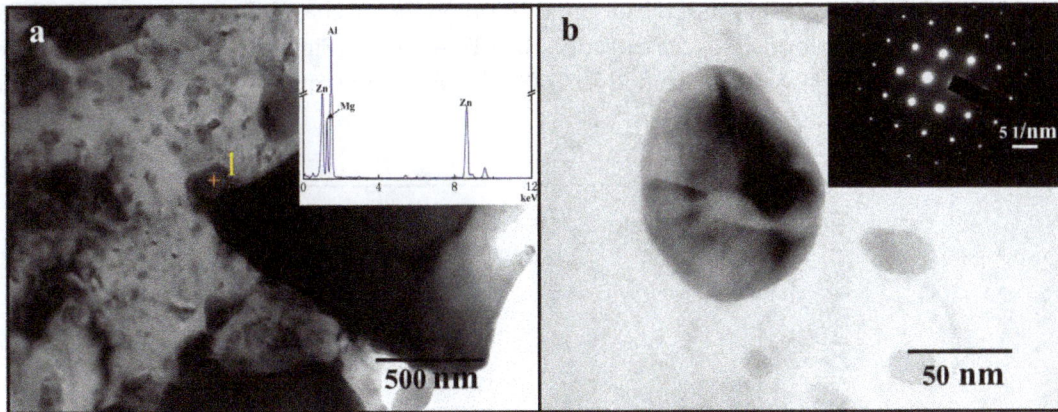

Figure 5. The TEM images of 7075 Al alloys without (**a**) and with 0.1 wt. % Sr (**b**).

Figure 6 shows the DSC curves of the 7075 T6 aluminum sample without and with 0.05 wt. %, 0.1 wt. %, and 0.2 wt. % Sr. Based on the DSC curves, the solidification temperatures of α-Al were 634, 631, 630, and 629 °C, respectively, which may indicate an increase in undercooling with the addition of Sr. Barrirero et al. has proved that the Sr promote the formation of ternary compound nanometre-sized clusters at the Si/liquid interface near the binary eutectic phase by APT method. They observed that, ahead of the growing Si crystal, a diffusion profile is formed by segregation leading to constitutional undercooling, thus altering the microstructure and obtained finer grain sizes [24]. The microstructural refinement observed in the present study can be attributed to the fact that Sr increased undercooling of the alloys and interacted with the growing α-Al.

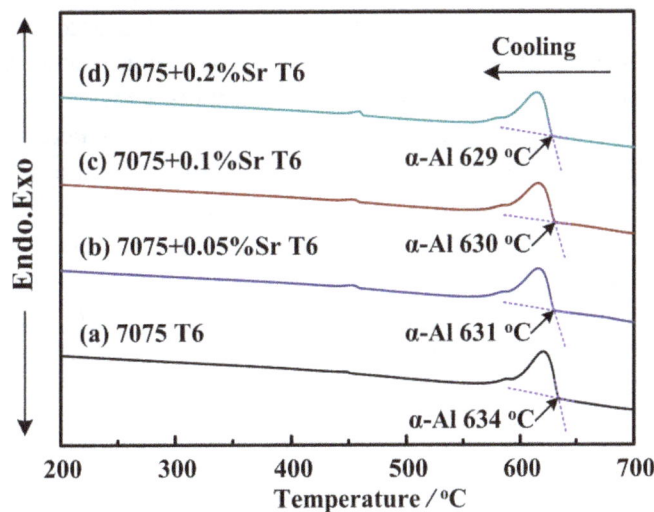

Figure 6. DSC curves for 7075Al T6 heat treated samples with various contents of Sr: (**a**) 0 wt. %; (**b**) 0.05 wt. %; (**c**) 0.1 wt. %; and (**d**) 0.2 wt. % Sr.

The constitutional undercooling usually promotes structural refinement [25]. The growth of α-Al is accompanied by the adsorption of Sr to the steps of a solid-liquid interface. Sr prevents Al atoms from attaching to their crystallographic sites and, thus, hinders the growth of the preferential direction, namely the <100> crystal orientation. As a consequence, the grain size is refined and the mechanical properties are improved. The effect of Sr contents on the grain size of extruded 7075 T6 alloy agrees well with the results of the as-cast alloys, even though the grain of 7075 Al changed from nearly globular to lamellar after extrusion.

In order to elucidate the effect of minor Sr addition on the mechanical properties of 7075 T6 Al alloys, tensile tests are performed for the extruded 7075 T6 Al alloys. Figure 7 presents the engineering stress–engineering strain curves of extruded 7075 T6 Al alloy without and with different Sr additions at room temperature.

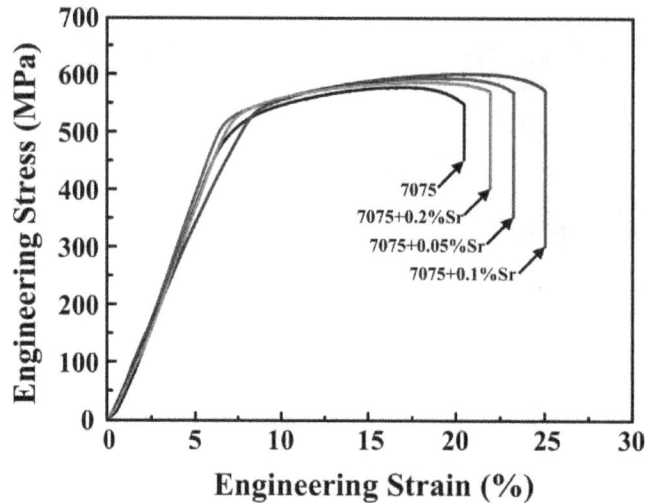

Figure 7. Engineering stress–stain curves of 7075 T6 alloy without and with different contents of Sr (wt. %).

Other mechanical properties such as average yield strength, ultimate tensile strengths, elongation, elongations-to-fracture, and the microhardness are shown in Table 1. From Table 1 we can see that the mechanical properties of the alloy are improved when Sr addition increased from 0 wt. % to 0.1 wt. %, but the improvement is subsequently degraded as the Sr addition reaches 0.2 wt. %. The tensile yield strengths, tensile strengths, elongation, and microhardness achieve their maximum value with 0.1 wt. % Sr addition. The yield strength and ultimate tensile strength increase from 490 to 526 MPa and from 573 to 598 MPa, respectively. Elongation and fracture elongation increase from 11.4% to 11.7% and from 19.5% to 24.9%, respectively. Microhardness improves from 182 to 195 Hv. In a word, 0.1 wt. % Sr can achieve the optimal modification effect for 7075 Al alloys. Our tensile strength is higher than the result reported by Chen et al. [26], as they gave a true stress–strain curve in their research with a true stress of about 600 MPa. The microhardness of 195 HV is the same value with the research reported by M. Tajally et al. [27], which is supplied by Alcoa, USA. However, their tensile strength is only 370 MPa.

Table 1. Mechanical properties of 7075 T6 alloys without and with different contents of Sr (wt. %).

Sample	σ_s/MPa	σ_b/MPa	δ/%	δ_f/%	Hardness/Hv
7075	490^{+9}_{-7}	573^{+3}_{-1}	$11.4^{+0.1}_{-0.1}$	$19.5^{+0.9}_{-0.9}$	182^{+2}_{-7}
7075 + 0.05%Sr	516^{+8}_{-3}	590^{+1}_{-2}	$11.6^{+0.1}_{-0.2}$	$23.2^{+0.2}_{-0.9}$	193^{+1}_{-0}
7075 + 0.1%Sr	526^{+4}_{-7}	598^{+1}_{-2}	$11.7^{+0.1}_{-0.2}$	$24.9^{+0.4}_{-0.8}$	195^{+1}_{-2}
7075 + 0.2%Sr	514^{+5}_{-9}	582^{+3}_{-4}	$11.5^{+0.2}_{-0.2}$	$21.0^{+1.0}_{-0.1}$	189^{+1}_{-1}

The XRD patterns of the ND-TD surface for extruded 7075 T6 Al alloys without and with different Sr addition are shown in Figure 8a–d. According to XRD results in Figure 8, only Al is identified by XRD in alloys without and with Sr addition. No $MgZn_2$ (η phase) and Al_4Sr are found after adding different Sr to 7075 Al alloys. The result reveals that the addition of different contents of Sr has no obvious influence on phase compositions of the alloy. The possible reason may be that the XRD technique is not sensitive enough for studying the low content of $MgZn_2$ and Al_4Sr intermetallic phases. Some Zn and Mg atoms dissolving in the Al matrix, thus, the content of nanosized $MgZn_2$ was too small to be detected.

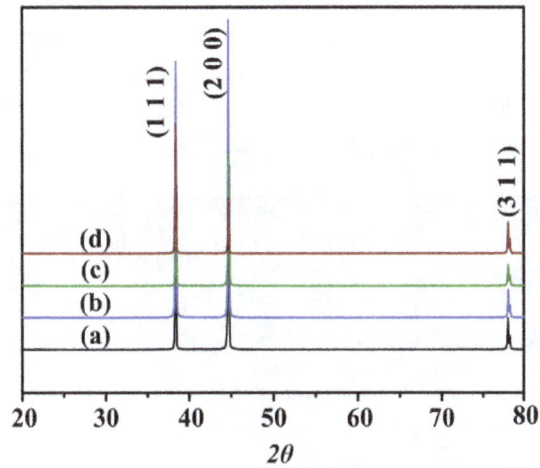

Figure 8. XRD results of ND-TD surface for 7075 T6 alloy without and with different contents of Sr addition: (**a**) 0 wt. %; (**b**) 0.05 wt. %; (**c**) 0.1 wt. %; and (**d**) 0.2 wt. % Sr.

To clarify the mechanism of Sr addition improving the mechanical properties of extruded 7075 T6 Al alloy, the sample with 0.1 wt. % Sr addition is analyzed by EBSD.

Figure 9 shows the EBSD image of as-cast 7075 T6 Al alloys without and with 0.1 wt. % Sr. Al_4Sr (yellow dot in the image) is detected both in grain boundary and grain interior (seen in Figure 9b) in the sample contained 0.1 wt. % Sr. This result indicate that minor Al_4Sr is formed when adding 0.1 wt. % Sr in 7075 Al alloy.

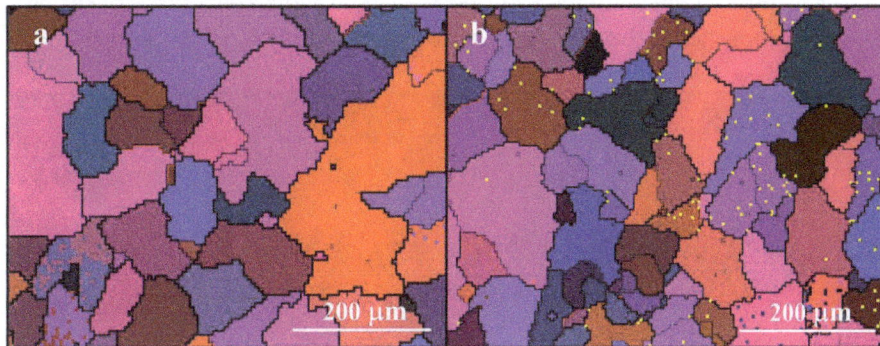

Figure 9. EBSD images of 7075 T6 Al alloy without (**a**) and with 0.1 wt. % Sr additions (**b**) (yellow dots in the image are set as Al_4Sr).

The distinctive feature of the tensile properties of the alloys obeys the Hall–Petch law qualitatively, as Equation (1) shows:

$$\sigma = \sigma_0 + kd^{-\frac{1}{2}} \tag{1}$$

where σ_0 and k are constants that are related to the crystal type and d is the average grain size. Thus, the finer the grain size, the better the mechanical properties. Unfortunately, the grain of 7075 Al changed from nearly globular to lamellar after extrusion; thus, a quantitative statistic of the grain size is difficult to obtain.

When the Sr addition increases to 0.2 wt. % in extruded 7075 Al alloy, the mechanical properties are inferior than the alloy with 0.1 wt. % Sr addition. SEM microstructures of as-cast 7075 Al alloy without and with 0.2 wt. % Sr modification is showed in Figure 6. Microporosity in the alloy with 0.2 wt. % Sr modification in show in Figure 10b, which is the main reason for the reduction of mechanical properties after extrusion and T6 heat treatment. While the grain boundary of 7075 Al alloy without Sr addition was very clear (Figure 10a). The increase in the Al_4Sr volume fraction increases the overall porosity area of the gas pores from 6.2% to 9.6%, compared with the sample without Sr. This porosity decreases both the yield and ultimate tensile strength values of the produced samples, as Tekman et al. have reported [28,29]. Porosity parameters, namely, the total porosity area is analyzed using Pixcavator IA 4.3 software (Marshall University, Huntington, WV, USA).

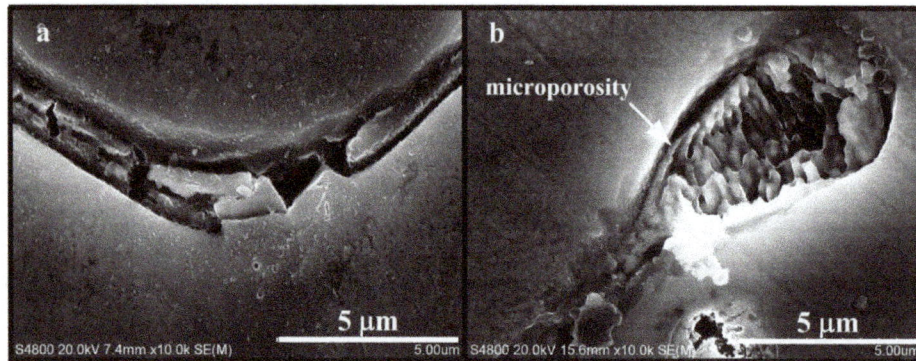

Figure 10. High-magnification SEM images of as-cast 7075 Al alloy without (**a**) and with 0.2 wt. % Sr (**b**).

The typical SEM images of the fracture surfaces in Figure 11 reveal a transition from brittle to ductile fracture mode by adding different contents of Sr. The alloy before modification has high fragility, which may cause low tensile strength and elongation. By contrast, the fracture surface of Sr-modified alloy (Figure 11c) shows more and finer dimples, which is to say the rupture has a ductile nature, indicating that the cracks hardly propagated through these precipitates. The morphology of $MgZn_2$ has a critical effect on the mechanical properties of the alloy. The $MgZn_2$ particles become finer, and the mechanical properties of the alloy are improved.

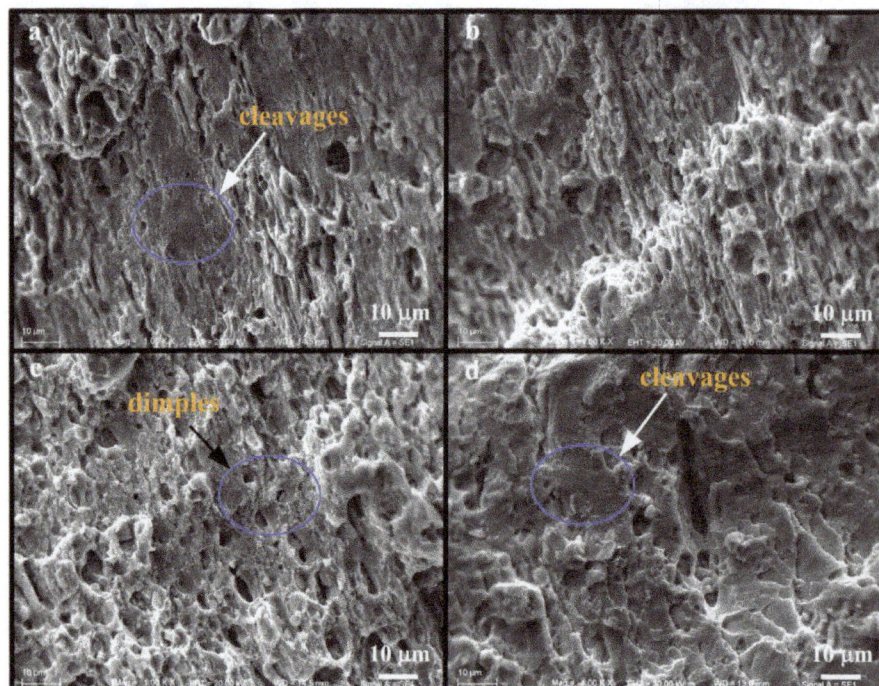

Figure 11. The SEM fracture morphology of 7075 T6 alloy without and with different contents of Sr: (**a**) 0 wt. %; (**b**) 0.05 wt. %; (**c**) 0.1 wt. %; and (**d**) 0.2 wt. % Sr.

4. Conclusions

Minor Sr additions have effects on the microstructures of as-cast and 7075 T6 Al alloy. The grain size of both cast and extruded T6 treated 7075 Al alloy was refined by different degrees after adding 0, 0.05, 0.1 and 0.2 wt. % Sr. The growth of α-Al was accompanied by the adsorption of Sr atom to the steps of a solid-liquid interface. The strength phase $MgZn_2$ in extruded 7075 T6 alloy was also refined and well-distributed after different Sr addition.

Minor Sr addition have effects on the mechanical properties of extruded 7075 T6 Al alloy. The mechanical properties of extruded 7075 T6 Al alloy were improved at first and then decreased as the Sr addition increased from 0.05 wt. % to 0.2 wt. %. By adding 0.1 wt. % Sr, the ultimate tensile strength of extruded 7075 T6 increased from 573 to 598 MPa. Fracture elongation increased from 19.5% to 24.9%. Microhardness improved from 182 to 195 Hv. The fracture mode revealed a transition from brittle fracture to ductile fracture as Sr addition increased from 0.05 wt. % to 0.2 wt. %. The improvement of the mechanical properties was mainly ascribed to the reduction of the grain size and the formation of high melting point phase Al_4Sr, which could act as barriers for dislocation movement.

Acknowledgments: This work is supported by SinoProbe-09-05 (Project No. 201011082), International S&T Cooperation Program of China (Grant No. 2013DFR70490).

Author Contributions: Youhong Sun conceived and designed the experiments; Shaoming Ma and Chi Zhang performed the experiments; Huiyuan Wang and Yinlong Ma analyzed the data; Ming Qian, Baochang Liu and Xiaoshu Lv contributed reagents/materials/analysis tools; Shaoming Ma wrote the paper.

Conflicts of Interest: The authors declare no conflict of interest.

References

1. Liang, J.; Sun, J.H.; Li, X.; Zhang, Y.Q.; Li, P. Development and Application of Aluminum Alloy Drill Rod in Geologic Drilling. *Process. Eng.* **2014**, *73*, 84–90.

2. Ziomek-Moroz, M. Environmentally Assisted Cracking of Drill Pipes in Deep Drilling Oil and Natural Gas Wells. *Mater. J. Eng. Perform.* **2012**, *21*, 1061–1069. [CrossRef]

3. Panigrahi, S.K.; Jayaganthan, R. Effect of ageing on microstructure and mechanical properties of bulk, cryorolled, and room temperature rolled Al 7075 alloy. *J. Alloy. Compd.* **2011**, *509*, 9609–9616. [CrossRef]

4. Xu, X.F.; Zhao, Y.G.; Ma, B.D.; Zhang, M. Electropulsing induced evolution of grain-boundary precipitates without loss of strength in the 7075 Al alloy. *Mater. Charact.* **2015**, *105*, 90–94. [CrossRef]

5. Fard, R.R.; Akhlaghi, F. Effect of extrusion temperature on the microstructure and porosity of A356-SiC p composites. *J. Mater. Process. Technol.* **2007**, *187–188*, 433–436. [CrossRef]

6. Lü, S.L.; Wu, S.S.; Dai, W.; Lin, C.; An, P. The indirect ultrasonic vibration process for rheo-squeeze casting of A356 aluminum alloy. *J. Alloy. Compd.* **2012**, *212*, 1281–1287. [CrossRef]

7. Haghayehi, R.; Heydari, A.; Kapranos, P. The effect of ultrasonic vibrations prior to high pressure die-casting of AA7075. *Mater. Lett.* **2015**, *153*, 175–178. [CrossRef]

8. Mapelli, C.; Gruttadauria, A.; Peroni, M. Application of electromagnetic stirring for the homogenization of aluminium billet cast in a semi-continuous machine. *J. Mater. Process. Technol.* **2010**, *210*, 306–314. [CrossRef]

9. Pacz, A. Alloy. U.S. Patent 1387900 [P/OL], 16 August 1921.

10. Kyziol, K.; Koper, K.; Sroda, M.; Klich, M.; Kaczmarek, L. Influence of gas mixture during N^+ ion modification under plasma conditions on surface structure and mechanical properties of Al–Zn alloy. *Surf. Coat. Technol.* **2015**, *278*, 30–37. [CrossRef]

11. Liao, C.W.; Chen, J.C.; Li, Y.L.; Chen, H.; Pan, C.X. Modification performance on 4032 Al alloy by using Al–10Sr master alloys manufactured from different processes. *Prog. Nat. Sci. Mater. Int.* **2014**, *24*, 87–96. [CrossRef]

12. Liu, G.L.; Si, N.C.; Sun, S.C.; Wu, Q.F. Influence of heat treatment on microstructure and friction and wear properties of multicomponent Al–7·5Si–4Cu alloy. *Trans. Nonferr. Met. Soc. China* **2014**, *24*, 946–953. [CrossRef]

13. Shivaprasad, C.G.; Narendranath, S.; Desai, V.; Swami, S.; Granesha, M.S. Influence of Combined Grain Refinement and Modification on the Microstructure and Mechanical Properties of Al-12Si, Al-12Si-4.5Cu Alloys. *Procedia Mater. Sci.* **2014**, *5*, 1368–1375. [CrossRef]

14. Srirangam, P.; Chattopadhyay, S.; Bhattacharya, A.; Nag, S.; Kaduk, J.; Shankar, S.; Shankar, R.; Banerjee, R.; Shibata, T. Probing the local atomic structure of Sr-modified Al–Si alloys. *Acta Mater.* **2014**, *65*, 186–193. [CrossRef]

15. Barriero, J.; Engstler, M.; Ghafoor, N.; Jonge, N.; Oden, M.; Mucklich, F. Comparison of segregations formed in unmodified and Sr-modified Al–Si alloys studied by atom probe tomography and transmission electron microscopy. *J. Alloy. Compd.* **2014**, *611*, 410–421. [CrossRef]

16. Timpel, M.; Wanderka, N.; Kumar, G.S.V.; Banhart, J. Microstructural investigation of Sr-modified Al-15 wt. % Si alloys in the range from micrometer to atomic scale. *Ultramicroscopy* **2011**, *111*, 695–700. [CrossRef] [PubMed]

17. Bai, J.; Sun, Y.S.; Xun, S.; Xue, F.; Zhu, T.B. Microstructure and tensile creep behavior of Mg–4Al based magnesium alloys with alkaline-earth elements Sr and Ca additions. *Mater. Sci. Eng. A* **2006**, *419*, 181–188.

18. Yang, M.B.; Pan, F.S.; Cheng, R.J.; Tang, A.T. Effects of various Mg-Sr master alloys on microstructural refinement of ZK60 magnesium alloy. *J. Alloy. Compd.* **2008**, *461*, 298–303. [CrossRef]

19. Binesh, B.; Aghaie-Khafri, M. Phase Evolution and Mechanical Behavior of the Semi-Solid SIMA Processed 7075 Aluminum Alloy. *Metals* **2016**. [CrossRef]

20. Tavighi, K.; Emamy, M.; Emami, A.R. Effects of extrusion temperature on the microstructure and tensile properties of Al–16 wt. % Al 4 Sr metal matrix composite. *Mater. Des.* **2013**, *46*, 598–604. [CrossRef]

21. Ma, K.; Wen, H.; Hu, T.; Topping, T.D.; Isheim, D.; Seidman, D.N.; Lavernia, E.J.; Schoenung, J.M. Mechanical behavior and strengthening mechanisms in ultrafine grain precipitation-strengthened aluminum alloy. *Acta Mater.* **2014**, *62*, 141–155. [CrossRef]

22. Sha, G.; Cerezo, A. Early-stage precipitation in Al–Zn–Mg–Cu alloy (7050). *Acta Mater.* **2004**, *52*, 4503–4516. [CrossRef]

23. Nicolas, M.; Deschamps, A. Characterisation and modelling of precipitate evolution in an Al–Zn–Mg alloy during non-isothermal heat treatments. *Acta Mater.* **2003**, *51*, 6077–6094. [CrossRef]

24. Barrirero, J.; Li, J.; Engstler, M.; Ghafoor, N.; Schumacher, P.; Odén, M.; Mücklich, F. Cluster formation at the Si/liquid interface in Sr and Na modified Al–Si alloys. *Scr. Mater.* **2016**, *117*, 16–19. [CrossRef]

25. SShin, S.; Kim, E.S.; Yeom, G.Y.; Lee, J.C. Modification effect of Sr on the microstructures and mechanical properties of Al–10.5Si–2.0Cu recycled alloy for die casting. *Mater. Sci. Eng. A* **2012**, *532*, 151–157. [CrossRef]

26. Chen, D.C.; You, C.S.; Gao, F.Y. Analysis and Experiment of 7075 Aluminum Alloy Tensile Test. *Procedia Eng.* **2014**, *81*, 1252–1258. [CrossRef]

27. Tajally, M.; Huda, Z.; Masjuki, H.H. A comparative analysis of tensile and impact-toughness behavior of cold-worked and annealed 7075 aluminum alloy. *Int. J. Impact Eng.* **2010**, *37*, 425–432. [CrossRef]

28. Tekmen, C.; Ozdemir, I.; Cocen, U.; Onel, K. The mechanical response of Al–Si–Mg/SiC p composite: Influence of porosity. *Mater. Sci. Eng. A* **2003**, *360*, 365–371. [CrossRef]

29. Farhoodi, B.; Raiszadeh, R.; Ghanaatian, M.H. Role of Double Oxide Film Defects in the Formation of Gas Porosity in Commercial Purity and Sr-containing Al Alloys. *J. Mater. Sci. Technol.* **2014**, *30*, 154–162. [CrossRef]

4

The Pseudo-Eutectic Microstructure and Enhanced Properties in Laser-Cladded Hypereutectic Ti–20%Si Coatings

Hui Zhang [1,2,*], **Zhonghong Zhang** [1] **and T. M. Yue** [2]

[1] School of Materials Science and Engineering, Anhui University of Technology, Ma'anshan 243002, China; zrzhhzrzhh@tom.com

[2] The Advanced Manufacturing Technology Research Centre, Department of Industrial and Systems Engineering, Hong Kong Polytechnic University, Hung Hom, Hong Kong, China; tm.yue@polyu.edu.hk

* Correspondence: huizhang@ahut.du.cn

Academic Editor: Hugo F. Lopez

Abstract: Ti_5Si_3 is an attractive light weight reinforcement phase in hypereutectic Ti–Si-based alloys, however, the proeutectic Ti_5Si_3 phase is brittle and is easily coarsened when the alloy is prepared under normal solidification conditions, thereby limiting its engineering applications in the aviation and biological industries. In this study, a hypereutectic Ti–20%Si coating with a pseudo-eutectic α-Ti + Ti_5Si_3 microstructure was successfully fabricated on a commercially available Ti alloy by laser cladding under non-equilibrium rapid solidification conditions. The fine, rod-like and well-dispersed eutectic Ti_5Si_3 phase, without the primary Ti_5Si_3 phase, that was produced resulted in a considerable improvement in hardness, corrosion resistance, and fracture resistance when compared to the same compositional alloy prepared by the conventional arc melting technique.

Keywords: laser cladding; hypereutectic Ti–Si alloy; corrosion; fracture resistance

1. Introduction

Ti–Si alloys are currently attracting wide academic interest due to their attractive characteristics, such as lightweight, high specific modulus, hardness, corrosion resistance, and good biocompatibility [1–3]. The Ti_5Si_3 intermetallic compound phase, which has a complex D88 hexagonal structure (Mn_5Si_3-type, a1/40.7444 nm, c1/40.5143 nm), has the highest melting temperature (2130 °C), highest hardness (11.3 GPa), and relatively high Young's modulus (225 GPa) among the five silicide phases, $TiSi_2$, $TiSi$, Ti_5Si_4, Ti_5Si_3, and Ti_3Si, that exist in the binary Ti–Si alloy system [4]. Thus, it could act as a desired reinforcement phase to significantly increase the hardness of the ductile titanium matrix. Evidently, the higher the Ti_5Si_3 phase content, the higher the hardness of the Ti–Si alloy. Indeed, hypereutectic Ti–Si alloys dominated by Ti_5Si_3 have received much attention in recent years, including applications as coating materials [2–5]. Nevertheless, the proeutectic Ti_5Si_3 phase nucleated from the liquid metal may develop into coarse particles, and result in low toughness of the alloy, which could restrict its applications [2,4]. It is believed that, if one can refine the Ti_5Si_3 phase in hypereutectic alloys and avoid the formation of coarse primary Ti_5Si_3 crystals, then the toughness of the alloy can be improved.

Based on the Ti–Si phase diagram [4,6], the eutectic reaction of the alloy occurs approximately at 13.5 atom % Si (atomic). This study aims to achieve the refinement of the Ti_5Si_3 eutectic phase as well as to suppress the nucleation of the primary Ti_5Si_3 phase of a Ti–20 atom % Si hypereutectic alloy using laser cladding. It is envisaged that, due to the nature of laser processing, i.e., under rapid solidification conditions, this aim can be achieved. This study also compares the hardness, fracture toughness, and corrosion resistance of the coating to those of the same compositional alloy prepared by the conventional arc melting and metal mold casting method.

2. Materials and Methods

The Ti–20 atom % Si alloy was prepared by both laser cladding and arc melting methods. For laser cladding, a continuous-wave CO_2 laser system was used to prepare the coating on a commercially available titanium Ti alloy (TC21) substrate. The Ti substrate was a square block (100×100 mm^2) with a thickness of 10 mm, of the nominal composition Ti–6Al–2Sn–2Zr–3Mo–1Cr–2Nb–0.1Si. It is an $\alpha + \beta$ titanium alloy with high strength and toughness, as well as high fatigue tolerance [7]. The Ti and Si powders used in the laser cladding had a purity of 99.7 wt % with particle sizes ranging between 50 and 120 µm. They were mechanically mixed, dried in a vacuum oven for 12 h, and delivered to the laser processing zone using a lateral powder feeder. High-purity argon gas flowing through a coaxial nozzle created a shielding region on the substrate to prevent oxidation of the powder. The laser cladding parameters used were as follows: a laser power of 2 kW, a laser beam diameter of 4 mm, a laser scan speed of 400 mm·min^{-1}, a laser track overlap ratio of 40%, and a powder feed rate of 200 mg/s. A single-layer coating was produced using one deposition pass by overlapping the laser tracks, resulting in a coating thickness of about 1.1 mm. Meanwhile, for the preparation of the cast material, high purity elements (Fe 99.9% and Si 99.9%) were used to obtain 25 g castings. The Ti–20 atom % Si alloy was prepared by means of Ti-gettered arc melting, where the molten alloy, produced using a non-consumable vacuum arc melting furnace (Shenyang Vacuum Technology Institute, Shenyang, China), was drop-cast into a 20 mm diameter copper mold. The melting was repeated four times to improve the chemical homogeneity of the alloy.

The phases and microstructure were characterized using a X-ray diffractometer (Rigaku, Tokyo, Japan) with Cu Kα radiation operated at 40 kV and 30 mA, and a JSM-6490 scanning electron microscope (JEOL, Tokyo, Japan) equipped with an energy dispersive spectrometer (EDAX, Mahwah, NJ, USA). The Vickers indentation technique was used, with an applied load of 4.9 N, to measure the hardness of the specimens, which were annealed for 2 h at 300 °C to relieve the residual stresses prior to the test. The indentation cracking-based method was used to evaluating the crack extension resistance (toughness) of the cast and laser-cladded specimens [8]. The test was performed using a loading force of 294 N to provoke crack nucleation. The electrochemical polarization measurements were carried out using a three-electrode cell system in Tyrode's artificially simulated body fluid solution at room temperature.

3. Results and Discussion

3.1. Characterization of Microstructure and Phases

An examination of the microstructure of the metal mold cast material showed that it consists of large grains with a coarse Ti$_5$Si$_3$ phase. Figure 1a shows two of such large columnar grains with a width of about 3 mm. An etched specimen revealed that it consists of coarse primary Ti$_5$Si$_3$ particles with sharp edges distributed in a α-Ti + Ti$_5$Si$_3$ eutectic matrix (Figure 1b). The primary Ti$_5$Si$_3$-phase, as marked by arrows, are about 15 µm in width, and more than 100 µm in length was nucleated from the liquid phase as the alloy experienced a eutectic reaction: Liquid $\rightarrow$ Ti$_5$Si$_3$ + β-Ti at 1330 °C [4,6]. The β-Ti subsequently transformed to α-Ti at a lower temperature. Figure 1c shows that the size of the Ti$_5$Si$_3$-phase in the eutectic matrix is about 3 µm in width and 10 µm in length.

Turning to the laser-cladded coating. Figure 2a shows the overall morphology of the microstructure of a cross section of a laser-cladded specimen. Here, the Ti–20%Si coating, the laser melted zone (LMZ), the heat affected zone (HAZ), and the unaffected substrate can be identified. The coating is free from porosity and cracks, and a metallurgical bonded interface was formed between the LMZ and the coating. The microstructure of the LMZ changed from equiaxed to columnar crystal growth towards the coating layer. This is thought to be due to the fact that, during the solidification of the LMZ, the crystal growth velocity increases with increasing distance from the bottom of the LMZ. A very narrow HAZ (about 400 μm) was observed between the substrate and the LMZ; this agreed with the hardness measurements obtained across the section of a coated specimen, and the results are presented in the next section. It was also observed that a continuous interface was formed between the overlapped regions between two laser tracks, and no cracks were found (Figure 2b).

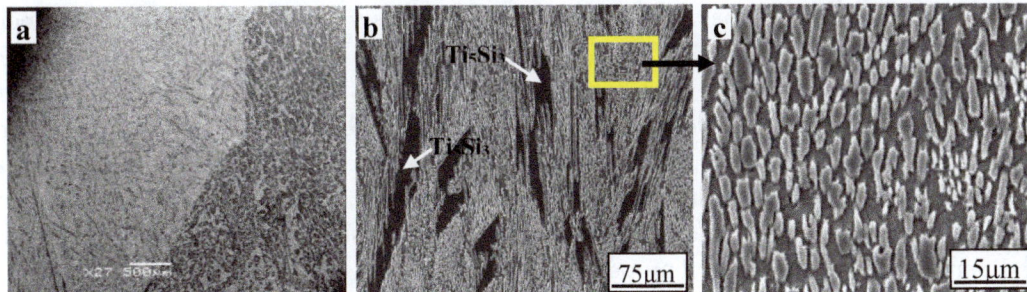

Figure 1. Microstructure of the Ti–20%Si alloy cast in metal mold showing (**a**) large grains; (**b**) a coarse primary Ti_5Si_3 phase; and (**c**) the eutectic matrix.

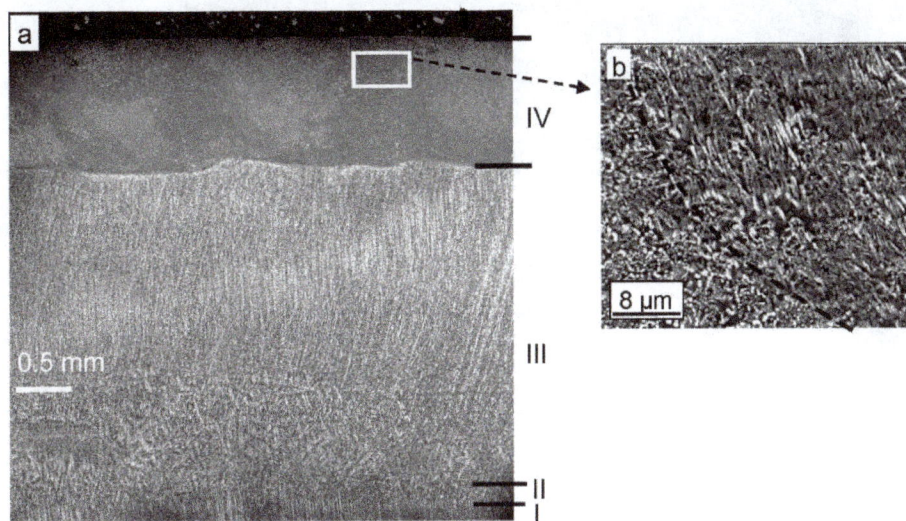

Figure 2. Cross-sectional images of the laser-cladded specimen showing (**a**) the unaffected substrate (I), the heat affected zone (II), the laser melted zone (III), the Ti–20% Si coating (IV); and (**b**) a laser overlapped region.

A closer examination of the coating microstructure shows that it exhibits a fine pseudo-eutectic microstructure. When compared to that of the metal mold casting, the laser-cladded material has a well-dispersed eutectic Ti_5Si_3 phase, which was greatly refined to the submicron level and has a rod-like growth morphology, suggesting that the formation of the primary Ti_5Si_3 phase was suppressed (Figure 3a,b). The eutectic phase has a kind of cellular structure having a size of about 10 μm. The formation of a eutectic structure, without the coarse primary Ti_5Si_3 phase, is considered to be due to non-equilibrium solidification conditions. During laser cladding, the outgrowth speed of the solid–liquid (S/L) interface is high and the local solidification time is limited, with large undercooling in front of the S/L interface. Thus, the sluggish diffused solute atoms accumulated at the liquid zone ahead of the S/L interface cannot be redistributed to the liquid region fast enough, causing the nucleation rate to decrease and suppress the nucleation of the Ti_5Si_3 phase [9]. Moreover, as the Ti–Si alloy belongs to a metal–nonmetal system, the non-faceted interface of the α-Ti grain grew more quickly than that of the faceted interface of the Ti_5Si_3 phase under the condition of large undercooling [9]. Therefore, crystallization occurs by diffusion decomposition into a mixture of crystals of α-Ti + Ti_5Si_3 phases that preceded the growth of the primary Ti_5Si_3 phase and resulted in the growth of a pseudo-eutectic structure in the Ti–20%Si alloy. The X-ray diffraction (XRD) patterns shown in Figure 4 indicate that both the metal mold cast and laser-cladded materials, primarily composed of α-Ti and Ti_5Si_3 phases without the Ti_3Si phase, suggests that the peritectoid reaction, Ti_5Si_3 + β-Ti → Ti_3Si, was supressed.

It is also worth noting that the rapidly solidified Ti–20%Si laser clad layer is metallurgically bonded to the TC21 alloy substrate (Figure 3c). The microstructure close to the substrate-coating interface has a typical hypoeutectic microstructure with a proeutectic α-Ti phase. The energy dispersive spectrometer (EDS) analysis shows that the Ti content in this area reached 89 atom %, which is higher than the nominal composition of the coating alloy. This is due to the remelting of the substrate and the alloy dilution effect. Such dilution at the interface is desirable, because the proeutectic α-Ti phase is relatively ductile and can decrease the tendency of cracking at the coating-substrate interface that may be caused by the differences in thermophysical properties.

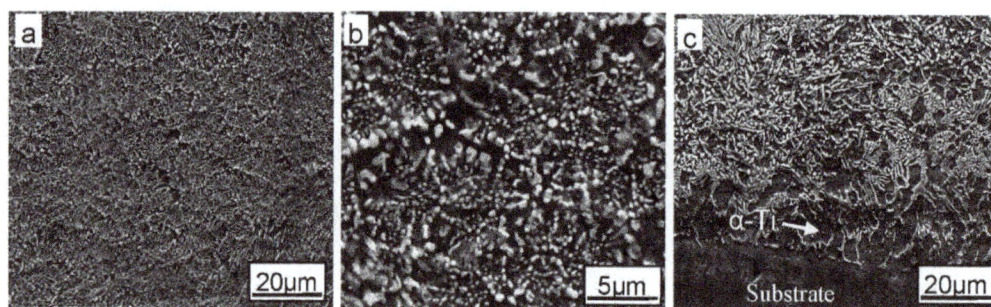

Figure 3. (**a**) Microstructure of the Ti–20%Si laser clad coating; (**b**) a higher magnification of (**a**) showing the fine eutectic structure; (**c**) the interface between the coating and the substrate.

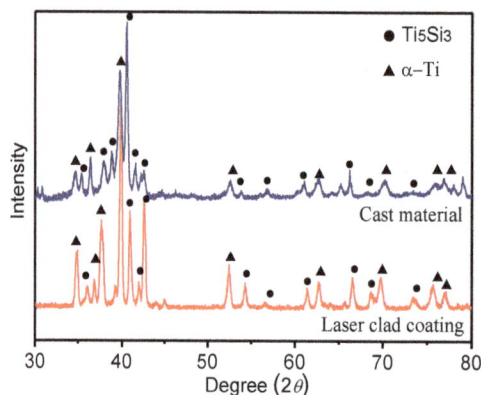

Figure 4. XRD patterns of the Ti–20%Si alloys.

3.2. Hardness and Fracture Resistance

Figure 5 presents the microhardness profile across a laser-cladded specimen, in which the hardness of the coating reached 672 $HV_{0.5}$. This hardness value is double that of the TC21 substrate (322 $HV_{0.5}$) and was approximately 30% higher than that of the metal mold cast material (504 $HV_{0.5}$). The hardness values of the remelted and HAZ zones were 389 $HV_{0.5}$ and 352 $HV_{0.5}$, respectively, which are actually higher than that of the substrate. This could be due to the refinement of the microstructure. Further, the high hardness obtained for the coating was a consequence of the rapid solidification rate in laser cladding (10^3–10^5 °C/s), which is much higher than that normally obtained in the arc melting-metal mold process (10^2 °C/s) [10,11]. Rapid solidification results in a much refined and well-distributed Ti_5Si_3 hard phase in the eutectic microstructure.

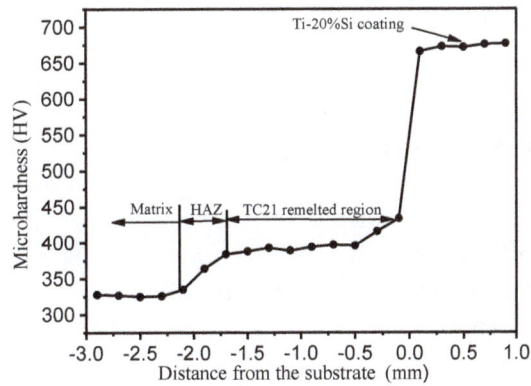

Figure 5. Hardness profile of a laser-cladded specimen.

Figure 6 shows the results of the indentation cracking test for the metal cast and laser-cladded materials. For the former, many microcracks developed at the tip of the indentation in the primary Ti_5Si_3 particles (Figure 6a,b), as well as across the coarse eutectic phase, within and close to the indentation (Figure 6c). It is believed that the stress concentration at the corner of the indentation can readily lead to nucleation of the cracks in the coarse primary Ti_5Si_3-phase. This indicates the brittle nature of coarse Ti_5Si_3 particles; this is consistent with findings of previous studies that the fracture toughness of Ti_5Si_3 particles decreased from 5 to 2.1 MPa·m$^{1/2}$ when the particle size increased from 6 μm to 20–50 μm [12–14].

On the other hand, no macrocracks, as were found in the cast material, were observed at the corner of the indentation of the laser-cladded material (Figure 6d), suggesting that the absence of coarse primary Ti_5Si_3 particles and the refined submicron size of the pseudo-eutectic Ti_5Si_3 phase effectively suppress nucleation of the cracks. This could improve the toughness of the hypereutectic Ti–20%Si alloy. Figure 6e shows a higher magnification image of the marked area in Figure 6d, where a microcrack developed along a diagonal edge of the indentation where a high concentration of stresses is present. However, the crack did not cut through the fine Ti_5Si_3 phase, but propagated along its interface, which resulted in a tortuous crack propagation path. It is believed that this kind of crack growth phenomenon can give rise to a higher toughness for the laser-cladded material.

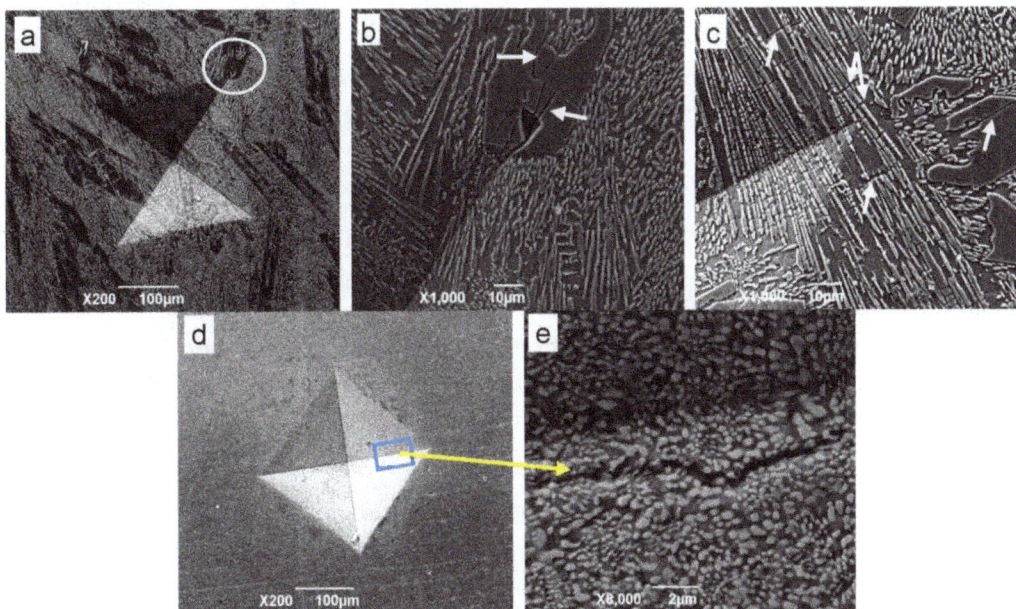

Figure 6. Cracks developed after the indentation test: (**a,b**) at the tip of the indentation, and (**c**) inside and around the indentation (cast material); (**d,e**) a microcrack developed along an edge (laser clad).

3.3. Corrosion Resistance in the Simulated Body Fluid Solution

Figure 7 presents the potentiodynamic polarization curves of the TC21 substrate and the Ti–20%Si specimen. The free corrosion potential (E_{corr}) and the corrosion current density (I_{corr}) of the coating material were -0.298 V and 0.31 μA·cm^{-2}, respectively, whereas, for the substrate materials, the values were -0.450 V and 0.86 μA·cm^{-2}. These figures show that the corrosion resistance of the former is much higher than that of the latter. The results are consistent with previous research work in which the formation of the Ti–Si compound contributed to the increase of E_{corr}, and a decrease in I_{corr} in Ti alloys corroded in a simulated body fluid [15–17]. In this study, it was also found that the laser-cladded material with a refined pseudo-eutectic microstructure has superior corrosion resistance to the cast material of the same composition. The E_{corr} and I_{corr} of the cast material were -0.327 V and 0.78 μA·cm^{-2}, respectively. When comparing these to those of the laser-cladded material, the corrosion potential of the cladded material was approximately 30 mV more noble than that of the cast material, and its current density was about half that of the cast material.

In addition, the linear polarization results also showed that the coating material exhibited a relatively large passive range when compared to the cast material. For the cast material, the passive range, as indicated by the arrows (from -0.25 to 0.0 V), is 0.25 V, while it is over 0.45 V (from -0.1 to 0.35 V) for the laser-cladded material. The extended passive range indicates that a more stable protective corrosion film formed on the laser clad coating, owing to the fine and well-distributed eutectic Ti$_5$Si$_3$ phase.

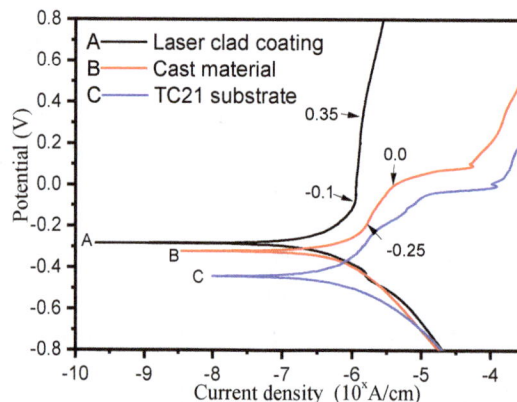

Figure 7. A comparison of the potentiodynamic polarization plots of the TC21 substrate and the Ti–20%Si alloys produced by laser cladding and metal mold casting.

4. Conclusions

(1) A pseudo-eutectic (α-Ti + Ti$_5$Si$_3$) microstructure was obtained in the hypereutectic Ti–20%Si coating produced by laser cladding. Unlike the metal cast material, coarse primary Ti$_5$Si$_3$ crystals are absent.

(2) The fine and well dispersed eutectic Ti$_5$Si$_3$ phase of the laser-cladded material greatly improves the hardness, corrosion resistance, and fracture resistance of the hypereutectic Ti–20%Si when compared with the same alloy prepared by the conventional metal mold casting method.

Acknowledgments: The work described in this paper was substantially supported by the research grant from the Hong Kong Polytechnic University (Project No. G-YM71), the authors also appreciate the support of the Joint Fund of Iron and Steel Research by the National Natural Science Foundation of China under Grant No. U1560105.

Author Contributions: T.M. Yue and Hui Zhang conceived and designed the experiments; Zhonghong Zhang and Hui Zhang performed the experiments; T.M. Yue and Hui Zhang analyzed the data; T.M. Yue and Hui Zhang wrote the paper.

Conflicts of Interest: The authors declare no conflict of interest.

References

1. Wang, X.; Wang, L.; Wang, Q.J.; Wu, Y.D.; Si, J.J.; Hui, X.D. Enhanced mechanical properties and structure stability induced by Si in Ti–8.5Al–1.5Mo alloys. *Mater. Eng. Sci. A* **2016**, *676*, 304–311. [CrossRef]

2. Alhammad, M.; Esmaeili, S.; Toyserkani, E. Surface modification of Ti–6Al–4V alloy using laser-assisted deposition of a Ti–Si compound. *Surf. Coat. Technol.* **2008**, *203*, 1–8. [CrossRef]

3. Hu, Z.H.; Zhan, Y.Z.; She, J. The role of Nd on the microstructural evolution and compressive behavior of Ti–Si alloys. *Mater. Eng. Sci. A* **2013**, *560*, 583–588. [CrossRef]

4. Kishida, K.; Fujiwara, M.; Adachi, H.; Tanaka, K.; Inui, H. Plastic deformation of single crystals of Ti$_5$Si$_3$ with the hexagonal D88 structure. *Acta Mater.* **2010**, *58*, 846–857. [CrossRef]

5. Riley, D.P. Synthesis and characterization of SHS bonded Ti$_5$Si$_3$ on Ti substrates. *Intermetallics* **2006**, *14*, 770–775. [CrossRef]

6. Murray, J.L. *Phase Diagrams of Binary Titanium Alloys*; ASM International: Materials Park, OH, USA, 1987; p. 289.

7. Zhang, Q.; Chen, J.; Zhao, Z.; Tan, H.; Lin, X.; Huang, W.D. Microstructure and anisotropic tensile behavior of laser additive manufactured TC21 titanium alloy. *Mater. Eng. Sci. A.* **2016**, *673*, 204–212. [CrossRef]

8. Tancret, F.; Osterstock, F. The Vickers indentation technique used to evaluate thermal shock resistance of brittle materials. *Scr. Mater.* **1997**, *37*, 443–447. [CrossRef]

9. Herlach, D.M. Non-equilibrium solidification of undercooled metallic melts. *Mater. Eng. Sci. R* **1994**, *12*, 177–272. [CrossRef]

10. Zhang, H.; He, Y.Z.; Pan, Y. Enhanced hardness and fracture toughness of the laser-solidified FeCoNiCrCuTiMoAlSiB$_{0.5}$ high-entropy alloy by martensite strengthening. *Scr. Mater.* **2013**, *69*, 342–345. [CrossRef]

11. Si, S.H.; Zhang, H.; He, Y.Z.; Li, M.X.; Guo, S. Liquid Phase Separation and the Aging Effect on Mechanical and Electrical Properties of Laser Rapidly Solidified Cu$_{100-x}$Cr$_x$ Alloys. *Metals* **2015**, *5*, 2119–2127. [CrossRef]

12. Zhang, L.; Wu, J. Ti$_5$Si$_3$ and Ti$_5$Si$_3$-based alloys: Alloying behavior, microstructure and mechanical property evaluation. *Acta Mater.* **1998**, *46*, 3535–3546. [CrossRef]

13. Frommeyer, G.; Rosenkranz, R.; Smarsly, W. Microstructure and properties of high melting point intermetallic Ti$_5$Si$_3$ and TiSi$_2$ compounds. *Mater. Sci. Eng. A* **1992**, *152*, 288–294.

14. Counihan, P.J.; Crawford, A.; Thadhani, N.N. Influence of dynamic densification on nanostructure formation in Ti$_5$Si$_3$ intermetallic alloy and its bulk properties. *Mater. Sci. Eng. A* **1999**, *267*, 26–35. [CrossRef]

15. Xu, J.; Liu, L.; Li, Z.; Munroe, P.; Xie, Z.H. Niobium addition enhancing the corrosion resistance of nanocrystalline Ti$_5$Si$_3$ coating in H$_2$SO$_4$ solution. *Acta Mater.* **2014**, *63*, 245–260. [CrossRef]

16. Xu, J.; Liu, L.L.; Xie, Z.H.; Munroe, P. Nanocomposite bilayer film for resisting wear and corrosion damage of a Ti–6Al–4V alloy. *Surf. Coat. Technol.* **2012**, *206*, 4156–4165. [CrossRef]

17. Wu, Y.; Wang, A.H.; Zhang, Z.; Zheng, R.R.; Xia, H.B.; Wang, Y.N. Laser alloying of Ti–Si compound coating on Ti–6Al–4V alloy for the improvement of bioactivity. *Appl. Surf. Sci.* **2014**, *305*, 16–23. [CrossRef]

5

The Effect of P on the Microstructure and Melting Temperature of Fe_2SiO_4 in Silicon-Containing Steels Investigated by In Situ Observation

Qing Yuan, Guang Xu *, Mingxing Zhou, Bei He and Haijiang Hu

The State Key Laboratory of Refractories and Metallurgy, Hubei Collaborative Innovation Center for Advanced Steels, Wuhan University of Science and Technology, 947 Heping Avenue, Qingshan District, Wuhan 430081, China; 15994235997@163.com (Q.Y.); kdmingxing@163.com (M.Z.); 15071412662@163.com (B.H.); hhjsunny@sina.com (H.H.)

* Correspondence: xuguang@wust.edu.cn

Academic Editor: Filippo Berto

Abstract: In this study, two silicon-containing steels with different P contents were used, and reheating tests were conducted in an industrial furnace in a hot strip plant. The effect of P on the microstructure and melting temperature of Fe_2SiO_4 in silicon-containing steels was investigated using a backscattered electron (BSE) detector and energy-dispersive spectroscopy (EDS). The melting process of Fe_2SiO_4 was also observed in situ for the two steels with different P contents. The results show that the addition of P could lower the melting point of the eutectic compound Fe_2SiO_4/FeO, which is helpful for descaling the oxide scale. The melting point decreases with the increasing P content, and the melting point of Fe_2SiO_4/FeO can reduce up to 954.2 °C when the content of P reaches 0.115 wt %. Furthermore, P-compounds form in the dispersive particles located in the iron matrix near the interface between the matrix and inner oxide scale when the P content is relatively high. In addition, a method of in situ observation was proposed to study the effect of P on the melting point of Fe_2SiO_4/FeO in silicon-containing steel. The results are of more practical significance for the descaling of oxide scale in silicon-containing steel.

Keywords: in situ observation; Fe_2SiO_4; $Fe_3(PO_4)_2$; melting temperature; dispersive particles

1. Introduction

The eutectic compound Fe_2SiO_4/FeO is the main inducement of red scale in silicon-containing steel. In general, the formation of the red scale is related not only to the content of Fe_2SiO_4, but also to its morphology and distribution. The formation mechanism of red scale has been reported in several studies [1–8]. Silicon reacts with oxygen diffusing into steel and precipitates as SiO_2, which combines with FeO and then forms a separate phase called fayalite (Fe_2SiO_4). The theoretical melting point of Fe_2SiO_4/FeO is approximately 1173 °C and liquid Fe_2SiO_4 irregularly penetrates into FeO and the matrix. It is difficult to completely wipe off the FeO layer after descaling due to the very high strength of the eutectic compound Fe_2SiO_4/FeO. Following the cooling process, the remaining FeO scale is oxidized into red Fe_2O_3.

To eliminate the red scale defect, many studies on Fe_2SiO_4 of silicon-containing steels have been conducted. Suarez et al. [9] reported that the amount of Fe_2SiO_4 increases with the silicon content and the liquid Fe_2SiO_4/FeO is distributed in the net-like form between the iron matrix and the inner oxide scale when the temperature is higher than the melting temperature of Fe_2SiO_4/FeO, resulting in more red scale. Mouayd et al. [10] reported that the penetrative depth of the eutectic FeO/Fe_2SiO_4 in the scale increases with the silicon content. In the present authors' previous study [11] on the relationship between silicon content and morphology of Fe_2SiO_4, it was shown that Fe_2SiO_4 appears in a net-like form in the innermost layer of oxide scale close to the iron matrix when the silicon content is 1.21 wt %. However, no obvious net-like Fe_2SiO_4 is observed when the silicon content is less than 0.25 wt %. In addition, the effects of chemical compositions on the scale were investigated by Kizu et al. [12]. They found that the blistering of the scale was promoted by increasing C, Mn and P contents and suppressed by increasing the S content at any temperature. Furthermore, Yu et al. [13] took the chemical compositions into account while investigating the tertiary scale characteristics of hot rolled strips. They pointed out that P can be enriched at the interface between the substrate and tertiary scale, which is beneficial to decrease the adhesion of the tertiary scale. However, Si enrichment at the interface between the substrate and tertiary scale increases the adhesion.

It is noteworthy that Fukagawa et al. [14] suggested a new method to decrease the red scale from the perspective of the effect of chemical compositions on the Fe_2SiO_4. They demonstrated that the addition of P lowers the binary eutectic temperature of FeO/Fe_2SiO_4 oxides (1167 °C, close to the theoretical melting point temperature of FeO/Fe_2SiO_4) formed during slab soaking, because the sarcopside ($Fe_3(PO_4)_2$) crystallizes in oxide scale when the content of P reaches a certain value. This leads to the formation of a ternary eutectic $FeO/Fe_2SiO_4/Fe_3(PO_4)_2$ having a low melting point of 890 °C. The appearance of $Fe_3(PO_4)_2$ lowers the binary eutectic temperature of FeO/Fe_2SiO_4. Therefore, the liquid eutectic compound in the scale/steel interface during the descaling improves the hydraulic-descaling-ability. However, in their studies, the measurement of melting points of FeO/Fe_2SiO_4 and other analyses on the oxide scale were based on differential scanning calorimetry (DSC) on the different reagent powders of Fe, Fe_3O_4, SiO_2 and P_2O_5. Meanwhile, the research about the effect of P on the oxide scale is scarce. Therefore, the present study focused on the effect of P on the microstructure and melting temperature of Fe_2SiO_4. First, two low-carbon steels with different P contents were used and heated in an industrial furnace. The chemical constitutions and experimental conditions of oxide scale conformed to the industrial scenario. Secondly, an in situ observation method was used to determine the melting point of FeO/Fe_2SiO_4; the melting process of FeO/Fe_2SiO_4 was completely recorded. The results are of more practical significance and provide a useful reference for the control of red scale defect in silicon-containing steels.

2. Materials and Methods

2.1. Industrial Oxidation Experiment

Tested materials were taken from two silicon-containing steels commercially produced in a hot strip plant (Baosteel, Shanghai, China). The chemical compositions of the two steels with different P contents were determined by a carbon sulfur analyzer (CS-8800) and infrared spectrometer (IS50) and are presented in Table 1. The samples were polished to remove the scale before heating in the furnace. The industrial reheating procedure is presented in Figure 1a. The samples were heated to 1260 °C by a segment heating route and held for 40 min, followed by air cooling to ambient temperature. The heating atmosphere in the furnace contained approximately 2% oxygen, 13% carbon dioxide, 11% water vapor and 74% nitrogen (vol %). After the oxidation experiment, three or more specimens of each steel were cut using a wire-electrode cutting device (HFang, Taizhou, China) whose flow velocity of lubricant was controlled in a relatively lower value for keeping the integrity of oxide scale. Moreover, a cold mounting method was used in the preparation of the samples for microscopic observation. The conventional hot mounting method was not utilized in the present study due to the strong compressive stress during the mounting process which is harmful for the protection of oxide scale. The cold mounting material is composed of 60% acrylic powder and 40% liquid hardener. The cross-sections of mounted samples were ground and polished. Subsequently, the microstructures were observed using a backscattered electron (BSE) detector on a Nova 400 Nano scanning electron microscope (SEM) (Hillsboro, OR, USA), operated at an accelerating voltage of 20 kV. Energy-dispersive spectroscopy (EDS) was applied to analyze the compositions of the oxide scale.

Two acquisition techniques, including point analysis and selected area elemental mapping, were employed. To obtain a better intensity of spectral lines, the acceleration voltage was selected to be a relatively low value as 18 kV. Appropriate long acquisition time was used to ensure sufficient acquisition counts (cts). For example, the acquisition time and counts are 155 s and 272,926, respectively, according to the following result of selected-area elemental mapping.

Table 1. The chemical compositions of the tested steels (wt %).

Steels	C	Si	Mn	P	S	Ni	Als	Ti	Fe
SiP-1	0.069	1.310	1.240	0.020	0.001	0.012	0.037	<0.004	Bal.
SiP-2	0.070	1.300	1.260	0.115	0.002	0.010	0.035	<0.004	Bal.

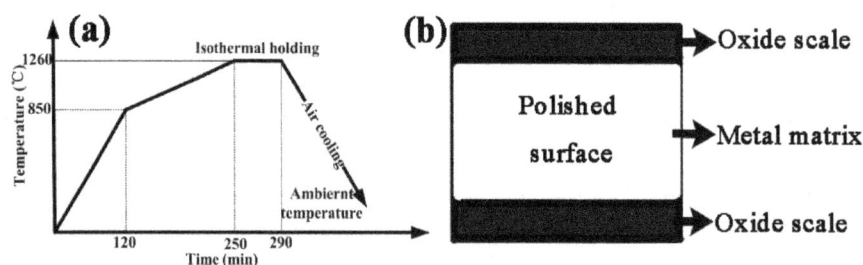

Figure 1. (a) The industrial heating procedure; (b) The schematic diagram of a sample for in situ observation.

2.2. In Situ Observation of the Melting Process of Fe_2SiO_4

The melting process of Fe_2SiO_4 was observed in situ using a VL2000DX laser scanning confocal microscope (LSCM) (Lasertec, Yokohama, Japan). The samples for in situ observations were prepared after the industrial oxidation experiments. The specimens were machined into a cube-shaped structure with dimensions of 4 mm × 4 mm × 4 mm. To maintain a level observation surface, the cross-section of the oxide scale and its opposite sides of the samples were conventionally polished. The polished surfaces were perpendicular to the observation direction. The Fe_2SiO_4 layer still adhered to the polished sample (Figure 1b). The specimen chamber was initially evacuated to 6×10^{-3} Pa before heating, and argon was used to protect the specimens from surface oxidation. The specimens were heated at a fast heating rate of 5 °C·s^{-1} to 800 °C. Furthermore, to clearly observe the melting process of Fe_2SiO_4, the specimens were then heated at a slow heating rate of 0.15 °C·s^{-1} to 1200 °C followed by 0.2 °C·s^{-1} to 1260 °C. Finally, the specimens were maintained at 1260 °C for 2 min and then cooled to an ambient temperature at 2 °C·s^{-1}. The LSCM images were continuously recorded at 15 frames·s^{-1} at a magnification of 500× during the whole heating treatment. A video showing the melting process of Fe_2SiO_4 was simultaneously recorded.

3. Results and Discussion

3.1. Microstructural Characterization of the Oxide Scale

Figure 2 shows the SEM images of the oxide scale in the two tested steels obtained with the backscattered electron (BSE) detector. As depicted in previous studies [15,16], the oxide scale mainly consists of three different layers, i.e., the upper layer Fe_2O_3, middle layer $FeO + Fe_3O_4$, and inner layer FeO/Fe_2SiO_4. The outer layers of Fe_2O_3 and $FeO + Fe_3O_4$ are brittle and easy to wipe off, but the inner layer of FeO/Fe_2SiO_4 strongly adheres to the metal matrix and is retained completely. The dark phase close to the scale–metal interface is a mixture of Fe_2SiO_4 and FeO. The red area in thumbnails located at the bottom left of Figure 2a,b is obtained by an image-processing software, Image-Pro plus 6.0 (Media Cybernetics, Rockville, MD, USA). Initially, the total area of Fe_2SiO_4 in the inner layers is isolated and measured by the auto-discernment function of color aberration in the software. Subsequently, to obtain the areas of Fe_2SiO_4 in unit widths, the total areas are divided by the width of the measured images [11]. The red area represents the distributions of solid Fe_2SiO_4 and the measured area can represent the amount of Fe_2SiO_4. Figure 2c shows the contents and penetration depths of Fe_2SiO_4 in the two specimens. Only the penetration depth of Fe_2SiO_4 into the FeO layer is considered because the formation mechanism of red scale is mainly related with the "anchor effect" between the upper FeO and lower Fe_2SiO_4. The result indicates that the contents of Fe_2SiO_4 in both the two tested steels have no significant difference. In addition, the penetration depths of Fe_2SiO_4 are almost the same. This is because the content and penetration depth of Fe_2SiO_4 are related to the silicon content [17–21], and the silicon contents in two tested steels are basically the same. Furthermore, the reasons for the net-like distribution of Fe_2SiO_4 can be explained by compressive stress at the oxide layer/metal interface [22]. The compressive stress at the oxide layer/metal interface is larger than that at the outer position when the Pilling–Bedworth Ratio (PBR) is more than 1 [23], resulting in the pressure differences between the interface and outside oxidation layer. The pressure difference always exists along with the displacement of the oxidation reaction interface [11,16,22]. The pressure difference in the liquefied Fe_2SiO_4 phase at a temperature higher than the melting point of Fe_2SiO_4 forces a part of Fe_2SiO_4 to permeate into the inner scale. The liquid Fe_2SiO_4 phase distributes along the FeO grain boundaries and the net-like Fe_2SiO_4 phase forms after its solidification. Moreover, the net-like Fe_2SiO_4 phase may also be related to the possibility of Si diffusion through FeO grain boundaries and possible oxidation there.

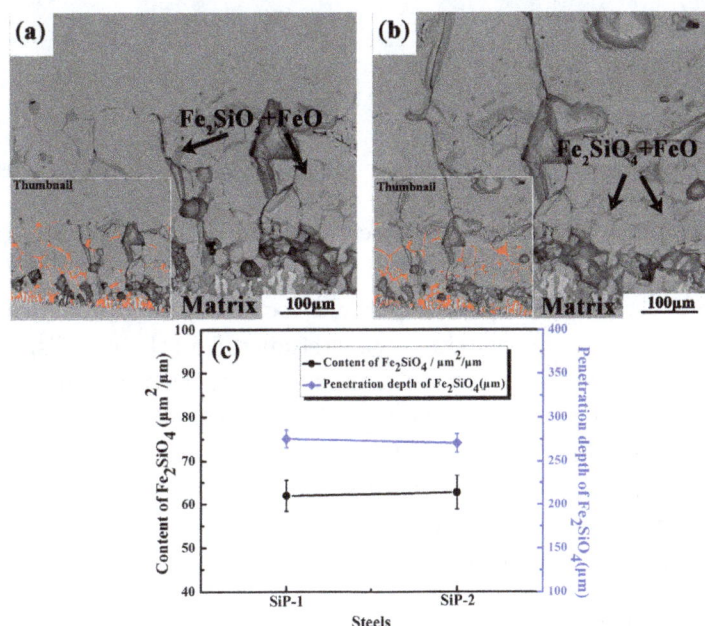

Figure 2. SEM images of the oxide scale obtained with the backscattered electron (BSE) detector. (a) SiP-1; (b) SiP-2; (c) The contents and penetration depths of Fe_2SiO_4 in the two specimens.

3.2. Distribution of Chemical Elements in the Fe_2SiO_4 Phase

The distributions of chemical elements in the Fe_2SiO_4 phase are illustrated in Figure 3. To observe the phase of spectrum such as the area of spectrum 1 in Figure 3a with greater clarity, the magnification of Figure 3b is larger than that of Figure 3a. Figure 3a shows that the oxygen content gradually decreases from the upper layer (spectrum 2) to the inner layer (spectrum 1). The oxygen content is relatively higher than the standard of iron oxide, probably because the oxygen sensitivity of the applied method (EDS) is not very good, resulting in a variable oxygen content. However, the Mn content increases from the outside to inside of the scale as a result of Mn diffusion from the metal to the oxide [24]. In addition, the contents of P in Fe_2SiO_4 of steel SiP-1 are much smaller, because of less P content in the steel SiP-1 (Table 1). The P content changes within a small range when it is quite small due to the limited sensitivity of the detectors in the EDS units. The maximum values for spectra 1 and 2 in Figure 3a are 0.02 and 0.03 (at %), respectively, demonstrating that the concentration of P element is very small. The atomic percentage of Fe:Si:O in spectra 1 and 2 (Figure 3a) is calculated to be approximately 2:1:4. Therefore, the dark gray district (net-like area) in Figure 3a is mainly the Fe_2SiO_4 phase and no $Fe_3(PO_4)_2$ phase is detected. On the other hand, the elemental distribution of spectrum 2 in Figure 3b is similar to that of Figure 3a. There is more P element in the spectrum 1, seen as a bright gray district in Figure 3b. The atomic percentage of Fe:P:O in spectrum 1 is calculated to be approximately 2.5:2.0:7.0, which is near to the standard atomic percentage of the $Fe_3(PO_4)_2$ phase. According to previous studies [12–14], P only exists as the $Fe_3(PO_4)_2$ phase which has an orthorhombic system as the Fe_2SiO_4 phase and aggregates at the interface between the scale and substrate. Therefore, it can be inferred that the phase in spectrum 1 of Figure 3b is mainly $Fe_3(PO_4)_2$. Therefore, there is only the Fe_2SiO_4 phase in the sample with less P content (Figure 3a), whereas there are two phases, Fe_2SiO_4 and $Fe_3(PO_4)_2$, in the sample with high P content (Figure 3b).

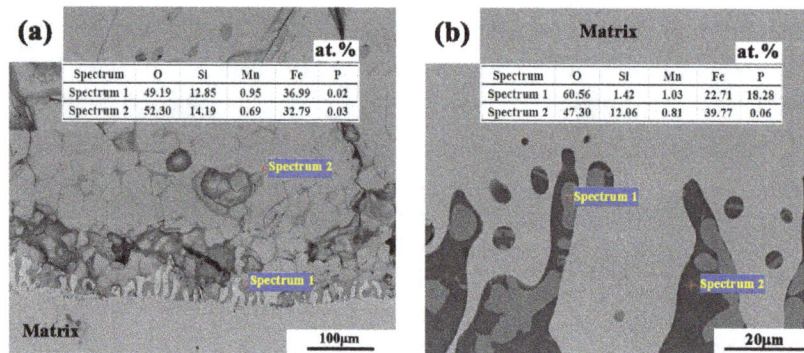

(a)

Spectrum	O	Si	Mn	Fe	P
Spectrum 1	49.19	12.85	0.95	36.99	0.02
Spectrum 2	52.30	14.19	0.69	32.79	0.03

(b)

Spectrum	O	Si	Mn	Fe	P
Spectrum 1	60.56	1.42	1.03	22.71	18.28
Spectrum 2	47.30	12.06	0.81	39.77	0.06

Figure 3. The distributions of chemical elements in the Fe_2SiO_4 phase (**a**): SiP-1; (**b**): SiP-2.

In addition, there were many dispersive particles located in the iron matrix near the interface between the matrix and inner oxide scale. In previous studies [5,6,11], these dispersive particles have been proven to be silicon dioxide in steels containing less P content. Figure 4 shows the results of selected-area elemental mapping on the dispersive particles located in the steel matrix near the interface between the iron matrix and inner oxide scale for steel SiP-2 containing 0.115 wt % P. P element can also be detected in these particles. Figure 5 gives the chemical compositions of the dispersive particles obtained by the point analysis method of EDS. It clarifies that not only silicon dioxide exists in these dispersive particles, but also P-compounds can be found when steel contains a higher P content.

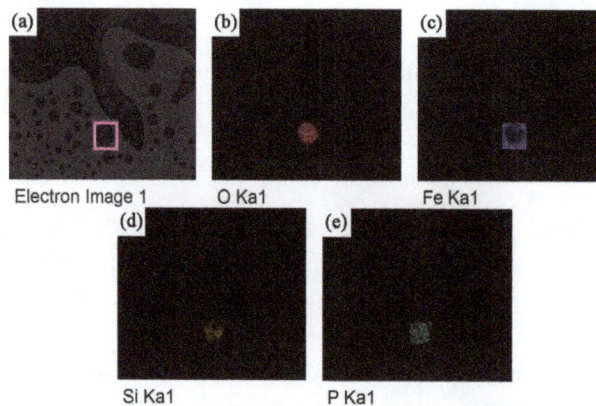

Figure 4. (**a–e**) The selected-area elemental mapping on dispersive particles located near the interface between the matrix and inner oxide scale.

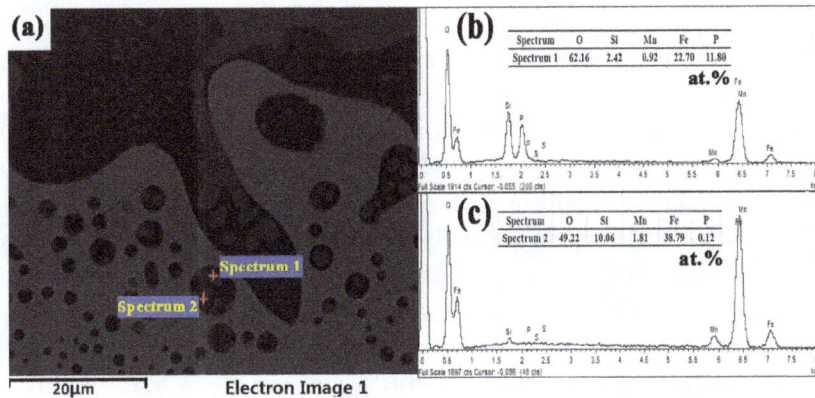

Figure 5. The chemical composition of the dispersive particles obtained by the point analysis method of EDS (**a**) the morphology of dispersive particles; (**b**) the elemental constitution of spectrum 1; (**c**) the elemental constitution of spectrum 2.

3.3. In Situ Observation of the Melting Process of Fe_2SiO_4

Figure 6 presents micrographs showing results from in situ observations of the melting process of Fe_2SiO_4 for steel SiP-1 (1.31 wt % Si; 0.020 wt % P). Figure 6a shows the morphology of oxide scale before melting. Fe_2SiO_4 is distributed in the net-like form. Figure 6b illustrates that the Fe_2SiO_4 phase begins to melt at 1170.6 °C, which is close to the theoretical melting temperature of Fe_2SiO_4 (1173 °C). It demonstrates that the melting point of Fe_2SiO_4 is close to the theoretical melting temperature when P content is relatively low. In earlier studies [1,3,5,9,25], the melting temperature of Fe_2SiO_4 was approximately 1173 °C because the P content was less than 0.051 wt %. Figure 6c shows the morphology of liquid Fe_2SiO_4 during melting; the FeO phase is still in the solid state due to its higher melting point (1370 °C). The end of the melting process of Fe_2SiO_4 is presented in Figure 6d. The time taken for the entire melting process of Fe_2SiO_4 is approximately 136.4 s.

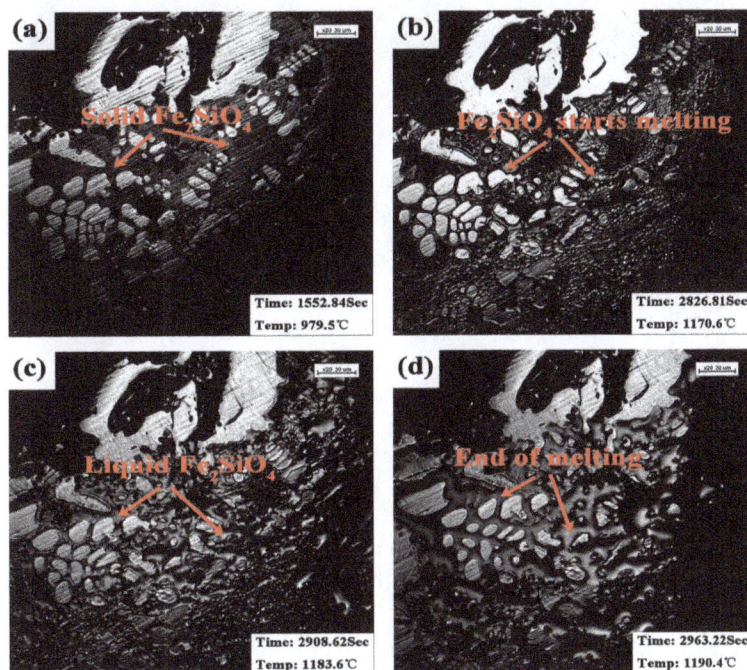

Figure 6. Micrographs showing results from in situ observations of the melting process of Fe_2SiO_4 for steel SiP-1 (**a**) before melting; (**b**) at the start of melting; (**c**) during melting; (**d**) at the end of melting.

The micrographs showing results from in situ observations of the melting process of Fe_2SiO_4 for steel SiP-2 (1.30 wt % Si; 0.115 wt % P) are given in Figure 7. Figure 7a shows the morphology of oxide scale before melting, and Figure 7b shows that the Fe_2SiO_4 phase begins to melt at 954.2 °C. The melting point of Fe_2SiO_4/FeO of steel SiP-2 is obviously lower than that of steel SiP-1 as well as the theoretical melting temperature of Fe_2SiO_4/FeO (1173 °C). It demonstrates that the melting point of Fe_2SiO_4/FeO decreases with the increasing P content. Therefore, this is an effective and practical way to decrease the Fe_2SiO_4 melting temperature with the addition of P element. By this way, the Fe_2SiO_4 phase is in a liquid state and can be removed easily when the descaling temperature is higher than the melting temperature of Fe_2SiO_4. Figure 7c illustrates the morphology of liquid Fe_2SiO_4 during melting. The end of the melting process of Fe_2SiO_4 is shown in Figure 7d. The time taken for the entire melting process of Fe_2SiO_4 is approximately 1458.5 s.

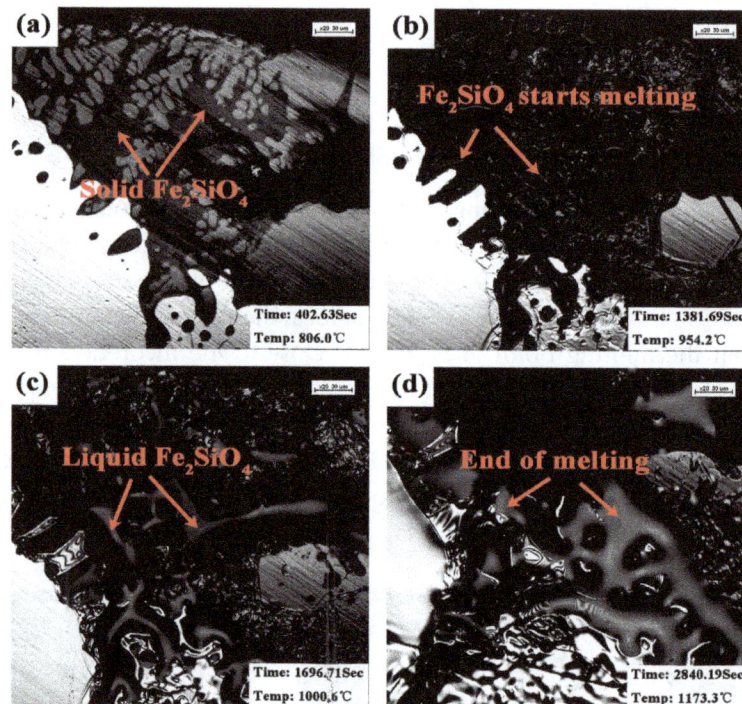

Figure 7. Micrographs showing results from in situ observations of the melting process of Fe_2SiO_4 for steel SiP-2 (**a**) before melting; (**b**) at the start of melting; (**c**) during melting; (**d**) at the end of melting.

Figure 8 shows the image of the FeO-SiO_2-P_2O_5 ternary eutectic system, which is reported by Fukagawa et al. [14]. This FeO-SiO_2-P_2O_5 ternary eutectic system is a regional image of the whole FeO-SiO_2-P_2O_5 ternary eutectic system which is obtained by the combination of three different binary eutectic systems. There are three temperature scales in the whole ternary eutectic system. For easier understanding of the ternary eutectic system, a regional image without temperature scales is used in the present study. Moreover, the three characteristic temperatures are given in order to show the relationship between the temperature and different phases. There are three different stages during the cooling process. First, the FeO phase crystallizes when the temperature is relatively high (point M_I (~1200 °C)). Then, the Fe_2SiO_4 phase starts crystallizing when the temperature decreases to point E_{II} (binary eutectic point (~954–1173 °C)). At last, the $Fe_3(PO_4)_2$ phase crystallizes when the temperature decreases to the point E_{III} (ternary eutectic (~890 °C)). Therefore, it can be concluded that the existence of $Fe_3(PO_4)_2$ phase can lower the binary eutectic point of compound Fe_2SiO_4/FeO. The results of this study also demonstrate that the addition of P could lower the melting point of the eutectic compound Fe_2SiO_4/FeO (Steel SiP-2), and a ternary eutectic compound $Fe_3(PO_4)_2$/Fe_2SiO_4/FeO forms upon the Fe_2SiO_4 phase area. It is easy to wipe off the red scale when the descaling temperature is higher than the melting point of Fe_2SiO_4/FeO. The melting point of the eutectic compound Fe_2SiO_4/FeO decreases with the increasing P content, and it could reduce to 954.2 °C when the P content is 0.115 wt %. In addition, the $Fe_3(PO_4)_2$ phase is detected in the net-like Fe_2SiO_4 area, and P-compounds are in the dispersive particles located in the iron matrix near the interface between the matrix and inner oxide scale. Furthermore, the silicon-containing steels in the present study are commercially produced and their oxide scale is formed during the reheating process in the industrial production. Therefore, the results represent oxidation behavior of steel and oxide microstructures in actual production and can provide some valuable reference from the viewpoint of industrial application.

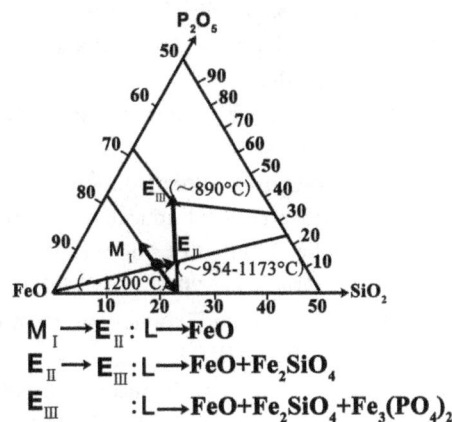

Figure 8. FeO-SiO_2-P_2O_5 ternary eutectic system.

4. Conclusions

In this study, two silicon-containing steels with different P contents were tested. High temperature oxidation tests were conducted in an industrial furnace in a hot strip plant. The effect of P on the microstructure and melting temperature of Fe_2SiO_4 in the two silicon-containing steel were investigated using a backscattered electron (BSE) detector on a Nova 400 Nano scanning electron microscope (SEM) and energy-dispersive spectroscopy (EDS). In situ observations of the melting process of Fe_2SiO_4 were conducted. The results indicate that the $Fe_3(PO_4)_2$ phase forms mainly at the interface between the iron matrix and inner oxide layer with the addition of P which could lower the melting point of the eutectic compound Fe_2SiO_4/FeO and the melting point of Fe_2SiO_4/FeO decreases with the increasing P content. The melting point of Fe_2SiO_4/FeO could reduce to 954.2 °C when the content of P is 0.115 wt %. In addition, when the P content is relatively high, P-compounds are formed in the dispersive particles located in the iron matrix near the interface between the matrix and inner oxide scale. Furthermore, the effect of P on the microstructure and melting temperature of Fe_2SiO_4/FeO in both the silicon-containing steels was investigated by in situ observation. This method may offer new insights for the research on metal oxidation and other fields of corrosion science.

Acknowledgments: The authors gratefully acknowledge the financial support from the National Natural Science Foundation of China (NSFC) (No. 51274154), the State Key Laboratory of Development and Application Technology of Automotive Steels (Baosteel Group).

Author Contributions: Guang Xu, supervisor, conceived and designed the experiments; Qing Yuan, doctoral student, conducted experiments, analyzed the data and wrote the paper; Mingxing Zhou, doctoral students, conducted experiments; Bei He, master students, conducted experiments; Haijiang Hu, doctoral students, conducted experiments.

Conflicts of Interest: The authors declare no conflict of interest.

References

1. Okada, H.; Fukagawa, T.; Ishihara, H. Prevention of red scale formation during hot rolling of steels. *ISIJ Int.* **1995**, *35*, 886–891. [CrossRef]

2.   Okada, H.; Fukagawa, T.; Ishihara, H.; Okamoto, A.; Azuma, M.; Matsuda, Y. Effects of hot-rolling and descaling condition on red scale defects formation. *ISIJ Int.* **1994**, *80*, 849–854.

3.   Fukagawa, T.; Okada, H.; Maeharara, Y. Mechanical of red scale defect formation in Si-added hot-rolled steels. *ISIJ Int.* **1994**, *11*, 906–911. [CrossRef]

4.   Fukagawa, T.; Okada, H.; Maehara, Y.; Fujikawa, H. Effect of small amount of Ni on Hydraulic-descaling-ability in Si-added Hot-rolled Steel Sheets. *ISIJ Int.* **1996**, *82*, 63–68.

5.   Liu, X.J.; Cao, G.M.; He, Y.Q.; Jia, T.; Liu, Z.Y. Effect of temperature on scale morphology of Fe-1.5Si alloy. *J. Iron Steel Res. Int.* **2013**, *20*, 73–78. [CrossRef]

6.   Yang, Y.L.; Yang, C.H.; Lin, S.N.; Chen, C.H.; Tsai, W.T. Effects of Si and its content on the scale formation on hot-rolled steel strips. *Mater. Chem. Phys.* **2008**, *112*, 566–571. [CrossRef]

7.   Cao, G.M.; Liu, X.J.; Sun, B.; Liu, Z.Y. Morphology of oxide scale and oxidation kinetics of low carbon steel. *J. Iron Steel Res. Int.* **2014**, *21*, 335–341. [CrossRef]

8.   Zhou, M.X.; Xu, G.; Hu, H.J.; Yuan, Q.; Tian, J.Y. The morphologies of different types of Fe_2SiO_4–FeO in Si-containing steel. *Metals.* **2017**, *7*, 8–15. [CrossRef]

9.   Suarez, L.; Schneider, J.; Houbaert, Y. High-temperature oxidation of Fe-Si alloys in the temperature range 900–1250 °C. In *Defect and Diffusion Forum*; Trans Tech Publications: Zurich, Switzerland, 2008; pp. 661–666.

10.   Mouayd, A.A.; Koltsov, E.; Sutter, B.; Tribollet, B. Effect of silicon content in steel and oxidation temperature on scale growth and morphology. *Mater. Chem. Phys.* **2014**, *143*, 996–1004. [CrossRef]

11.   Yuan, Q.; Xu, G.; Zhou, M.X.; He, B. The effect of the Si content on the morphology and amount of Fe_2SiO_4 in low carbon steels. *Metals* **2016**, *6*, 94–102. [CrossRef]

12.   Taro, K.; Yasunobu, N.; Toru, I.; Yoshihiro, H. Effects of chemical composition and oxidation temperature on the adhesion of scale in plain carbon steels. *ISIJ Int.* **2001**, *41*, 1494–1501.

13.   Yu, Y.; Wang, C.; Wang, L.; Chen, J.; Hui, Y.J.; Sun, C.K. Combination effect of Si and P on tertiary scale characteristic of hot rolled strip. *J. Iron Steel Res. Int.* **2015**, *22*, 232–237. [CrossRef]

14.   Fukagawa, T.; Okada, H.; Fujikawa, H. Effect of P on hydraulic-descaling-ability in Si-added hot-rolled steel sheets. *Steels* **1997**, *83*, 305–311. (In Japanese)

15.   Yuan, Q.; Xu, G.; Zhou, M.X.; He, B. New insights into the effects of silicon content on the oxidation process in silicon-containing steels. *Int. J. Miner. Metall. Mater.* **2016**, *23*, 1–8. [CrossRef]

16.   He, B.; Xu, G.; Zhou, M.X.; Yuan, Q. Effect of oxidation temperature on the oxidation process of silicon-containing steel. *Metals* **2016**, *6*, 137–145. [CrossRef]

17.   Kusabiraki, K.; Watanabe, R.; Ikehata, T. High-temperature oxidation behavior and scale morphology of Si-containing steels. *ISIJ Int.* **2007**, *9*, 1329–1334. [CrossRef]

18.   Suarez, L.; Schneider, J.; Houbaert, Y. *Effect of Si on high-temperature oxidation of steel during hot rolling In Defect and Diffusion Forum*; Trans Tech Publications: Zurich, Switzerland, 2008; pp. 655–660.

19.   Chen, R.Y.; Yuen, W.Y.D. Review of the high-temperature oxidation of iron and carbon steels in air or oxygen. *Oxid. Met.* **2003**, *59*, 433–468. [CrossRef]

20.   Taniguchi, S.; Yamamoto, K.; Megumi, D.; Shibata, T. Characteristics of scale/substrate interface area of Si-containing low-carbon steels at high temperatures. *Mater. Sci. Eng. A* **2001**, *1*, 250–257. [CrossRef]

21.   Li, S.J.; Liu, Y.B.; Zhang, W.; Sun, Q.S.; Wang, L.P. Effects of silicon on spring steel oxidation rate under 2% residual oxygen atmosphere. *J. Iron Steel Res. Int.* **2015**, *5*, 55–60. (In Chinese)

22.   Garnaud, G.; Rapp, R.A. Thickness of the oxide layers formed during the oxidation of iron. *Oxid. Met.* **1977**, *11*, 193–198. [CrossRef]

23.   Staettle, R.W.; Fontana, M.G. *Advances in Corrosion Science and Technology*; Springer-Verlag: New York, NY, USA, 1974; pp. 239–356.

24.   Zhang, Y.S.; Rapp, R.A. Solubilities of CeO_2, HfO_2 and Y_2O_3 in Fused Na_2SO_4-30 mol% $NaVO_3$ and CeO_2 in Pure Na_2SO_4 at 900 °C. *Corrosion* **1987**, *43*, 348–352. [CrossRef]

25.   Asai, T.; Soshiroda, T.; MiyaharaI, M. Influence of Ni impurity in steel on the removability of primary scale in hydraulic descaling. *ISIJ Int.* **1997**, *37*, 272–277. [CrossRef]

6

Mathematical Modeling of the Concentrated Energy Flow Effect on Metallic Materials

Sergey Konovalov [1,2,3], **Xizhang Chen** [2,*], **Vladimir Sarychev** [1], **Sergey Nevskii** [1], **Victor Gromov** [1] and **Milan Trtica** [2,4]

[1]   Physics Department, Siberian State Industrial University, 42 Kirova str., Novokuznetsk 654007, Russia; konovserg@mail.ru (S.K.); sarychev_vd@mail.ru (V.S.); nevskiy.sergei@yandex.ru (S.N.); gromov@physics.sibsiu.ru (V.G.)

[2]   School of Mechanical and Electrical Engineering, Wenzhou University, Wenzhou 325035, China

[3]   Department of Metals Technology and Aviation Materials, Samara National Research University, Moskovskoye Shosse 34, Samara 443086, Russia

[4]   VINCA Institute of Nuclear Sciences, University of Belgrade, P.O. Box 522, Belgrade 11001, Serbia; etrtica@vin.bg.ac.rs

*   Correspondence: kernel.chen@gmail.com or chenxizhang@wzu.edu.cn

Academic Editor: Hugo F. Lopez

Abstract: Numerous processes take place in materials under the action of concentrated energy flows. The most important ones include heating together with the temperature misdistribution throughout the depth, probable vaporization on the surface layer, melting to a definite depth, and hydrodynamic flotation; generation of thermo-elastic waves; dissolution of heterogeneous matrix particles; and formation of nanolayers. The heat-based model is presented in an enthalpy statement involving changes in the boundary conditions, which makes it possible to consider melting and vaporization on the material surface. As a result, a linear dependence of penetration depth vs. energy density has been derived. The model of thermo-elastic wave generation is based on the system of equations on the uncoupled one-dimensional problem of dynamic thermo-elasticity for a layer with the finite thickness. This problem was solved analytically by the symbolic method. It has been revealed for the first time that the generated stress pulse comprises tension and compression zones, which are caused by increases and decreases in temperature on the boundary. The dissolution of alloying elements is modeled on the example of a titanium-carbon system in the process of electron beam action. The mathematical model is proposed to describe it, and a procedure is suggested to solve the problem of carbon distribution in titanium carbide and liquid titanium-carbide solution in terms of the state diagram and temperature changes caused by phase transitions. Carbon concentration vs. spatial values were calculated for various points of time at diverse initial temperatures of the cell. The dependence of carbon particle dissolution on initial temperature and radius of the particle were derived. A hydrodynamic model based on the evolution of Kelvin-Helmholtz instability in shear viscous flows has been proposed to specify the formation of nanostructures in materials subjected to the action of concentrated energy flows. It has been pointed out for the first time that, for certain parameters of the problem, that there are two micro- and nanoscale peaks in the relation of the decrement to the wavelength of the interface disturbance.

Keywords: electron-beam treatment; Kelvin-Helmholtz instability; thermoelastic waves; nanostructures

1. Introduction

At present the flows of concentrated energy find wide application in processing surfaces of various materials [1–5]. For instance, low-energy, high-current electron beams with durations up to milliseconds, which is used for strengthening the surface layer [2], smoothing the surface relief [3,5], and forming micro- and nano-crystalline grains [4–6]. The advantages of electron beam treatment mainly include [7]: extensive control over modes and fine adjustment of heat flow; suitable for both metals and non-metals; high efficiency coefficient (up to 98%); and automation of process. Combined treatment is a promising procedure of surface modification. This implies the use of heterogeneous plasma flows generated by electric explosion of conductors with the subsequent exposure to electron beams [8–10]. The surfaces modified by above method have quite unique characteristics, for example, the wear resistance properties after surfaces modification increase by 7.5 times [8–10]. Wide-scale research is required to determine the relationship of processing parameters and properties. To reduce the time needed for this kind of research some mechanisms are to be developed, as well as some mathematical models are necessary to specify the processes of combined treatment: heat transfer to the target, generation of stress waves, structural and phase transformations in liquid and solid states, and hydrodynamic flows in the molten layer.

The first mathematical models describing heat and thermo-elastic processes under the action of electron beams were proposed almost 20 years ago [11,12]. Leyvi et al. [13] analyzed the present day experimental and theoretical studies focused on modification of engineering materials by intense flows of charged particles and plasma. They comprise: kinetic equation to simulate the interaction of concentrated energy flows with the material, continuum mechanics equations (variation of mass, pulse, and energy) for target material and the wide-range equation of state.

Researchers [14–16] took a close look at the works linked with simulations of heat processes with evaporation. Models and software complexes are developed, which can provide comprehensive data for working out certain technologies. The problems of heat impact on laser emission target were investigated [14], taking into account evaporation and vapor dissemination. It is a complex problem of gas-dynamic flow in the flame of evaporated substances and thermal processes. However, the study was unable to demonstrate how to monitor the melt front. Bleykher et al. [15] modeled the field of temperatures and evaporation rates under magnetron action on the target. The field of temperatures and evaporation rates are calculated when finding solution of boundary problem, thermal conductivity equation, and boundary conditions, with the evaporation rate considered. The rate of evaporation here is determined relying on Hertz-Knudsen equation, although processes occurring in the target were not analyzed. The authors [16] dealt with the effect of charged particles on the target and suggested a model, which rests on kinetic equation for particles, continuum mechanics equation, and the wide-range equation of state. However, no correlation between penetration depth and beam parameters was found. Therefore, it makes sense to use a simplified model [17], which provides a quite correct description of temperature fields in the target under the action of electron beams and allows the determination of the penetration depth.

Some authors [13,18,19] have attempted to develop models for the stress state determination under concentrated energy flows. They referred to complex behavioral models of the continuum, and dynamic problems for layers with finite thickness have a computational solution. It is possible to model the definite dissemination of stress waves, but wave reflection can be studied in detail only on the basis of the analytical solution of the simplified model of disconnected thermo-elasticity [20].

Combined treatment implies, first, the injection of carbon powder into molten titanium by means of the heterogeneous plasma flow, and second, the impact of electron beam on the heterogeneous titanium-carbon particles system. Under the action of electron beams on the heterogeneous titanium mixture, melting occurs, but carbon particles are in a solid state, as the fusion temperature of carbon is 3550 °C. Whether carbon particles dissolve over the time of the electron beam impact (100 μm), a homogenous layer of titanium carbide will be formed. As a consequence, a mathematical model, which allows the determination of the electron beam parameters, is required to further synthesize a homogenous layer of carbide. The problem of iso-thermal dissolution of a new phase spherical center within the diffusion model has been solved [21], and was confined to the Stefan problem for the diffusion equation in the zone, on the flexible boundary of which the conditions of mass balance are satisfied. When carbon reacts with titanium, the temperature changes as heating and cooling occur in the process of electron beam impact and exothermic reaction. Diffusion and temperature problems are solved simultaneously in [22,23] in accordance with diverse spatial and temporal scales. This approach is quite complex since it necessitates consideration of heat release in the thermo-conductivity equation. The method stated in combustion theory [24], and applied further in [25,26], was used for consideration of the changing temperature. It relies on the heat-balance equation for a cell on the whole; that is, there is no change in temperature along the spatial value. The first phase of experimental interaction in the system Ti-C and physical and chemical processes taking place in carbon fusion have been investigated [27], and the double-wave structure of wave transformation was proved experimentally. A mathematical model of carbon distribution in titanium carbide and liquid titanium-carbon solution was suggested [28], which, taking into account the diagram of states and phase transitions, result from temperature changes. The authors of this work suggested a framework implying the determination of such process characteristics, which allow the understanding of the controlled process parameters (e.g., energy density of electron beam, pulse duration, pulse repetition, etc.) on the outcomes of experiments. For instance, whether a particle of carbon dissolves in liquid titanium exposed to electron beam. In this case it is quite reasonable to address only to diffusion problem for homogenous temperature field, which can vary over time due to heat of chemical reactions and heat removal. Solution of this problem helps to realize the reason for generation of homogenous titanium layers exposed to heterogeneous plasma flows and processed further by low-energy high-current electron beam [8,9].

Numerous publications emphasize the interest of researchers in development of mathematical models describing nano-structures under the action of concentrated energy flows [29–39]. These models are based mainly on computer simulation of processes, going with the development of nano-structural states. The researchers [29,30] speculated that nano-structural state evolves together with the growing hydrodynamic instability in shear-related motion of liquid layers, one of them is viscous. In this case the dispersion equation is considered as a quartic algebraic equation with complex variables. It is rather complicated to analyze the parameters, so decrement vs. wavenumber was assessed numerically. The wavelength agreeing with the peak of the decrement is the most instable; therefore, this wave preserves its length, making its observation possible in the experiment. Two peaks were revealed in decrement vs. wavelength correlation. Approximation of short waves is used in [39], and a simplified dispersion equation is derived. The values of wavelengths were determined for two peaks on the base of this equation. This fact is of significant importance for parameterization of experimental outcomes. This work is focused on derivation and analysis of the dispersion equation for finite layers.

2. Mathematical Models of Electron-Beam Effect on Metallic Materials

2.1. Thermal Model

Let us consider the impact of the electron beam with the surface energy density E_S on a flat plate with the thickness l. As we are interested in the distribution of temperature over the depth of the sample, we confine ourselves to solving the one-dimensional problem of thermal conductivity. The x axis is directed inside the plate. The surface $x = 0$ is affected by the electron current during the period of time t_0, and there is no heat flow on the back side of the plate $x = l$. The equation of thermal conductivity is written in an enthalpy statement in order to take into consideration the phase transformations on the boundary:

$$\frac{\partial H}{\partial t} = \frac{\partial}{\partial x}\left(\lambda \frac{\partial T}{\partial x}\right)$$

(1)

where enthalpy is:

$$H(T) = \int_0^T C(T)\rho(T)dT, \quad C(T)\rho(T) = \begin{cases} C_S\rho_S, \ T < T_S \\ L_L\rho_L/\Delta T_1, \ T_S < T < T_L \\ C_L\rho_L, \ T_L < T < T_V \\ L_V\rho_L/\Delta T_2, \ T_V < T \end{cases} \tag{2}$$

where C is specific heat; ρ—density; T—temperature; t—time; λ—thermal conductivity, for which:

$$\lambda = \begin{cases} \lambda_S, \ T < T_S \\ \lambda_S + \frac{\lambda_L - \lambda_S}{\Delta T_1}(T - T_L), \ T_S < T < T_L \\ \lambda_L, \ T_L < T < T_V \\ \lambda_L + \frac{\lambda_V - \lambda_L}{\Delta T_2}(T - T_V), \ T_V < T \end{cases} \tag{3}$$

Here indices S, L, and V refer to solid, liquid, and gas phases; L_L and L_V are latent heats of melting and vaporization. We suppose $\Delta T_1 = 100$ K; $\Delta T_2 = 100$ K. The solution of Equation (1) requires setting initial and boundary conditions. The initial condition is dependent on the constant temperature, which equals to 300 K.

$$T(0,x) = T_0, \ 0 < x < l \tag{4}$$

The boundary condition at $x = l$ is written as follows:

$$\frac{\partial T(t,l)}{\partial x} = 0, \ t > 0 \tag{5}$$

Mixed boundary conditions at $x = 0$ are used to take into account the effect of vaporization, and a heat flow is set till the start of vaporization:

$$\lambda\frac{\partial T(0,t)}{\partial x} = -\frac{E_S}{t_0}, \ 0 < t < t_1 \tag{6}$$

where t_1—time when boundary conditions is changed, and it is determined on assumption that the enthalpy on the surface is similar to the vaporization enthalpy at normal pressure, i.e., it results from the equation $H(0,t_1) = H_V$. Over the period from t_1 to the completion of the electron beam impact (t_0) energy is supplied to the system and consumed for vaporization and emission of the evaporated substance. At these points the enthalpy on the vapor-liquid boundary is constant and equals to the heat of vaporization H_V. This boundary moves into the zone of the plate (up to 1 μm), but we disregard this shift and suppose the boundary condition is as follows:

$$H(0,t) = H_V, \ t_1 < t < t_0 \tag{7}$$

As soon as the pulse impact is over, vapor contacts with the plate surface; since the time of vapor emission is several hundreds of microseconds, the temperature of the sample surface is equated to that of vaporization:

$$T(0,t) = T_V, \ t_0 < t < t_2 \tag{8}$$

After completion of the vapor contact with the surface, its temperature can vary in the absence of heat exchange, therefore, the condition of heat exchange is stated as:

$$\frac{\partial T(t,0)}{\partial x} = 0, \ t > t_2 \tag{9}$$

Thermal and physical characteristics of commercially pure titanium [40–42]: $T_S = 1998$ K; $L_m = 304$ kJ/kg; $T_V = 3560$ K; $L_V = 8900$ kJ/kg; $\rho_S = 4.5 \times 10^3$ kg/m³; $\rho_L = 4.1 \times 10^3$ kg/m³; $C_S = 318$ J/(kg·K); $C_L = 400$ J/(kg·K); $\lambda_S = 25$ W/(m·K); $\lambda_L = 10$ W/(m·K). The problem was solved numerically (1)–(9) for λ_L, varying in the range from 20–40 W/(m·K). The coefficient of thermal conductivity of the liquid titanium was selected on assumption that the estimated and experimental values of the penetration depth are concurrent.

To calculate time t_2 we integrate the Equation (1) according to x from zero to l:

$$\frac{\partial W}{\partial t} = -\lambda \frac{\partial T(t,0)}{\partial x}, \quad W(t) = \int_0^l H(t,x)dx \tag{10}$$

The Equation (10) we integrate according to time from zero to t_2:

$$W(t_2) - W(0) = -\int_0^{t_1} \lambda \frac{\partial T(t,0)}{\partial x}dt - \int_{t_1}^{t_0} \lambda \frac{\partial T(t,0)}{\partial x}dt - \int_{t_0}^{t_2} \lambda \frac{\partial T(t,0)}{\partial x}dt \tag{11}$$

We obtain an expression for each summand in the right part of Equation (11):

$$-\int_0^{t_1} \lambda \frac{\partial T(t,0)}{\partial x}dt = \frac{E_S t_1}{t_0}$$

$$-\int_{t_1}^{t_0} \lambda \frac{\partial T(t,0)}{\partial x}dt = -(T_V - T_2)\int_{t_1}^{t_0} \frac{\lambda_V}{x_V}dt \tag{12}$$

$$-\int_{t_0}^{t_2} \lambda \frac{\partial T(t,0)}{\partial x}dt = -(T_1 - T_V)\int_{t_0}^{t_2} \frac{\lambda_L}{x_L}dt$$

Here $x_V(t)$ and $x_L(t)$ are computable dependencies of vaporization front coordinates and melting on time, which are given in Figure 1.

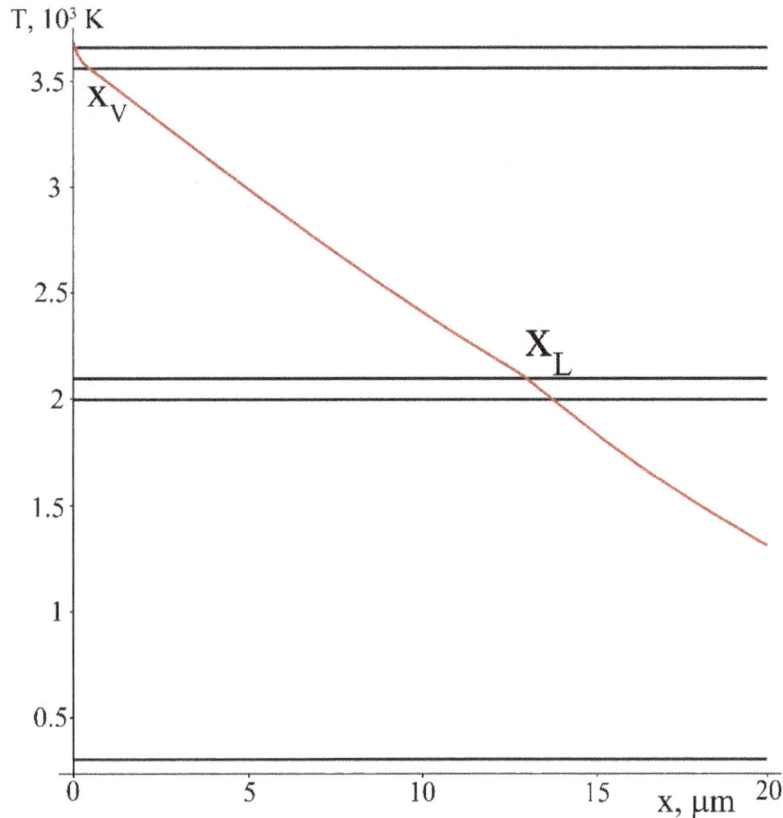

Figure 1. The temperature dependence of coordinates on time $t_1 = 31$ μs. Here are $x_V(t)$ and $x_L(t)$. The calculations are carried out for $E_S = 60$ J/sm², $t_0 = 100$ μs.

The system is supplied with the energy per a unit of area $W(t_2) - W(0) = E_S$; taking into consideration Equations (12). Equation (11) is written as follows:

$$E_S\left(1 - \frac{t_1}{t_0}\right) + (T_V - T_2)\int_{t_1}^{t_0}\frac{\lambda_V}{x_V}dt + (T_1 - T_V)\int_{t_0}^{t_2}\frac{\lambda_L}{x_L}dt = 0 \tag{13}$$

On the basis of Equation (13) we determine t_2. The temperature was calculated by means of the implicit first-order difference scheme with respect to time and by the second-order scheme with respect to space. The time increment is 1 µs, and the space increment is 0.1 µm. The obtained algebraic system of equations was solved by the double-sweep method. The thickness of the plate l is set to 600 µm; this quite large thickness can provide results similar to those of an infinitely thick plate over the time, up to 2000 µs [11–17,40]. The depths of penetration in diverse process conditions of electron beam treatment are given in Table 1. It is seen that estimated and experimental values are quite identical. In accordance to the data in Table 1 the dependence of penetration depth on the surface energy density is linear.

Table 1. The dependence of penetration depth on energy density.

E_S, J/cm^2	t_0, µs	h_{melt}, µm	
		Experimental	Estimated
18	50	6	7
21	50	8	8
25	50	13	15
45	100	31.1	31.2
50	100	36	35.0
60	100	50.1	50

The processes of vaporization are quite irrelevant at energy density ranging from 10–30 J/cm^2 because temperature on the surface does not reach that of vaporization. As soon as pulse action is completed there is a sharp drop in temperature on the surface, therefore, the penetration depth is small ~10 µm. At $Es > 30$ J/cm^2 temperature on the surface reaches the temperature of vaporization. A layer of vapor generated over this process creates a heat buffer, as stated above. As a result, temperature on the surface is quite constant from time $T = T_V$ to the time when pulse action is completed. This period of time is dependent on energy density. The higher energy density is the longer temperature on the surface is constant. As a consequence, penetration depth increases. The derived dependence of penetration depth on energy density can provide a basis for choosing proper modes of low-energy high-current electron beam treatment.

2.2. Model of Thermoelastic Waves Generation

Generation of the thermoelastic waves is a principal problem to be solved for modelling the processes caused by the concentrated energy flows. An attempt to model a thermal situation relying on heat transfer from the inner heat source and generation of thermoelastic waves has been made in [18–20,43], where the bipolarity of the thermoelastic wave was modeled numerically. However, the mechanism of origination, that is, the relationship arising from compression and tension to the zones of heating and cooling was not identified.

In this paper the causes and the mechanism of bipolar thermoelastic wave generation were revealed on the basis of an analytical solution of the thermoelastic problem for the layer with a finite thickness when heating the surface, the stress states close to the back and front surfaces were determined as well.

We consider the problem of thermal stresses for a stress-free flat sample with the thickness, which is subject to the heat flow impact. We set thermal conditions as temperature heating. The axis is perpendicular to the surface of the sample; the other two axes are in the plane of energy flow impact. The plane is referred to as a front surface, and the plane is a back one. We assume that the energy flow is homogenous in the cross-section, therefore, one-dimensional statement of problem is selected, that is, all functions are dependent on the only one spatial variable, where we consider the one-axial stress-strain state with non-zero components of the strain tensors. To state the mathematical problem we use the equations of motion and thermal conductivity, as well as the Duhamel-Neumann relation, the so called dynamic problem of the disconnected thermo-elastisity [31]:

$$\rho\frac{\partial^2 u}{\partial t^2} = \frac{\partial \sigma_x}{\partial x}, \ \frac{1}{\chi}\frac{\partial T}{\partial t} = \frac{\partial^2 T}{\partial x^2}, \ \sigma_x = \rho c^2\frac{\partial u}{\partial x} - \gamma T \tag{14}$$

Here, ρ, χ are the density and thermal diffusivity of the material, $c = \sqrt{(2\mu + \lambda)/\rho}$ is the rate of longitudinal wave propagation, $\gamma = (3\lambda + 2\mu)\alpha_t$ is the coefficient of the thermoelastic coupling, λ, μ are the Lame coefficients, α_t is the coefficient of linear expansion, $u = u(t,x)$ is the component of the displacement vector, and $T = T(t,x)$ is the temperature distribution.

A thermal situation at the front surface under the action of energy flows can be written as temperature heating:

$$T(t,0) = \Psi_0(t) \tag{15}$$

We set $\Psi_0(t)$ as a trapezoid. In this case the surface temperature increases in the range from zero to T_0 over the period of time t_1, it is constant for the period of time t_2, and goes down from T_0 to zero over the period of time t_3. The right part Equation (15) is written as follows:

$$\Psi_0(t) = \tfrac{t}{t_1}(H(t) - H(t-t_1)) + (H(t-t_1) - H(t-t_1-t_2)) - \tfrac{t-t_1-t_2}{t_3}(H(t-t_1-t_2) - H(t-t_1-t_2-t_3)) \tag{16}$$

where $H(t)$ is the Heaviside step function.

In Equation (15) the initial conditions are generalized by taking Equation (16) into consideration, and the following problems are solved for these conditions: at $l_0 = \infty$: Danilevskaya problem—momentary growth of temperature ($t_1 = 0$, $t_2 = \infty$, $t_3 = 0$), the problem with the finite time of temperature increase up to the certain value (t_1—finite, $t_2 = \infty$), and the case of the increase and decline of temperature over the same time ($t_1 = t_3$, $t_2 = 0$) [44].

Without any regard to the energy flow we assume that a sample is not fixed in the plane of the flow action. The boundary conditions for the stress equal to zero:

$$\sigma_x(t,0) = \sigma_x(t,l_0) = 0 \tag{17}$$

The initial conditions of the problem:

$$\sigma_x(0,x) = 0, \ T(0,x) = T_2, \ \frac{\partial \sigma_x}{\partial t}(0,x) = 0 \tag{18}$$

The stated problem has two typical spatial scales: dynamic, equal to the thickness of the sample $l_0 = \sim 1$ mm, and thermal—$l_T = \sim 10$ μm. Therefore, the equation of dynamics is considered in the range from zero to l_0, whereas the equation of thermal conductivity is considered on the semi-finite line with a zero temperature at infinity.

The mathematical problem in non-dimensional variables Equations (14)–(18) is written as follows:

$$\ddot{\sigma} + \ddot{\theta} = \sigma'', \ 0 < \xi < 1, \ \tau > 0; \ \alpha\dot{\theta} = \theta'', \ \xi > 0, \ \tau > 0$$
$$\theta(\tau,0) = \Psi_0(\tau), \ \theta(\tau, \infty) = 0; \tag{19}$$
$$\sigma(\tau,0) = \sigma(\tau,1) = 0; \sigma(0,\xi) = \sigma(0,\xi) = \dot{\sigma}(0,\xi) = 0.$$

The non-dimensional variables are set according to formulae:

$$\xi = \frac{x}{l_0}, \tau = \frac{tc}{l_0}, \sigma = \frac{\sigma_x}{\gamma T_0}, \theta = \frac{T}{T_0}, \alpha = \frac{l_0 c}{\chi}, \tau_1 = \frac{t_1 c}{l_0}, \tau_2 = \frac{t_2 c}{l_0}, \tau_3 = \frac{t_3 c}{l_0} \tag{20}$$

To solve the problem Equation (19) the Laplace transform is used. For the images a system of differential equations is written with the corresponding boundary conditions:

$$\sigma_L''(p,\xi) - p^2\sigma_L(p,\xi) = p^2\theta_L(p,\xi), \ \sigma_L(p,0) = \sigma_L(p,1) = 0$$
$$\theta_L''(p,\xi) - b^2\theta_L(p,\xi) = 0, \ \theta(p,0) = \Psi(p), \ \theta_L(p,\infty) = 0 \tag{21}$$

where $b^2 = \alpha p$ and $\Psi_0(p)$ is the Laplace transform of the function $\Psi_0(\tau)$ set according to Equation (16). The solution of the problem Equation (21) is written as follows:

$$\theta_L(p,\xi) = \frac{\Psi(p)}{b}\exp(-b\,\xi) \tag{22}$$

$$\sigma_L(p,\xi) = \frac{p\,\Psi(p)}{(p-\alpha)}\left(W(\xi) - W(1)\frac{\sinh(p\xi)}{\sinh(p)}\right) \tag{23}$$

where:

$$\Psi(p) = \frac{1}{p^2}\left(\frac{1-\exp(-p\tau_1)}{\tau_1} - \exp(-p(\tau_1+\tau_2))\frac{1-\exp(-p\tau_3)}{\tau_3}\right)$$
$$W(\xi) = \exp(-p\xi) - \exp(-b\,\xi) \tag{24}$$

The first summand in Equation (23) is an image of the stress in the wave, generated via temperature heating, and the second summand represents the superposition of the straight wave and the wave reflected from the back and front surfaces:

$$\sigma_L(p,\xi) = \sigma_{L\infty}(p,\xi) + \sigma_{L\leftarrow}(p,\xi) + \sigma_{L\rightarrow}(p,\xi) \tag{25}$$

The expression below is obtained according to the sum of infinite geometric sequence formula:

$$\sigma_{L\infty}(p,\xi) = \frac{p\,\Psi(p)W(\xi)}{(p-\alpha)}$$
$$\sigma_{L\leftarrow}(p,\xi) = -\frac{p\,\Psi(p)W(1)}{(p-\alpha)}\sum_{n=0}^{\infty}\exp(-p(1-\xi+2n))$$
$$\sigma_{L\rightarrow}(p,\xi) = \frac{p\,\Psi(p)W(1)}{(p-\alpha)}\sum_{n=0}^{\infty}\exp(-p(1+\xi+2n)) \tag{26}$$

The second summand in Equation (25) describes the waves moving from the back surface to the front one (backward waves), and the third summand in Equation (25) is for forward waves. We calculate the stress originals:

$$\sigma_{L\infty} = \frac{(\exp(-p\xi)-\exp(-\xi\sqrt{ap}))-(\exp(-p(\xi+\tau_1))-\exp(-\xi\sqrt{ap}-p\tau_1))}{\tau_1 p(p-\alpha)} +$$
$$\frac{(\exp(-p(\xi+\tau_1+\tau_2+\tau_3))-\exp(-\xi\sqrt{ap}-p(\tau_1+\tau_2)))}{\tau_3 p(p-\alpha)} \rightarrow \tag{27}$$
$$\sigma_\infty(\tau,\xi) = \frac{F(\tau,\xi)-H(\tau-\tau_1)F(\tau-\tau_1,\xi)}{\alpha\tau_1} + \frac{H(\tau-\tau_1-\tau_2-\tau_3)F(\tau-\tau_1-\tau_2-\tau_3,\xi)}{\alpha\tau_3}$$

The designations are used here:

$$F(\tau,\xi) = H(\tau-\xi)(\exp(\alpha(\tau-\xi))-1) + Erfc\left(\frac{\xi\sqrt{\alpha}}{2\sqrt{\tau}}\right) -$$
$$\frac{1}{2}\left(\exp(\alpha(\tau+\xi)Erfc\left(\frac{\xi\sqrt{\alpha}}{2\sqrt{\tau}} + \sqrt{\alpha\tau}\right) + \exp(\alpha(\tau-\xi)Erfc\left(\frac{\xi\sqrt{\alpha}}{2\sqrt{\tau}} - \sqrt{\alpha\tau}\right)\right)$$
$$Erfc(z) = \frac{2}{\sqrt{\pi}}\int_z^\infty \exp(-y^2)dy \tag{28}$$

The obtained Equations (27) and (28) are in line with those written for a semi-finite layer, Equation (28) confirms the formula given in [45]. Using the theorem of multiplication of originals we obtain the images for the forward and backward waves:

$$\sigma_{n\rightarrow}(\tau,\xi) = \frac{\Phi(\tau-\xi-(2n+1))-\Phi(\tau-\xi-(2n+1+\tau_1))}{\alpha\tau_1} +$$
$$\frac{\Phi(\tau-\xi-(2n+1+\tau_1+\tau_2+\tau_3))-\Phi(\tau-\xi-(2n+1+\tau_1+\tau_2))}{\alpha\tau_3} \qquad (29)$$

$$\sigma_{n\leftarrow}(\tau,\xi) = \frac{\Phi(\tau+\xi-(2n+1))-\Phi(\tau+\xi-(2n+1+\tau_1))}{\alpha\tau_1} +$$
$$\frac{\Phi(\tau+\xi-(2n+1+\tau_1+\tau_2+\tau_3))-\Phi(\tau+\xi-(2n+1+\tau_1+\tau_2))}{\alpha\tau_3} \qquad (30)$$

The designation used here is:

$$\Phi(z) = H(z-1)(\exp(\alpha(z-1))-1) + Erfc\left(\frac{\sqrt{\alpha}(2z+1)}{2\sqrt{z}}\right) +$$
$$\exp(\alpha(z-1))Erfc\left(\frac{\sqrt{\alpha}(1-2z)}{2\sqrt{z}}\right) \qquad (31)$$

The results of stress distribution computation according to the coordinate for different instants of time are depicted in Figure 2a,b. This distribution makes it evident that the thermoelastic wave is a bipolar one and zones of pressure and extension are located symmetrically (Figure 2a). Provided that temperature heating is a non-symmetrical one, zones of pressure and extension will be distributed non-symmetrically, too. When the wave approaches the back surface, reflection occurs: the zone of pressure in an incident wave is compensated by the reflected tensile wave; the stress is equal to zero for the instant of time up to 1.2 (0.17 μs), if the distance from the back surface does not exceed 0.2 (120 μm) (Figure 2b, curve 2). From the instant of time 1.1 the reflected tensile wave is added to the zone of extension in an incident wave and, as a consequence, tension gets doubled (Figure 2b, curve 1). Splitting off arises because of such an increase of stresses. A detailed investigation of the zone, where splitting off arises, preconditions taking into consideration the period of active tensile stresses and application of the time strength model. Therefore, it is a separate problem to detect the place of splitting off.

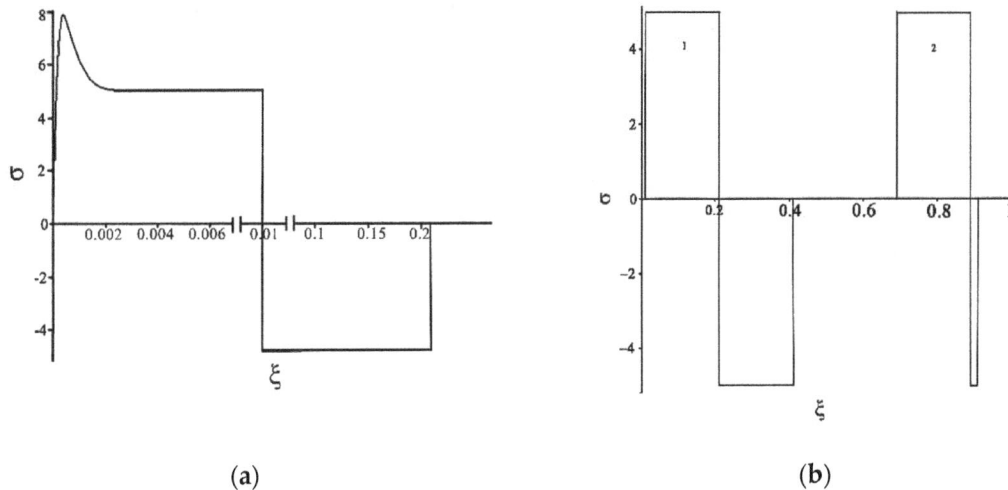

(a)   (b)

Figure 2. The dependence of stress on the coordinates: in the instants of non-dimensional time: (**a**) 0.21; and (**b**) 0.41.

At the instant of time $\tau = 1.4$ (0.2 μs) the wave turns completely around and moves to the face, where it is reflected; the maximum tensile stresses are recorded at the definite distance (up to 150 μm, approximately) (Figure 2b, curve 2). It facilitates, however, strengthening processes, associated with microhardness peak. In experiments, carried out in [46], microhardness peak occurs at the depth of about 50–100 μm from the face of thin plates. As experimental and estimated values agree with each other, the selected model can be considered a constituent one.

Hence, bipolarity of the thermoelastic wave is the result of change in heating and cooling modes, according to the solution of decoupled thermoelastic problem and parabolic equation of thermal conductivity for the set triangle temperature profile on the surface. The zones of doubled tension have been revealed at the back surface, as well as those of doubled pressure at the face. The developed stress field facilitates rearrangement of defect structure of material, as the consequence, furthers accelerated mass transfer. This fact can help to account for the fact, why the depth of the strengthened layer is increased more than that of thermal impact.

The obtained results can be used for discuss probable mechanisms how to improve mechanical properties of materials at the depth exceeding dimensions of the zone affected by plasma flows.

2.3. The Diffusion Model of Dissolution Refractory Inclusions in Metals under Concentrated Flows of Energy Action

Let us consider the dissolution of a high-melting element in liquid titanium, taking carbon as an example. A mathematical model and a method to solve the problem of carbon distribution in titanium carbide and in liquid titanium and carbon solution are considered below in terms of the state diagram and temperature changes caused by phase transitions and heat exchange with the ambient environment. The dependence of temperature of the surrounding cell environment with respect to time is the key feature of this model. This problem can be reduced to a non-autonomous singular differential second-degree equation, but it is difficult to analyze it in terms of quality [21,26,47]. Therefore, a numerical method is selected to investigate some definite models with particular parameters.

In the model below a cell is considered, which contains a r_0 size carbon particle, placed into liquid titanium with a definite volume of R_0 size. On the C-Ti interface a chemical reaction takes place, as the result of which titanium carbide generates and heat releases. A wave of titanium carbide generation with the coordinate $r_1(t)$ starts spreading towards carbon. Carbon, which is diffusing through the layer TiC, reacts with titanium, as the result, carbide is generated, therefore, front $r_2(t)$ is moving. Carbon also dissolves in liquid titanium. Hence, a three-layer structure is generated in the cell: hard carbon—$0 < r < r_1$, titanium carbide—$r_1 < r < r_2$, carbon solution in liquid titanium—$r_2 < r < R_0$. If temperature is assumed to be constant, mathematical statement of the diffusive problem contains the level of diffusion, edge, and initial conditions. Let us consider $T_e < T < T_m$ in more detail: here T_e, T_m are the temperatures of eutectic titanium-titanium carbide and titanium carbide-graphite.

Equations of diffusion (sphere—$n = 2$, cylinder—$n = 1$, flat plate—$n = 0$) are as follows:

$$\frac{\partial c_V(r,t)}{\partial t} = \frac{1}{r^n}\frac{\partial}{\partial r}\left(D_1 r^n \frac{\partial c_V(r,t)}{\partial r}\right),\ r_1(t) < r < r_2(t)$$
$$\frac{\partial c_V(r,t)}{\partial t} = \frac{1}{r^n}\frac{\partial}{\partial r}\left(D_2 r^n \frac{\partial c_V(r,t)}{\partial r}\right),\ r_2(t) < r < R_0 \tag{32}$$

Edge conditions on the extremities of the interval:

$$c_V(r_1(t)+0,t) = c_{V1},\ c_V(r_2(t)-0,t) = c_{V2}$$
$$c_V(r_2(t)+0,t) = c_{V3},\ \left.\frac{\partial c_V(r,t)}{\partial r}\right|_{R_0} = 0 \tag{33}$$

Conditions to determine the coordinates of boundaries:

$$(1-c_{V1})\frac{dr_1}{dt} = D_1 \left.\frac{\partial c_V(r,t)}{\partial r}\right|_{r_1+0},\ r_1(0) = r_0$$
$$(c_{V2}-c_{V3})\frac{dr_2}{dt} = -D_1 \left.\frac{\partial c_V(r,t)}{\partial r}\right|_{r_2-0} + D_2 \left.\frac{\partial c_V(r,t)}{\partial r}\right|_{r_2+0},\ r_2(0) = r_0 \tag{34}$$

Initial conditions:

$$c_V(r,0) = \begin{cases} 1,\ 0 < r < r_0 \\ 0,\ r_0 < r < R_0 \end{cases} \tag{35}$$

Here $c_V = V_C/(V_C + V_{Ti})$ is the inclusion volume fraction of carbon in phases, which depends on the spatial values r and time t; V_J are the volumes of corresponding pure components ($J = C, Ti$). Figure 3 shows the graph of atomic fraction $c = N_C/(N_C + N_{Ti})$ dependent on the coordinate. Atomic fracture can be used to determine the chemical formulae of phases. However, equations of diffusion are written for inclusion volume fractions or mass fractions, mole or atomic concentrations, which are proportional to them. Atomic-inclusion volume fraction conversion and vice versa are described in Equation (36):

$$c_V = \frac{1}{\left(\frac{1}{c} - 1\right) \frac{\rho_C}{\rho_{Ti}} \frac{\mu_{Ti}}{\mu_C} + 1}, \quad c = \frac{1}{\left(\frac{1}{c_V} - 1\right) \frac{\rho_{Ti}}{]\rho_C} \frac{\mu_C}{\mu_{Ti}} + 1} \tag{36}$$

To calculate the temperature in the cell we use the heat-balance equation:

$$S_0 \nu_0 \frac{dT}{dt} = Q_1 \frac{d\nu_1}{dt} + Q_2 \frac{d\nu_2}{dt} - a(T - T_a(t))$$
$$\nu_1(t) = \int_{r_1}^{r_2} c_V(r,t) r^n dr, \quad \nu_2(t) = \int_{r_2}^{R_0} c_V(r,t) r^n dr \tag{37}$$
$$\nu_0 = \frac{1}{n+1} R_0{}^{n+1}\left(c_{V0} + \frac{\rho_{Ti}}{\mu_{Ti}} \frac{\mu_C}{\rho_C}(1 - c_{V0})\right)$$

Here ($i = 1, 2$). C_0 is the mean molar capacity of the cell, ν_1, ν_2, ν_0 are values, proportional to the moles of carbon in titanium carbide and in solution, as well as to the total mole concentration at the initial instant of time, the proportionality factor is similar. Q_i is the molar heat of titanium carbide generation and carbon dissolution in liquid titanium, respectively; and c_V is the bulk concentration of carbon in titanium carbide or in solution, $c_{V0} = (r_0/R_0)^{n+1}$ is the initial bulk concentration of carbon in the system, ρ_J and μ_J are the density and molar mass of the component J ($J = C, Ti$), a is the coefficient of heat exchange, T is the temperature of the cell, and T_a is the ambient temperature. We admit that the dependence of ambient temperature on time is pre-set.

The stated mathematical problem Equations (32)–(37) are solved numerically by the method of scalar marching. Time is assumed to be discrete at the moments $t_n = k\tau$, τ is the time interval of discretization, and k is the time step number. Let us consider the values c_V^k, r_1^k, r_2^k, T^k to be known for the k time step, we calculate the values for $k + 1$ time step. An initial-boundary value problem is numerically solved with edge conditions and coefficients of diffusion, taken from k time step.

Calculations were done for a one-dimensional case ($n = 0$, flat plate), half-thickness of the cell to be calculated R_0, the initial half-thickness of graphite impurity r_0. Parameters of calculations are as follows: $D_{10} = 45.4 \times 10^{-4}$ m^2/s, $E_1 = 447.5 \times 10^3$ J/mol, $D_{20} = 38.5 \times 10^{-4}$ m^2/s, $E_2 = 22.4 \times 10^3$ J/mol, $c_{2e} = 0.31$, $T_e = 1926$ K, $T_m = 3346$ K, and $Q_1 = 231.7$ kJ/mol, $\delta = 0.1$ [48].

Ambient temperature was decreasing according to the linear law $T_a = T_{00} - \beta t$. Here β is the rate of cooling. Parameters T_{00}, α, and β were varied. Initial temperature was selected equal to T_{00}. This approach helped to analyze whether carbon could dissolve in titanium, to determine its dissolution time according to initial temperature and the rate of cooling.

Figures 3 and 4 show the results of calculations when the rate of cooling is equal to zero and there is no heat exchange. Values for r_0 were selected equal to either 10 μm or 10 nm. This was done to compare the time of carbon dissolution under conditions of various scales of cells.

The thickness of titanium carbide layer increases first, then it gets smaller (Figure 3). The time of dissolution is getting reduced as the initial temperature grows (Figure 4a,b). Here, if initial thickness of graphite layer reduces from 10 μm to 10 nm, time of dissolution decreases 10^6 times. The time of graphite dissolution is on the order of microseconds; this conforms to the experiments on the electron-beam treatment of the surface after electro-explosive doping [10].

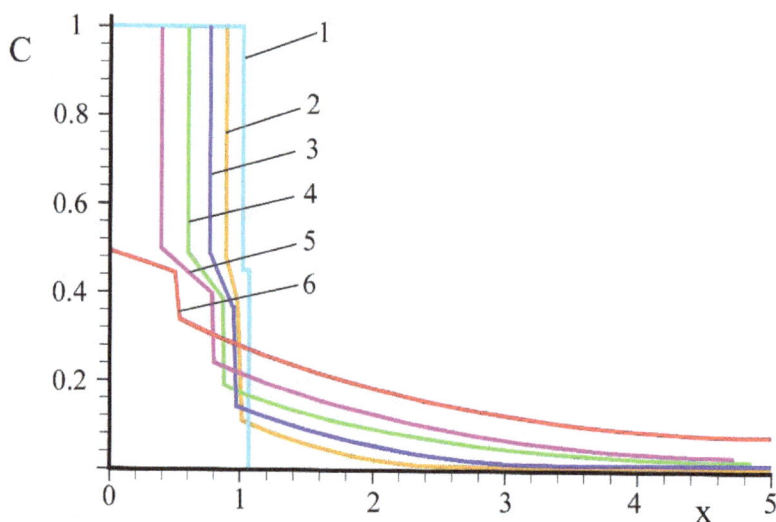

Figure 3. The dependencies of atomic fraction of carbon (C) according to the value $x = r/r_0$. 1: 0, 2: 1.2, 3: 2.5, 4: 3.8, 5: 5.0, 6: 6.136 (μs) (initial temperature 2626 K, $r_0 = 10$ nm).

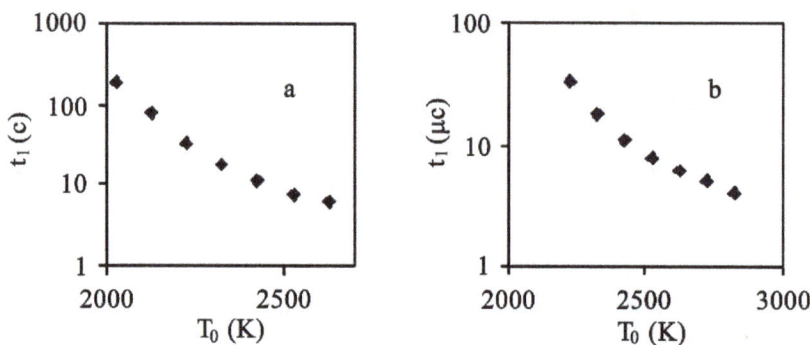

Figure 4. Time dependence of graphite dissolution on initial temperature of the cell: (**a**) $r_0 = 10$ μm; (**b**) $r_0 = 10$ nm.

Therefore, the submitted model and calculations made on its ground, made it possible to demonstrate that micrometer-order carbon particles dissolve for about 10 s. This time goes far beyond the time of impact on material by concentrated flow of energy. If the dimensions of the particles are of nanometer-order ones, the time of dissolution gets equal to 10 μs in order of magnitude. In this case carbon particle can be dissolved as long as it is treated by electron-beam. We can say that the process of combined treatment requires for carbon-graphite fibers, micro-fibrille diameter of them is about tens of nanometers.

3. Modeling of Subsurface Nanostructure Formation

The research into the impact of the electric conductor, explosive-produced, heterogeneous plasma flows on the structure and properties of materials has demonstrated that a multi-layer gradient structure forms at a distance from the treated surface (Figure 5). The first ~1-μm thick nano-structural layer (I) reacts with plasma of incident flow [49]. The dimensions of vertically-located cells in the second layer (II) with the structure of cellular-type crystallization are much larger than those located horizontally. The third layer (III) has approximately equal longitudinal and crosscut dimensions of the grain. The forth inner nano-structural layer (IV) is ~1 μm thick. These experimental facts, two nano-structural layers in particular, have become the issue of theoretical investigations. The external nano-layer is possible to account for the significant heat removal and, as a consequence, for the high rate of cooling. However, specific conditions are necessary in order to obtain such significant rates of cooling, which is hardly ever possible. The internal nano-layer could not be modeled in the thermal model. Moreover, this layer was considered an artifact. Nevertheless, in products hardened quickly from the melt on a rapidly-moving disk (hardening by spinning) there are four structural zones, described in [50]. The zone located near to the surface to be hardened, or the zone of freezing, has a fine, disperse structure. It is thought to be generated as the result of multiple-crystal nucleation. The second zone is one of columnar crystal—highly extended grains. A dendrite or dendrite-cellular structure is formed in the third zone. The fourth zone is one of equiaxed randomly-oriented nano-grains. The situation here is much aike that one described above—an internal nano-layer arises under the action of heterogeneous plasma. Furthermore, nano-dimensional structures, which form under significant plastic deformations, arise, if two materials are sheared. Nanostructures were detected in shear bands, when detailed microscopic research was carried out [29]. The aforementioned experimental data emphasizes nanostructures to appear often in shear flows.

In [51] one explains why the first nano-structural layer arises involving the mechanism of Kelvin-Helmholtz instability [30] in the nano-dimensional range of wavelengths in conditions of tangentiall- flowing plasma and a layer of molten metal. If two media are moving relative to each other, waves appear (like ripples on water's surface when it is windy). A dispersion evolution is obtained in linear approximation of the evolving boundary surface which is, as a rule, a transcendental algebraic equation attributing increments to the wavelength. This depends on a number of parameters; hence, analytical parameterization can rarely be accomplished. Therefore, a dispersion equation is solved numerically to obtain the dependence of the decrement on the wavelength for definite values of parameters.

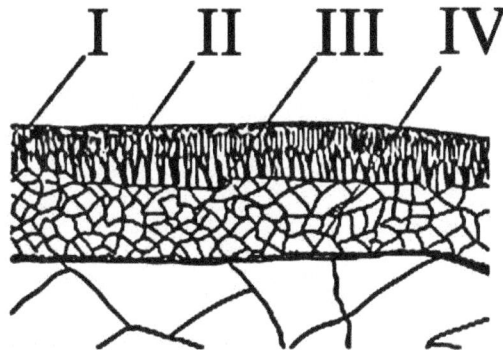

Figure 5. The arrangement of four structural zones.

The maximums in this dependence are obtained at particular wavelengths λ_{max} under conditions of positive increments. Hence, $\sim\lambda_{max}$-dimensioned waves, and vortices generated by them, are developed, and waves of other lengths, other scales, respectively, cannot be formed. This approach is a conventional one for investigations into instabilities. The idea of using this mechanism to explain the generation of the second internal nano-layer (zone IV) was offered and elaborated in [30]. In this work the following dispersion relation is obtained in the approximation of viscous and viscous-potential fluid in terms of Navier-Stokes and Euler equations.

To derive a dispersion equation we deal with a double-layer incompressible liquid (Figure 6), like in [30]. The underlayer is assumed to be fixed and viscous. The upper layer is modeled as a perfect liquid moving at speed u_0 parallel to the underlayer. For each layer Navier-Stocks and Euler linearized equations are written:

$$\frac{\partial U_1}{\partial t} = -\frac{1}{\rho_1}\frac{\partial P_1}{\partial x} + \nu_1\left(\frac{\partial^2 U_1}{\partial x^2} + \frac{\partial^2 U_1}{\partial y^2}\right)$$
$$\frac{\partial V_1}{\partial t} = -\frac{1}{\rho_1}\frac{\partial P_1}{\partial y} + \nu_1\left(\frac{\partial^2 V_1}{\partial x^2} + \frac{\partial^2 V_1}{\partial y^2}\right),\ \frac{\partial U_1}{\partial x} + \frac{\partial V_1}{\partial y} = 0 \tag{38}$$

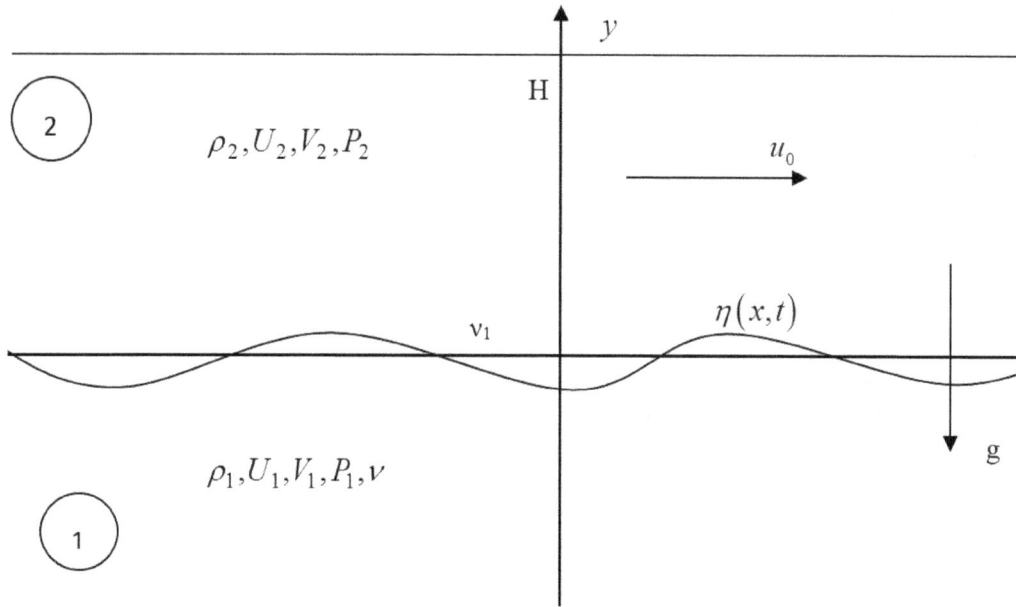

Figure 6. Computation scheme.

For the upper layer:

$$\frac{\partial U_2}{\partial t} + u_0 \frac{\partial U_2}{\partial x} = -\frac{1}{\rho_2}\frac{\partial P_2}{\partial x}, \ \frac{\partial V_2}{\partial t} + u_0 \frac{\partial V_2}{\partial x} = -\frac{1}{\rho_2}\frac{\partial P_2}{\partial y}, \ \frac{\partial U_2}{\partial x} + \frac{\partial V_2}{\partial y} = 0 \tag{39}$$

Kinematic and dynamic conditions on the boundary of layers:

$$y = H: V_2 = 0, \ y = -h: U_1 = 0, \ V_1 = 0$$

$$y = 0: \frac{\partial \eta}{\partial t} + u_0 \frac{\partial \eta}{\partial x} = V_2, \ \frac{\partial \eta}{\partial t} = V_1 \tag{40}$$

$$-P_1 + P_2 + 2\rho_1 \nu_1 \frac{\partial V_1}{\partial y} = \sigma \frac{\partial^2 \eta}{\partial x^2} + g(\rho_2 - \rho_1)\eta$$

We note that index 1 is for the underlayer, whereas index 2 is for the upper one. Equations (38) and (39) are solved the following way:

$$U_1(x,y,t) = u_1(y)\exp(\omega t - ikx), \ U_2(x,y,t) = u_2(y)\exp(\omega t - ikx)$$
$$V_1(x,y,t) = v_1(y)\exp(\omega t - ikx), \ V_2(x,y,t) = v_2(y)\exp(\omega t - ikx)$$
$$P_1(x,y,t) = p_1(y)\exp(\omega t - ikx), \ P_2(x,y,t) = p_2(y)\exp(\omega t - ikx) \tag{41}$$
$$\eta = \eta_0 \exp(\omega t - ikx)$$

Substituting Equation (41) into Equations (38)–(40) and carrying out further transformations we obtain a dispersion equation:

$$G(z,d) + \mu\,\Omega_1^2 \mathrm{cth}(kH) + W = 0$$

$$G(z,d) = \frac{\left((z^4-1)(z-\mathrm{th}(zd)\mathrm{th}(d))\right)}{(\mathrm{th}(zd)-z\mathrm{th}(d))}\nu^2 + \frac{2z(1-z^2)\nu^2 - 2z(z^2+3)\nu\nu_1}{\mathrm{sh}(zd)\mathrm{ch}(d)-z\mathrm{sh}(d)\mathrm{ch}(zd)} \tag{42}$$

$$+ \frac{(2z(z^2+3)-2(1+3z^2)\mathrm{th}(d)\mathrm{th}(zd))}{(\mathrm{th}(zd)-z\mathrm{th}(d))}\nu\nu_1$$

where $\Omega_1 = \frac{\Omega}{k^2} = \nu(z^2-1)i - \frac{u_0}{k}$, $\mu = \frac{\rho_2}{\rho_1}$, $W = \frac{A}{\rho_1 k^3} = \frac{g(\rho_2-\rho_1)-\sigma_0 k^2}{\rho_1 k^3}$, and ν_1 is the viscosity on the boundary of layers. We consider cases when the underlayer is viscous or viscous-potential liquid. In the first case $\nu_1 = \nu$, so Equation (42) is written as follows:

$$\nu^2 G_1(z,d) + \mu \Omega_1^2 \coth(kH) + W = 0$$
$$G_1(z,d) = \frac{-4z(z^2+1)}{\sinh(zd)\cosh(d) - z\sinh(d)\cosh(zd)} + \frac{z(z^4+6z^2+5) - (z^4+6z^2+1)\tanh(d)\tanh(zd)}{\tanh(zd) - z\tanh(d)}$$

(43)

For viscous-potential liquid $\nu = 0$ and $\nu_1 \neq 0$, so Equation (42) is written:

$$G_2(z,d) + \mu\left(\frac{\nu-ku_0}{k^2}\right)^2 \coth(kH) + W = 0$$
$$G_2(z,d) = \frac{2i\omega}{\tanh(d)}\nu_1 + \frac{\omega^2}{k^4\tanh(d)}$$

(44)

The analysis of this relation has demonstrated that increment dependencies on the wavelength for viscous and viscous-potential liquids are nearly the same in the wide range of parameters (Figure 7). Therefore, a less complicated dispersion equation of viscous-potential liquid can be applied to analysis in terms of the quality of Kelvin-Helmholtz instability of a perfect and viscous liquid boundary. Numerical computations of increment dependencies on the wavelength at definite values of parameters, which are like those of internal nano-layer formations under the action of heterogeneous plasma flows on ferrum, have revealed that two maximums in the increment-wavelength dependence—in the micro- and nano-range—are possible, if a moving layer is thin ($\rho_1 = \rho_2 = 6.3 \times 10^3$ kg/m^3, $\nu = 6 \times 10^{-7}$ m^2/s, $\sigma_0 = 1.2$ N/m [52], $h = 10^{-6}$ m, $H = 10^{-6}$ m, $u_0 = 30$ m/s). This conforms to the development of two-mode instability.

The microwave mode corresponds with the interaction of perfect liquid layers. The nano-wave mode is formed due to viscosity. In general, the offered interpretation has the following results. In $H = h = 1$-μm thick layers, if the rates exceed 40 m/s, $\lambda = 1$-μm long waves arise; for these values of layer thickness and wavelength the wave can be considered as a moving $H_c \approx 10$-nm thick boundary layer, where a nano-wave mode is developing; the wavelength is $\lambda_c = 100$ nm.

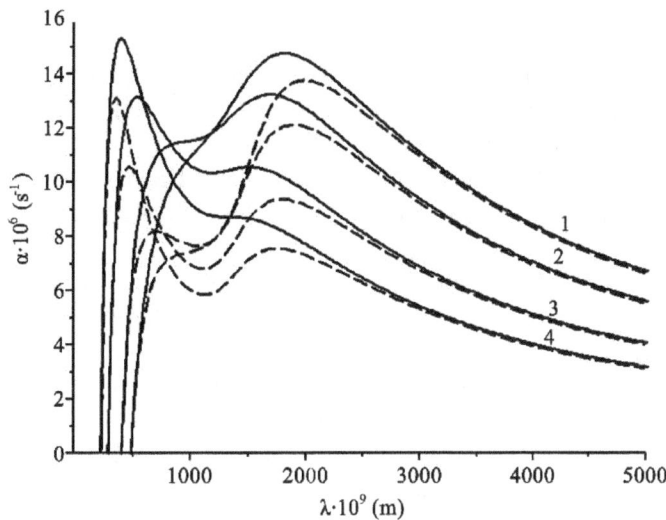

Figure 7. Increment α dependencies on the wave length λ for various H and $\sigma_0 = 1.2$ N/m. Full line—viscous model, dotted line—viscous-potential. (**1**) $H = 6$ nm; (**2**) $H = 10$ nm; (**3**) $H = 20$ nm; and (**4**) $H = 30$ nm.

4. Conclusions

1. A thermal mathematical model, which considers vaporization from the surface of the material, has been presented. The dependence of penetration depth vs. energy surface density is obtained. Its linear character has been shown. When comparing penetration depths determined computationally with the experimental data their satisfactory fit has been determined. The period of vaporization has been determined without solving the gas-dynamic problem.

2. The mechanism of bipolar thermo-elastic wave generation has been revealed on the basis of being solved analytically, not on the basis of the uncoupled thermo-elasticity problem. The matter of it is that tension and compression in the thermo-elastic wave are caused by an increase and a subsequent drop of temperature on the edge.

3. A model of the dissolution of carbon particles in titanium has been analyzed under the action of electron beams. It has been stated that micrometer-dimensional carbon particles get dissolved for about 10 s. This time significantly exceeds the time of concentrated energy flow impact on the material. If particles are nanometer-dimensional ones, the time of dissolution is 10 μs in order of magnitude. In this case carbon particles get dissolved as long as they are impacted by electron beams.

4. Formation of internal nano-structural layers has been analyzed under the action of heterogeneous plasma flows. Instability increment dependence on the wavelength with two maximums—in nano- and micro- ranges—have been developed.

Acknowledgments: The research has been carried out under financial support of the Russian Scientific Foundation (project No. 15-19-00065) (Sections 2.2 and 3), Russian Foundation for Basic Research (project No. 15-08-03411a) (Section 2.1). This work (Section 2.3) was partly sponsored by National Natural Science Foundation of China (No. 51575401), Zhejiang Provincial Natural Science Foundation (No. LY16E050007).

Author Contributions: Sergey Konovalov, model building and carrying out. Xizhang Chen, design the model and analyze the simulation results. Vladimir Sarychev, the calculations of thermal and thermoelastic models. Sergey Nevskii, performing calculations on models of carbon dissolved in the titanium and hydrodynamic model of the formation of nanostructures. Victor Gromov, the formulation of the problem of power influence on titanium alloys. Milan Trtica, main comments on the revision of model and analysis.

Conflicts of Interest: The authors declare no conflict of interest.

References

1. Ilyuschenko, A.P.; Shevtsov, A.I.; Astashynski, V.M.; Kuzmitski, A.M.; Gromyko, G.F.; Chumakov, A.N.; Bosak, N.A.; Buikus, K.V. Friction and wear of powder coatings produced by using high-energy pulsed flows. *High Temp. Mater. Process.* **2015**, *19*, 141–152. [CrossRef]

2. Huang, Z.-T.; Suo, H.-B.; Yang, G.; Yang, F.; Dong, W. Effect of heat treatment on microstructure and property of TC18 titanium alloy prepared by electron beam rapid manufacturing. *Trans. Mater. Heat Treat.* **2015**, *36*, 50–54.

3. Da Silva, M.R.; Gargarella, P.; Gustmann, T.; Filho, W.J.B.; Kiminami, C.S.; Eckert, J.; Pauly, S.; Bolfarini, C. Laser surface remelting of a Cu-Al-Ni-Mn shape memory alloy. *Mater. Sci. Eng. A* **2016**, *66*, 161–167. [CrossRef]

4. Chen, X.; Wang, J.; Fang, Y.; Madigan, B.; Xu, G.; Zhou, J. Investigation of microstructures and residual stresses in laser peened Incoloy 800H weldments. *Opt. Laser Technol.* **2014**, *57*, 159–164. [CrossRef]

5. Hu, J.; Chai, L.; Xu, H.; Ma, C.; Deng, S. Microstructural modification of brush-plated nanocrystalline Cr by high current pulsed electron beam irradiation. *J. Nano Res.* **2016**, *41*, 87–95. [CrossRef]

6. Zhou, Z.; Chen, B.; Xiao, H.; Tu, J.; Chai, L.; Huang, W.; Hu, J. Microstructure and properties of CuFe10 alloys treated by high current pulsed electron beam. *High Power Laser Part. Beams* **2015**, *27*, 024105. [CrossRef]

7. Devyatkov, V.N.; Koval, N.N.; Schanin, P.M.; Grigoryev, V.P.; Koval, T.V. Generation and propagation of high-current low-energy electron beams. *Laser Part. Beams* **2003**, *21*, 243–248. [CrossRef]

8. Sosnin, K.V.; Raykov, S.V.; Vaschuk, E.S.; Budovskikh, E.A.; Gromov, V.E.; Ivanov, Y.F. Morphology of the surface of technically pure titanium VT1-0 after electroexplosive carbonization with a weighed zirconium oxide powder sample and electron beam treatment. In Proceedings of the International Conference on Physical Mesomechanics of Multilevel Systems, Tomsk, Russia, 3–5 September 2014.

9. Sosnin, K.V.; Raikov, S.V.; Gromov, V.E.; Ivanov, Y.F.; Budovskikh, E.A.; Vashchuk, E.S. Formation of a microcomposite structure in the surface layer of yttrium-doped titanium. *J. Surf. Investig. X-ray Synchrotron Neutron Tech.* **2015**, *9*, 377–382. [CrossRef]

10. Ivanov, Yu.F.; Teresov, A.D.; Petrikova, E.A.; Raikov, S.V.; Goryushkin, V.F.; Budovskikh, E.A. Surface layer of commercially pure VT1-0 titanium after electric-explosion alloying and subsequent treatment by a high-intensity pulsed electron beam. *Steel Transl.* **2013**, *43*, 798–802. [CrossRef]

11. Markov, A.B.; Rotshtein, V.P. Calculation and experimental determination of hardening and tempering zones in quenched U7A steel irradiated with a pulsed electron beam. *Nucl. Instrum. Methods Phys. Res. Sect. B* **1997**, *132*, 79–86. [CrossRef]

12. Markov, A.B.; Ivanov, Yu.F.; Proskurovsky, D.I.; Rotshtein, V.P. Mechanisms for hardening of carbon steel with a nanosecond high-energy, high-current electron beam. *Mater. Manuf. Process.* **1999**, *14*, 205–216. [CrossRef]

13. Leyvi, A.Ya.; Talala, K.A.; Krasnikov, V.S.; Yalovets, A.P. Modification of the Constructional Materials with the Intensive Charged Particle Beams and Plasma Flows. *Ser. Mech. Eng. Ind.* **2016**, *16*, 28–55. (In Russian)

14. Mazhukin, V.I.; Mazhukin, A.V.; Lobok, M.G. Mathematical modeling of dynamics of fast phase transitions and overheated metastable states during nano- and femtosecond laser treatment of metal targets. *Math. Model. Comput. Simul.* **2010**, *2*, 396–405. [CrossRef]

15. Bleykher, G.A.; Krivobokov, V.P.; Yuryeva, A.V. Thermal Processes and Emission of Atoms from the Liquid Phase Target Surface of a Magnetron Sputtering System. *Russ. Phys. J.* **2015**, *58*, 431–437. [CrossRef]

16. Yalovets, A.P. Calculation of flows of a medium induced by high-power beams of charged particles. *J. Appl. Mech. Tech. Phys.* **1997**, *38*, 137–150. [CrossRef]

17. Khaimzon, B.B.; Sarychev, V.D.; Gromov, V.E. Temperature distribution produced by pulsed energy fluxes, with evaporation of the target. *Steel Transl.* **2013**, *43*, 55–58. [CrossRef]

18. Maier, A.E.; Yalovets, A.P. Mechanical stresses in an irradiated target with a disturbed surface. *Tech. Phys.* **2006**, *51*, 459–465. [CrossRef]

19. Chumakov, Yu.A.; Knyazeva, A.G. Interrelated processes of heat mass transfer and stress evolution in a disk with an inclusion under the action of high density energy flow. *Phys. Mesomech.* **2013**, *16*, 85–91.

20. Sarychev, V.D.; Voloshina, M.S.; Gromov, V.E. Mathematical model of generation of the thermoplastic waves under action of concentrated energy fluxes at the materials. *Basic Probl. Mater. Sci.* **2011**, *8*, 71–76. (In Russian)

21. Lyubov, B.Ya. *Diffusion Processes in Nonhomogeneous Solid State*; Nauka: Moscow, Russia, 1981. (In Russian)

22. Bukrina, N.V.; Knyazeva, A.G. Numerical solution algorithm for non-isothermal diffusion problems in surface treatment processes. *Phys. Mesomech.* **2006**, *9*, 55–62. (In Russian)

23. Knyazeva, A.G.; Krukova, O.N.; Bukrina, O.V.; Sorokova, S.N. Simulation issues of surface treatment and coating materials using high energy sources. *Izv. TPU* **2010**, *317*, 93–101. (In Russian)

24. Merzhanov, A.G.; Averson, A.E. The present state of the thermal ignition theory: An invited review. *Combust. Flame* **1971**, *16*, 89–124. [CrossRef]

25. Nekrasov, E.A.; Smolyakov, V.K.; Maksimov, Yu.M. Adiabatic heating in the titanium-carbon system. *Combust. Explos. Shock Waves* **1981**, *17*, 305–311. [CrossRef]

26. Khina, B.B. *Combustion Synthesis of Advanced Materials*; Nova Science Publishers Inc.: New York, NY, USA, 2010.

27. Aleksandrov, V.V.; Korchagin, M.A. Mechanism and macrokinetics of reactions accompanying the combustion of SHS systems. *Combust. Explos. Shock Waves* **1987**, *23*, 557–564. [CrossRef]

28. Sarychev, V.D.; Khaimzon, B.B.; Gromov, V.E. Mathematical model of dissolution of particles of carbon in the titan at influence of the concentrated streams of energy. *Titan* **2012**, *1*, 4–8. (In Russian)

29. Sarychev, V.D.; Mochalov, S.P.; Budovskikh, E.A.; Vashchuk, E.S.; Gromov, V.E. Formation of convective structures in metals and alloys under the action of pulsed multiphase plasma jets. *Steel Transl.* **2010**, *40*, 531–536. [CrossRef]

30. Granovskii, A.Yu.; Sarychev, V.D.; Gromov, V.E. Model of formation of inner nanolayers in shear flows of material. *Tech. Phys.* **2013**, *58*, 1544–1547. [CrossRef]

31. Inogamov, N.A.; Zhakhovsky, V.V.; Khokhlov, V.A.; Petrov, Y.V.; Migdal, K.P. Solitary nanostructures produced by ultrashort laser pulse. *Nanoscale Res. Lett.* **2016**, *11*, 177. [CrossRef] [PubMed]

32. Suh, K.Y.; Jeong, H.E.; Kim, D.-H.; Singh, R.A.; Yoon, E.-S. Capillarity-assisted fabrication of nanostructures using a less permeable mold for nanotribological applications. *J. Appl. Phys.* **2006**, *100*, 034303. [CrossRef]

33. Prokoshev, V.G.; Parfionov, S.D.; Obgadze, T.A. Mathematical modelling of the temperature fields induced under the laser processing material. In Proceedings of the International conference on laser assisted net shape engineering LANE 2001, Erlangen, Germany, 28–31 March 2001; pp. 185–190.

34. Galkin, A.F.; Abramov, D.V.; Savina, L.D.; Fedotova, O.Yu.; Prokoshev, V.G.; Arakelian, S.M. Laser-induced hydrodynamics waves on the surface of melt. In Proceedings of the Laser-induced hydrodynamics waves on the surface of melt, Vladimir/Suzdal, Russia, 21 September 2000; Volume 4429, pp. 101–104.

35. Lee, P.D.; Quested, P.N.; McLean, M. Modelling of Marangoni effects in electron beam melting. *Philos. Trans. R. Soc. A Math. Phys. Eng. Sci.* **1998**, *356*, 1027–1043. [CrossRef]

36. Lukashov, E.A.; Radkevich, E.V.; Yakovlev, N.N. Structurization of the instability zone and crystallization. *J. Math. Sci.* **2011**, *179*, 491–514. [CrossRef]

37. Kuznetsov, V.P.; Smolin, I.Y.; Dmitriev, A.I.; Tarasov, S.Y.; Gorgots, V.G. Toward control of subsurface strain accumulation in nanostructuring burnishing on thermostrengthened steel. *Surf. Coat. Technol.* **2016**, *285*, 171–178. [CrossRef]

38. Sarychev, V.D.; Nevskii, S.A.; Konovalov, S.V.; Komissarova, I.A.; Cheremushkina, E.V. Thermocapillary model of formation of surface nanostructure in metals at electron beam treatment. *IOP Conf. Ser. Mater. Sci. Eng.* **2015**, *91*, 012028. [CrossRef]

39. Sarychev, V.D.; Nevskii, S.A.; Sarycheva, E.V.; Konovalov, S.V.; Gromov, V.E. Viscous Flow Analysis of the Kelvin–Helmholtz Instability for Short Waves. In Proceedings of the International Conference on Advanced Materials with Hierarchical Structure for New Technologies and Reliable Structures 2016, Tomsk, Russia, 19–23 September 2016; Volume 1783.

40. Samarskii, A.A.; Babishchevich, P.N. *Vychislitel'naya Teploperedacha (Computational Heat Transmission)*; Editorial URSS: Moscow, Russia, 2003.

41. Cheynet, B.; Dubois, J.-D.; Milesi, M. Données thermodynamiques des éléments chimiques. *Tech. Ing. Traité Mater. Metall.* **1993**, M64-1–M64-22.

42. Missenard, A. *Conductivite Thermique des Solides, Liquides, Gaz et Leurs Melanges*; Editions Eyrolles: Paris, France, 1965.

43. Iesan, D.; Antonio, S. *Thermoelastic Deformations*; Springer: Berlin, Germany, 1996.

44. Awrejcewicz, J.; Krysko, V.A. *Nonclassical Thermoelastic Problems in Nonlinear Dynamics of Shells*; Springer: Berlin, Germany, 2003.

45. *Three-Dimensional Problems of the Mathematical Theory of Elasticity and Thermoelasticity*; Kupradze, V.D. (Ed.) Elsevier: Amsterdam, The Netherlands, 1976.

46. Proskurovsky, D.I.; Rotshtein, V.P.; Ozur, G.E.; Ivanov, Y.F.; Markov, A.B. Physical foundations for surface treatment of materials with low energy, high current electron beams. *Surf. Coat. Technol.* **2000**, *125*, 49–56. [CrossRef]

47. Paul, T.; Vuorinen, V.; Divinski, S.V.; Laurila, T. *Thermodynamics, Diffusion and the Kirkendall Effect in Solids*; Springer: Berlin, Gemany, 2014.

48. Mehrer, H. *Diffusion in Solids*; Springer: Berlin, Germany, 2007.

49. Bagautdinov, A.Ya.; Tsvirkun, O.A.; Budovskikh, E.A.; Ivanov, Yu.F.; Gromov, V.E. Gradient state of the surface layers of iron and nickel after electro-explosive alloying. *Metallurgist* **2007**, *51*, 151–158. [CrossRef]

50. Glezer, A.M.; Permyakova, I.E. *Melt-Quenched Nanocrystals*; CRC Press: Boca Raton, FL, USA, 2013.

51. Bazylev, B.; Janeschitz, G.; Landman, I.; Loarte, A.; Klimov, N.S.; Podkovyrovd, V.L.; Safronov, V.M. Experimental and theoretical investigation of droplet emission from tungsten melt layer. *Fusion Eng. Des.* **2009**, *84*, 441–445. [CrossRef]

52. Forsythe, W.E. *Smithsonian Physical Tables*; Knovel: Norwich, NY, USA, 2003.

7

Effect of Batch Annealing Temperature on Microstructure and Resistance to Fish Scaling of Ultra-Low Carbon Enamel Steel

Zaiwang Liu [1,2]**, Yonglin Kang** [1,*]**, Zhimin Zhang** [2] **and Xiaojing Shao** [2]

[1] School of Materials Science and Engineering, University of Science and Technology Beijing, Beijing 100083, China; lzwbeijing2007@163.com

[2] Shougang Research Institute of Technology, Beijing 100043, China; 2001zhimin@163.com (Z.Z.); shaoxiaojing@shougang.com.cn (X.S.)

* Correspondence: kangylin@ustb.edu.cn

Academic Editor: Hugo F. Lopez

Abstract: In the present work, an ultra-low carbon enamel steel was batch annealed at different temperatures, and the effect of the batch annealing temperature on the microstructure and resistance to fish scaling was investigated by optical microscopy, transmission electron microscopy, and a hydrogen permeation test. The results show that the main precipitates in experimental steel are fine TiC and coarse $Ti_4C_2S_2$ particles. The average sizes of both TiC and $Ti_4C_2S_2$ increase with increasing the batch annealing temperature. The resistance to fish scaling decreases with increasing the annealing temperature, which is caused by the growth of ferrite grain and the coarsening of the TiC and $Ti_4C_2S_2$ particles.

Keywords: ultra-low carbon enamel steel; batch annealing temperature; microstructure; resistance to fish scaling; hydrogen permeation test

1. Introduction

Ultra-low carbon steels were produced to enamel products, such as bathtubs, kitchen utensils, and decorative panels due to their extraordinary deep drawability [1,2]. Enamel coatings have been widely applied for the protection of steel products due to their excellent engineering properties, such as corrosion protection, resistance to heat and abrasion, hygiene and ease of cleaning [3]. Fish scaling is one of the most dangerous defects in the production of enameled steel products. Studies have found that it is hydrogen which plays a key role in the formation of the fish scaling. The resistance to fish scaling of enamel steel is usually evaluated by the hydrogen permeation test, and the hydrogen permeation value (*TH* value) is an important parameter characterizing the resistance to fish scaling. High *TH* value means good resistance to fish scaling. *TH* value should be larger than 6.7 min/mm^2 to ensure satisfactory resistance to fish scaling [4].

The resistance to fish scaling can be improved by increasing the number of hydrogen traps. Hydrogen traps are generally classified as reversible and irreversible traps depending on their binding energy with hydrogen atoms [5,6]. A reversible trap is one from which a hydrogen atom can easily jump out of due to fluctuations in thermal energy [7]. It is known that grain boundaries, dislocations, vacancies, and microvoids have low binding energy with hydrogen atoms and are considered as reversible traps. Hydrogen atoms in these sites are diffusible, and these traps have an influence on the effective hydrogen diffusivity. Irreversible traps are sites with high binding energy, and thus the trapped hydrogen is considered as non-diffusible [8]. (Ti, Nb)(C, N), TiC, TiN, NbC, VC and non-metallic inclusions are considered as irreversible traps because of their high binding energy.

Ti is usually added to ultra-low carbon steel to improve the resistance to fish scaling. The main irreversible hydrogen traps in ultra-low carbon Ti-bearing steel are TiN, TiC, TiS and $Ti_4C_2S_2$

particles which influence hydrogen diffusivity obviously [2]. There are several factors which affect the precipitation behavior of titanium precipitates, i.e., chemical composition, finishing temperature, coiling temperature, annealing temperature and so on. It is reported that Ti and S content will affect the type and fraction of precipitates [9]. Mo can also influence the precipitation behavior of TiC particles [10]. It was found that low finishing temperature was beneficial to the occurrence of strain-induced precipitation of TiC [11]. Kim et al. [12] and Xu et al. [13] found that interphase precipitation took place at high coiling temperature, while dispersed precipitation was formed at low coiling temperature. The annealing temperature will influence the size, distribution and number of precipitates [14,15]. The characteristics of precipitates will influence their binding energy and hydrogen storage capacity, so the annealing temperature will affect the resistance to fish scaling of Ti-bearing steel. Much is still unknown about the effect of batch annealing temperature on microstructure and resistance to fish scaling of ultra-low carbon Ti-bearing enamel steel. It is essential to perform relevant research work to promote the application of ultra-low carbon Ti-bearing enamel steel.

2. Materials and Methods

The experimental steel used in this study was produced by Shougang Group, and the chemical composition is listed in Table 1. The slab was reheated to 1250 °C for 2 h, and then hot rolled to a sheet of 5 mm at finishing temperature of 900 °C. The sheet was water cooled to coiling temperature of 700 °C. After acid pickling, the sheet was cold rolled to 0.8 mm in thickness. The cold rolled sheets were batch annealed at 580, 630, 680, and 730 °C for 5 h, and the schematic of the batch annealing process is shown in Figure 1. These batch annealing temperatures were chosen to obtain different sized ferrite grains and precipitates, so the effects of ferrite grain boundaries and precipitates on resistance to fish scaling can be investigated by a hydrogen permeation test.

Table 1. Chemical composition of experimental steel (in wt %).

C	Si	Mn	P	S	Alt	Ti
$\leq$0.005	$\leq$0.05	0.12–0.25	$\leq$0.02	$\leq$0.05	$\leq$0.05	0.05–0.12

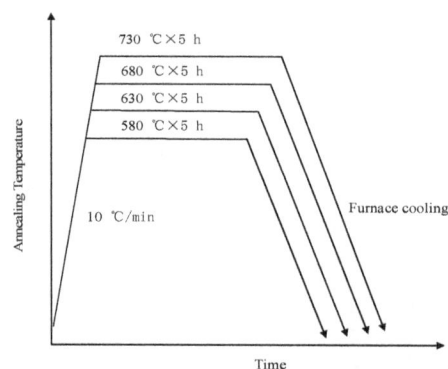

Figure 1. Schematic of the batch annealing process of ultra-low carbon enamel steel.

The metallographic specimens and tensile specimens were cut along the transverse direction. Microstructure observations were conducted on an optical microscope (Leica, Wetzlar, Hesse, Germany). The precipitates in specimens were extracted on carbon replicas and examined by Tecnai G^2 F20 transmission electron microscope (TEM) (FEI, Houston, TX, USA) with an energy dispersive spectrometer (EDS). The dog–bone tensile specimens have 12.5 mm width and 80 mm gauge length. A tensile test was performed on a computer material test (CMT) 5105 testing machine (MTS, Shenzhen, Guangdong, China) at room temperature, and the crosshead speed was 3 mm/min.

The hydrogen permeation specimens with 50 mm × 80 mm × 0.8 mm were prepared. Both sides of the specimens were ground by 1200 grit abrasive paper, and then the specimens were washed by acetone and rinsed by distilled water. The hydrogen permeation test was conducted at room temperature on a Devanathan–Stachursky type cell [16]. The hydrogen permeation setup was composed of two parts separated by the specimen into the cathodic cell and anodic cell. The anodic cell was filled with 0.2 N NaOH solution, and a constant anodic potential 200 mV was applied. Once the background current was less than 0.1 $\mu A/cm^2$, the hydrogen permeation test was started [17], and the cathodic cell was filled with 0.5 N H_2SO_4 + 0.22 g/L H_2NCSNH_2 solution immediately. The charging current density was maintained at 1 mA/cm^2, which was low enough to avoid damage to the steel sheet. In the test, hydrogen is produced electrolytically in the charging cell by cathodic polarization [18]. H_2NCSNH_2 facilitated hydrogen pick-up by promoting the breakdown of molecular hydrogen. High-purity nitrogen gas was purged prior to cathodic polarization and during the entire test to remove the dissolved oxygen which would contribute to the anodic current [19]. The measured anodic current was proportional to the hydrogen flow rate out of the specimen [20], and it was recorded by using an automatic data-acquisition system. Once the measured anodic current reached a steady-state, the hydrogen permeation test can be finished.

The hydrogen permeation time (t_b) was determined by plotting the cumulative anodic current through the specimen and extrapolating its asymptote to intercept with the horizontal axis [21]. The hydrogen permeation value can be calculated by the following formula:

$$TH = t_b/d^2 \tag{1}$$

where TH is the hydrogen permeation value, t_b is the hydrogen permeation time in minutes, d is the sheet thickness in mm.

3. Results and Discussion

3.1. Microstructure

The optical micrographs of experimental steel batch annealed at different temperatures are presented in Figure 2. As can be seen in Figure 2a, the shear bands (indicated by arrows) along the rolling direction and small recrystallized grains were observed at a batch annealing temperature of 580 °C, which reveals that partial recrystallization has occurred at this temperature. At an annealing temperature of 580 °C, there are a large number of dislocations in the shear bands due to the incomplete recrystallization. When the annealing temperature is higher than 630 °C, the microstructures are equiaxed ferrite grains, which mean that full recrystallization has occurred in the annealing process.

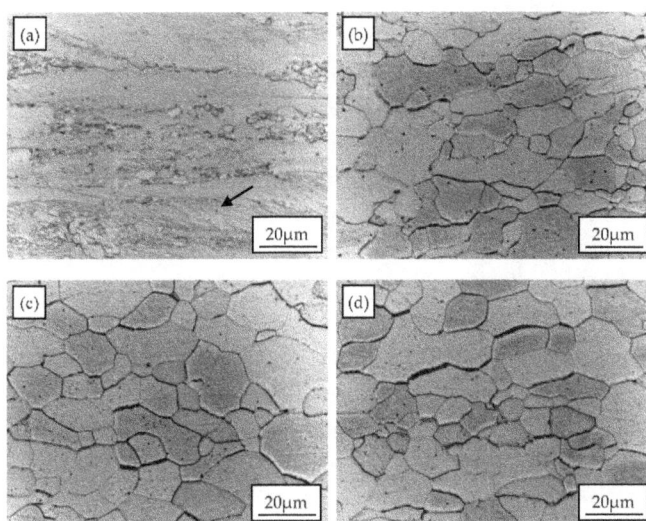

Figure 2. Optical micrographs of experimental steel batch annealed at different temperatures (**a**) 580 °C; (**b**) 630 °C; (**c**) 680 °C; (**d**) 730 °C.

The average size of ferrite grain is measured by the line intercept method, and the effect of the batch annealing temperature on the average size of ferrite grain is shown in Figure 3. The average size of ferrite grain increased when the batch annealing temperature increased from 630 to 730 °C. There is not a large number of dislocations in the specimens annealed at 630, 680 and 730 °C due to the full recrystallization.

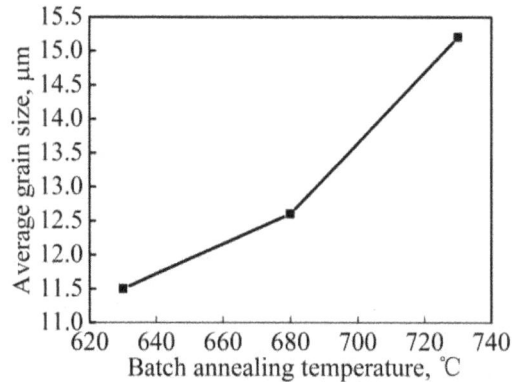

Figure 3. Effect of the batch annealing temperature on the average size of ferrite grain of ultra-low carbon enamel steel.

The precipitates of experimental steel annealed at different temperatures exhibit the same features by TEM observations. As shown in Figure 4, there are TiC, TiN and $Ti_4C_2S_2$ particles in all the specimens. Figure 4a shows typical TEM morphologies of TiN and TiC. A large cubic particle contains Ti, C and N (as shown in Figure 4c), which indicates that the particle is TiN due to its coarse and cubic shape; the peak characteristic of the C element can be ignored because of the carbon extraction replicas. Large TiN particles formed during the solidification process; they are rare in the specimens, and their sizes change a little at different annealing temperatures. A small sized elliptical particle contains Ti and C (as shown in Figure 4d), which is considered as TiC. In experimental steel, TiC is finer and has a denser distribution than TiN and $Ti_4C_2S_2$. The EDS spectrum of the particle in Figure 4b shows that the atomic ratio of Ti to S is close to 2 (as shown in Figure 4e), and the particle can be identified as $Ti_4C_2S_2$ [9]. The distribution of $Ti_4C_2S_2$ is very inhomogeneous, often in the shape of strings or clusters (as shown in Figure 4b).

The chemical composition of ultra-low carbon enamel steel is similar to that of ultra-low carbon Ti-bearing steel, but the former contains higher Ti and S contents to ensure that sufficient $Ti_4C_2S_2$ particles can be formed [22]. TiS was not observed in all the specimens. Figure 5 is a schematic illustration of the stability of various Ti compounds in Interstitial-Free (IF) steels as a function of the precipitation temperature [23]. The stability of the precipitates mainly depends on the temperature and the chemical composition. After the formation of TiN, TiS precipitation is likely to take place. However, the stability of TiS is quite low, and TiS decomposes during hot rolling and the coiling process [23]. TiS may also change to $Ti_4C_2S_2$ during the annealing process [9,24]. Because TiN particles are large and rare, their influence on hydrogen diffusion behavior can be ignored, and the main irreversible hydrogen traps are fine TiC and coarse $Ti_4C_2S_2$. TiC and $Ti_4C_2S_2$ are both considered as precipitating in the batch annealing process [25]. The effect of batch annealing temperature on average sizes of TiC and $Ti_4C_2S_2$ particles is shown in Figure 6. It can be seen that the average sizes of TiC and $Ti_4C_2S_2$ particles increased with increasing the batch annealing temperature, which means that TiC and $Ti_4C_2S_2$ particles coarsened in the annealing process.

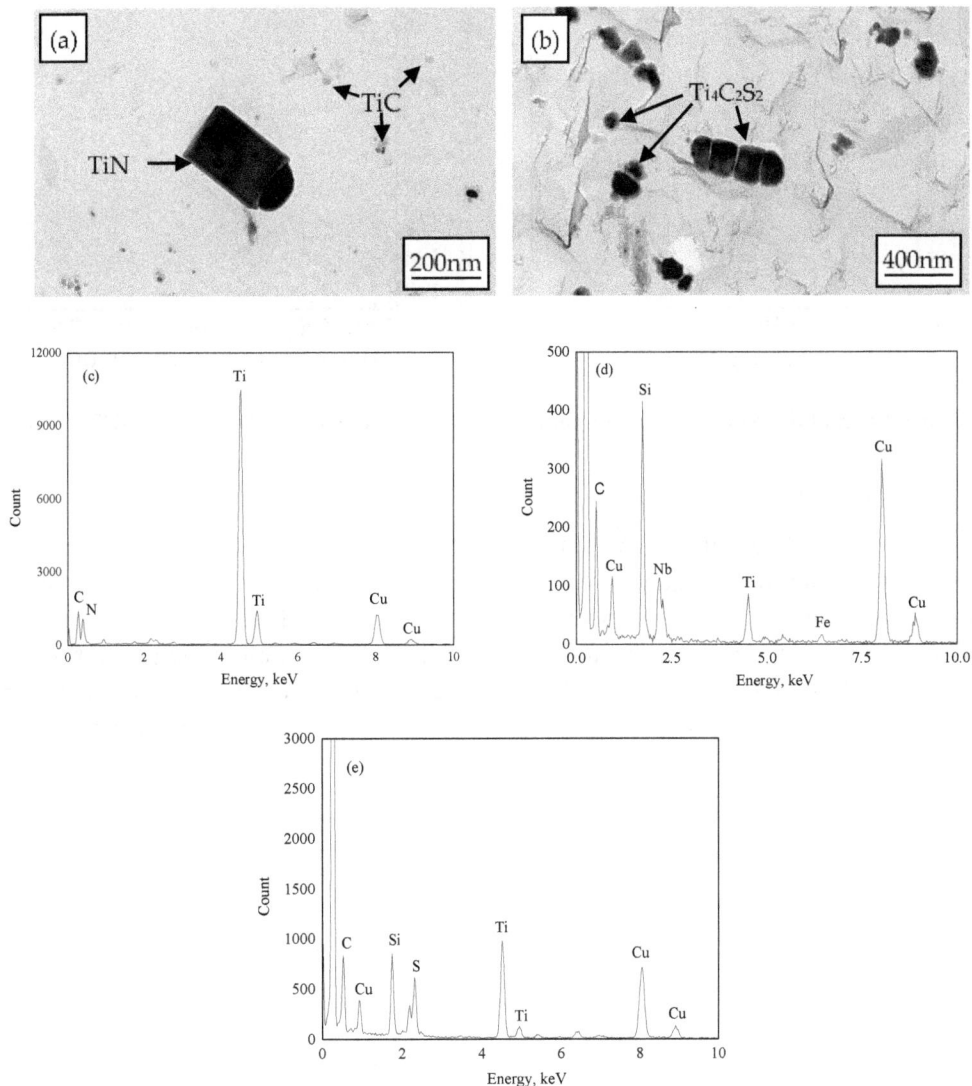

Figure 4. Transmission electron microscope (TEM) micrographs showing precipitates in experimental steel (**a**) TiN and TiC particles; (**b**) $Ti_4C_2S_2$ particles; (**c**) Energy dispersive spectrometer (EDS) spectrum of TiN; (**d**) EDS spectrum of TiC; (**e**) EDS spectrum of $Ti_4C_2S_2$.

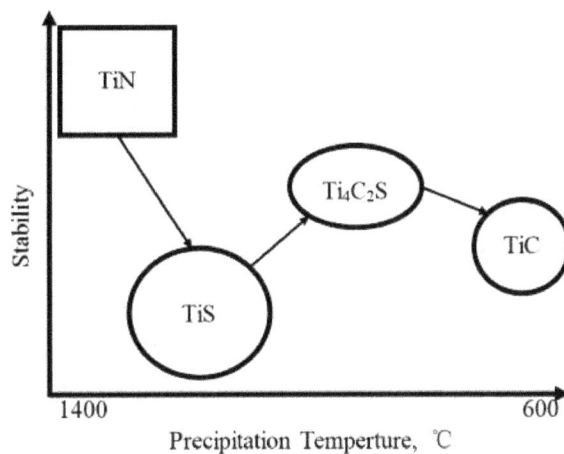

Figure 5. Schematic illustration on the stability of various Ti compounds in Interstitial-Free (IF) steels as a function of the precipitation temperature.

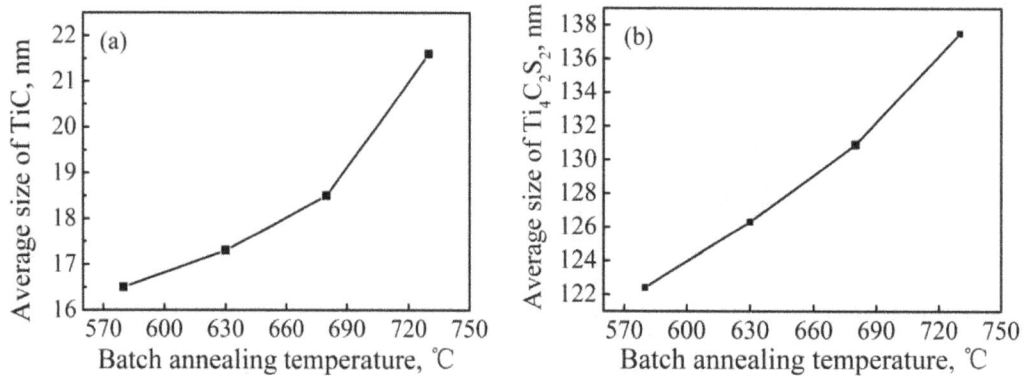

Figure 6. Effect of the batch annealing temperature on the average sizes of TiC (**a**) and $Ti_4C_2S_2$ (**b**).

3.2. Mechanical Properties and Resistance to Fish Scaling

The relationship between the mechanical properties and batch annealing temperature of ultra-low carbon enamel steel is shown in Figure 7. At the annealing temperature of 580 °C, the yield strength and tensile strength are obviously higher than those of experimental steel annealed at other temperatures; this is because of the incomplete recrystallization at 580 °C. At annealing temperatures of 630, 680, and 730 °C, the yield strength, tensile strength, elongation, n-value and r-value do not vary very much. That is to say, the mechanical properties of experimental steel change a little when recrystallization has finished.

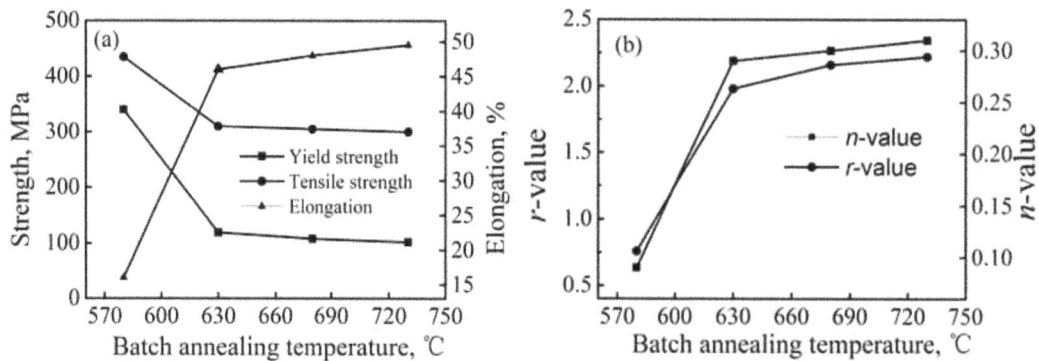

Figure 7. Relationship between the mechanical properties and batch annealing temperature of ultra-low carbon enamel steel. (**a**) Strength and Elongtation; (**b**) n and r value

The hydrogen permeation curves (charge quantity vs. time curves) are shown in Figure 8. The *TH* values of experimental steel annealed at 580, 630, 680 and 730 °C are calculated to be 30.0 ± 3.4, 18.6 ± 1.1, 13.5 ± 1.7 and 10.4 ± 1.6 min/mm^2, respectively. As shown in Figure 9, the *TH* value decreases with increasing batch annealing temperature.

The *TH* value is related to the hydrogen diffusion coefficient; high *TH* value means low hydrogen diffusivity. Both reversible traps and irreversible traps can reduce the hydrogen diffusion coefficient. At annealing temperatures of 630, 680 and 730 °C, there are a few dislocations in the specimen due to the recrystallization, so the main reversible hydrogen traps are ferrite grain boundaries, while the main irreversible hydrogen traps are TiC and $Ti_4C_2S_2$ particles. With the annealing temperature increasing, the average size of ferrite grain and the mean sizes of TiC and $Ti_4C_2S_2$ particles increase. The ferrite grain boundary is a kind of reversible trap; it traps hydrogen atoms in the hydrogen charging process and contributes significantly to hydrogen trapping (*TH* value) [26,27]. The hydrogen storage capacity of the grain boundaries is generally proportional to the grain boundary area [28]. The growth of ferrite grain leads to the reduction of the grain boundary area, which is one reason for the reduction of the resistance to fish scaling. Takahashi et al. [29] have observed the hydrogen trapping sites of nano-sized TiC by using three dimensional atom probe (3DAP) for the first time. They revealed that the broad interface between the matrix and TiC was the main trapping site. Wei et al. [30] found that incoherent TiC particles have higher binding energy than that of coherent TiC particles, but they are not able to trap hydrogen during cathodic charging at room temperature due to its high energy barrier for trapping. Small TiC and $Ti_4C_2S_2$ particles at low annealing temperature have more interfaces, which will result in the decrease of hydrogen diffusivity. The coarsening of precipitates is the other reason for the reduction of resistance to fish scaling. Yuan [2] also indicated that small precipitates can lead to low hydrogen diffusivity. Because irreversible traps always lead to a great decrease in hydrogen diffusivity [31], a great number of small precipitates are favorable to the enhancement of the resistance to fish scaling of enamel steel.

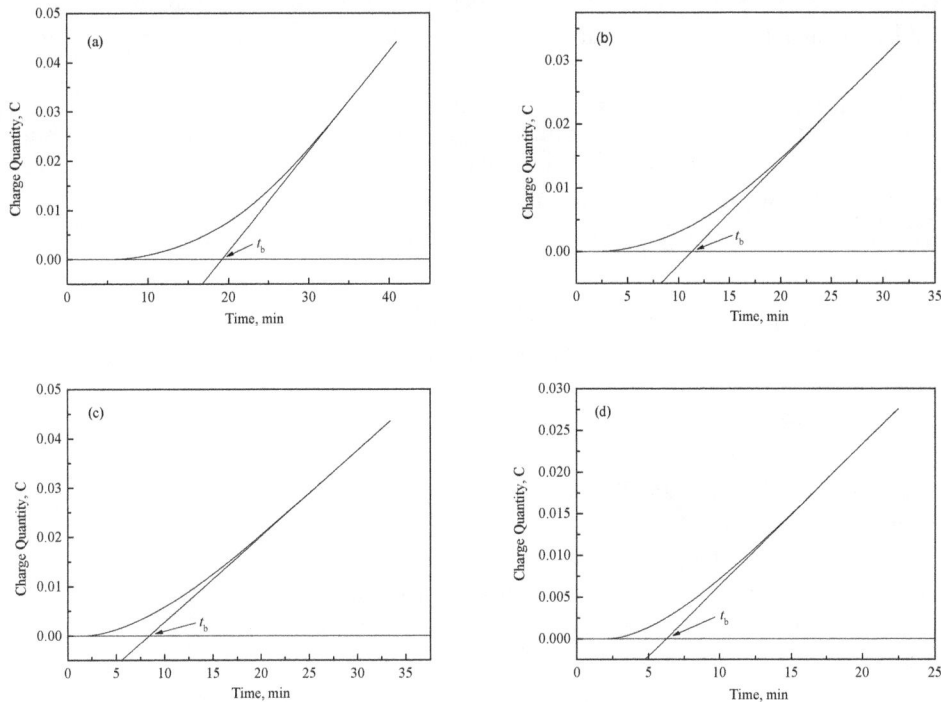

Figure 8. Hydrogen permeation curves (charge quantity vs. time curves) of ultra-low carbon enamel steel annealed at different temperatures (**a**) 580 °C; (**b**) 630 °C; (**c**) 680 °C; (**d**) 730 °C.

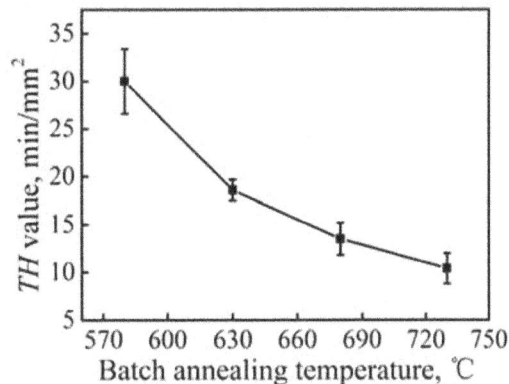

Figure 9. Effect of the batch annealing temperature on the *TH* value of ultra-low carbon enamel steel.

At an annealing temperature of 580 °C, there are much more dislocations and grain boundaries in the specimen due to the incomplete recrystallization. Dislocation is another kind of reversible trap, and it can reduce the hydrogen diffusion coefficient of steel [32]. The decrease of the hydrogen diffusion coefficient means the increase of the *TH* value. More dislocations can lead to higher *TH* value. The presence of a large number of dislocations is the reason why the *TH* value is higher than those of experimental steel annealed at other temperatures.

4. Conclusions

(1)   The main irreversible hydrogen traps of ultra-low carbon enamel steel are fine TiC and coarse $Ti_4C_2S_2$ particles. The reversible hydrogen traps are grain boundaries and dislocations. The mean sizes of TiC and $Ti_4C_2S_2$ particles increase with increasing the batch annealing temperature.

(2)   Both reversible and irreversible traps will influence the resistance to fish scaling. The resistance to fish scaling can be enhanced with increasing the number of reversible and irreversible traps. The resistance to fish scaling decreases with increasing the batch annealing temperature, which is caused by the growth of ferrite grain and the coarsening of TiC and $Ti_4C_2S_2$ particles.

Author Contributions: Zaiwang Liu analyzed the experimental data and wrote the manuscript. Yonglin Kang conceived and designed the experiments. Zhimin Zhang and Xiaojing Shao performed the experimental work. All authors have participated in the discussion of results.

Conflicts of Interest: The authors declare no conflict of interest.

References

1.   Dong, F.T.; Du, L.X.; Liu, X.H.; Hu, J.; Xue, F. Effect of Ti(C, N) precipitation on texture evolution and fish-scale resistance of ultra-low carbon Ti bearing enamel steel. *J. Iron Steel Res. Int.* **2013**, *20*, 39–45. [CrossRef]

2.   Yuan, X. Precipitates and hydrogen permeation behavior in ultra-low carbon steel. *Mater. Sci. Eng. A* **2007**, *452*, 116–120. [CrossRef]

3.   Son, Y.K.; Lee, C.J.; Lee, J.M.; Kim, B.M. Deformation prediction of porcelain-enameled steels with strain history by press forming and high-temperature behavior of coating layer. *Trans. Nonferr. Met. Soc. China* **2012**, *22*, s838–s844. [CrossRef]

4.   Okuyamas, T.; Nishimoto, A.; Kurokawa, T. New type cold rolled steel sheet for enamelling produced by the continuous casting method. *Vitreous Enamel.* **1990**, *41*, 49–60.

5.   Mohtadi-Bonab, M.A.; Szpunar, J.A.; Collins, L.; Stankievech, R. Evaluation of hydrogen induced cracking behavior of API X70 pipeline steel at different heat treatments. *Int. J. Hydrog. Energy* **2014**, *39*, 6076–6088. [CrossRef]

6.   Krom, A.H.; Bakker, A. Hydrogen trapping models in steel. *Metall. Mater. Trans. B* **2000**, *31*, 1475–1482. [CrossRef]

7.   Liu, Q.; Atrens, A. Reversible hydrogen trapping in a 3.5NiCrMoV medium strength steel. *Corros. Sci.* **2015**, *96*, 112–120. [CrossRef]

8.   Liu, Q.L.; Venezuela, J.; Zhang, M.X.; Zhou, Q.J.; Atrens, A. Hydrogen trapping in some advanced high strength steels. *Corros. Sci.* **2016**, *111*, 770–785. [CrossRef]

9.   Hua, M.; Garcia, C.I.; Eloot, K.; DeArdo, A.J. Identification of Ti-S-C-containing multi-phase precipitates in ultra-low carbon steels by analytical electron microscopy. *ISIJ Int.* **1997**, *37*, 1129–1132. [CrossRef]

10.   Funakawa, Y.; Shiozaki, T.; Tomita, K.; Yamamoto, T.; Maeda, E. Development of high strength hot-rolled sheet steel consisting of ferrite and nanometer-sized carbides. *ISIJ Int.* **2004**, *44*, 1945–1951. [CrossRef]

11.   Huo, X.; Li, L.; Peng, Z.; Chen, S. Effect of TMCP schedule on precipitation, microstructure and properties of Ti-microalloyed high strength steel. *J. Iron Steel Res. Int.* **2016**, *23*, 593–601. [CrossRef]

12.   Kim, Y.W.; Song, S.W.; Seo, S.J.; Hong, S.; Lee, C.S. Development of Ti and Mo micro-alloyed hot-rolled high strength sheet steel by controlling thermomechanical controlled processing schedule. *Mater. Sci. Eng. A* **2013**, *565*, 430–438. [CrossRef]

13.   Xu, Y.; Zhang, W.; Sun, M.; Yi, H.; Liu, Z. The blocking effects of interphase precipitation on dislocations' movement in Ti-bearing micro-alloyed steels. *Mater. Lett.* **2015**, *139*, 177–181. [CrossRef]

14.   Shi, J.; Wang, X. Comparison of precipitate behaviors in ultra-low carbon, titanium-stabilized interstitial free steel sheets under different annealing processes. *J. Mater. Eng. Perform.* **1999**, *8*, 641–648. [CrossRef]

15. Ghosh, P.; Ghosh, C.; Ray, R.K.; Bhattcharjee, D. Precipitation behavior and texture formation at different stages of processing in an interstitial free high strength steel. *Scr. Mater.* **2008**, *59*, 276–278. [CrossRef]

16. Devanathan, M.A.V.; Stachurski, Z. The mechanism of hydrogen evolution on iron in acid solutions by determination of permeation rates. *J. Electrochem. Soc.* **1964**, *111*, 619–623. [CrossRef]

17. International Organization for Standardization. *Method of Measure of Hydrogen Permeation and Determination of Hydrogen Uptake and Transport in Metals by an Electrochemical Technique*; ISO: Geneva, Switzerland, 2004.

18. Mohtadi-Bonab, M.A.; Szpunar, J.A.; Razavi-Tousi, S.S. A comparative study of hydrogen induced cracking behavior in API 5L X60 and X70 pipeline steels. *Eng. Fail. Anal.* **2013**, *33*, 163–175. [CrossRef]

19. Mohtadi-Bonab, M.A.; Karimdadashi, R.; Eskandari, M.; Szpunar, J.A. Hydrogen-induced cracking assessment in pipeline steels through permeation and crystallographic texture measurements. *J. Mater. Eng. Perform.* **2016**, *25*, 1781–1793. [CrossRef]

20. Haq, A.J.; Muzaka, K.; Dunne, D.P.; Calka, A.; Pereloma, E.V. Effect of microstructure and composition on hydrogen permeation in X70 pipeline steels. *Int. J. Hydroen Energy* **2013**, *38*, 2544–2556. [CrossRef]

21. Winzer, N.; Rott, O.; Thiessen, R.; Thomas, I.; Mraczek, K.; Höche, T.; Wright, L.; Mrovec, M. Hydrogen diffusion and trapping in Ti-modified advanced high strength steels. *Mater. Des.* **2016**, *92*, 450–461. [CrossRef]

22. Yoshinaga, N.; Ushioda, K.; Akamatsu, S.; Akisue, O. Precipitation behavior of sulfides in Ti-added ultra low-carbon steels in austenite. *ISIJ Int.* **1994**, *34*, 24–32. [CrossRef]

23. Ghosh, P.; Ghosh, C.; Ray, R.K. Thermodynamics of precipitation and textural development in batch-annealed interstitial-free high-strength steels. *Acta Mater.* **2010**, *58*, 3842–3850. [CrossRef]

24. Hua, M.; Garcia, C.I.; DeArdo, A.J. Multi-phase precipitates in interstitial-free steels. *Scr. Metall. Mater.* **1993**, *28*, 973–978. [CrossRef]

25. Carabajar, S.; Merlin, J.; Massardier, V.; Chabanet, S. Precipitation evolution during the annealing of an interstitial-free steel. *Mater. Sci. Eng. A* **2000**, *281*, 132–142. [CrossRef]

26. Ono, K.; Meshii, M. Hydrogen detrapping from grain boundaries and dislocations in high purity iron. *Acta Metall. Mater.* **1992**, *40*, 1357–1364. [CrossRef]

27. Choo, W.Y.; Lee, J.Y. Thermal analysis of trapped hydrogen in pure iron. *Metall. Trans. A* **1982**, *13*, 135–140. [CrossRef]

28. Valentini, R.; Solina, A.; Matera, S.; de Gregorio, P. Influence of titanium and carbon contents on the hydrogen trapping of microalloyed steels. *Metall. Mater. Trans. A* **1996**, *27*, 3773–3780. [CrossRef]

29. Takahashi, J.; Kawakami, K.; Kobayashi, Y.; Tarui, T. The first direct observation of hydrogen trapping sites in TiC precipitation-hardening steel through atom probe tomography. *Scr. Mater.* **2010**, *63*, 261–264. [CrossRef]

30. Wei, F.G.; Tsuzaki, K. Quantitative analysis on hydrogen trapping of TiC particles in steel. *Metall. Mater. Trans. A* **2006**, *37*, 331–353. [CrossRef]

31. Pressouyre, G.M.; Bernstein, I.M. A quantitative analysis of hydrogen trapping. *Metall. Trans. A* **1978**, *9*, 1571–1580. [CrossRef]

32. Jebaraj, J.J.M.; Morrison, D.J.; Suni, I.I. Hydrogen diffusion coefficients through Inconel 718 in different metallurgical conditions. *Corros. Sci.* **2014**, *80*, 517–522. [CrossRef]

8

Statistical Approach to Optimize the Process Parameters of HAZ of Tool Steel EN X32CrMoV12-28 after Die-Sinking EDM with SF-Cu Electrode

Ľuboslav Straka [1],*, Ivan Čorný [2], Ján Piteľ [3] and Slavomíra Hašová [1]

[1] Department of Manufacturing Processes Operation, The Technical University of Košice, Štúrova 31, 08001 Prešov, Slovakia; slavomira.hasova@tuke.sk

[2] Department of Science and Research, The Technical University of Košice, Bayerova 1, 08001 Prešov, Slovakia; ivan.corny@tuke.sk

[3] Department of Mathematics, Informatics and Cybernetics, The Technical University of Košice, Bayerova 1, 08001 Prešov, Slovakia; jan.pitel@tuke.sk

* Correspondence: luboslav.straka@tuke.sk

Academic Editor: Hugo F. Lopez

Abstract: The paper describes the results of the experimental research of the heat affected zone (HAZ) of an eroded surface after die-sinking electrical discharge machining (EDM). The research was carried out on chrome-molybdenum-vanadium alloyed tool steel EN X32CrMoV12-28 (W.-Nr. 1.2365) after die-sinking EDM with a SF-Cu electrode. The aim of the experimental measurements was to contribute to the database of knowledge that characterizes the significant impact of the main technological and process parameters on the eroded surface properties during die-sinking EDM. The quality of the eroded surface was assessed from the viewpoint of surface roughness, microhardness variation, and the total HAZ depth of the thin sub-surface layer adjacent to the eroded surface. On the basis of measurement results, mathematical models were established by statistical methods. These models can be applied for computer simulation and prediction of the resultant quality of the machined surface after die-sinking EDM. The results achieved by simulation were compared with the results of experimental measurements and high correlation indexes between the predicted and real values were achieved. Suggested mathematical models can be also applied for the determination of the optimal combination of significant technological parameters in order to minimize microhardness and total HAZ depth variations of tool steel EN X32CrMoV12-28 after die-sinking EDM with a SF-Cu electrode.

Keywords: electro-erosion; microhardness; affected zone; tool steel; electrode

1. Introduction

Electrical discharge machining (EDM) is in general characterized by thermal processes that take place directly on the eroded surface and proceed into the inner material. Therefore, it can be expected that certain microstructural changes occur in the thin sub-surface layer also called the heat affected zone (HAZ). This phenomenon has been mentioned by many researchers, to name a few: Kompella et al., Ťavodová, and Choudhary et al. [1–3]. The researchers claim in their papers, that the surfaces of the material before, and after EDM differ substantially. They put an emphasis mainly on the significance of qualitative parameters such as surface microhardness and the total HAZ depth. The microhardness variation in HAZ is also accompanied by a change of the microstructure, the fact mentioned in Shrestha's research [4]. The microhardness variation in HAZ is typically manifested by its decline, which has in many cases an adverse impact on the products' functional surfaces produced by die-sinking EDM technology. These assumptions have also been confirmed by the studies of many authors, of whom we can mention Čada, Švecová, and Abu Zeid [5–7]. The authors state in their articles that particularly in this regard, tool steel products such as shearing tools, and molds are the most susceptible. According to Marafona [8], there is occurrence of a so-called black layer (BL) directly on the surface of tool steels after die-sinking EDM. However, the black layer has not shown such an adverse impact on machined surface quality as the presence of a so-called white layer (WL) which is located just below the black layer. The creation of the white layer is caused by the action of high temperature on the machined surface followed by rapid cooling (quenching). WL represents the structure after secondary hardening as a consequence of the conversion of the residual austenite. The thickness of the WL falls into the μm range. According to Ekmekci and Zang [9,10], due to heating and subsequent rapid cooling by dielectric liquid, additional residual stresses are generated in the HAZ. The size and extent of residual stresses in the HAZ depends on the combination of electrical discharge intensity and the cooling effect of the dielectric fluid. These residual stresses often result in the formation of microcracks, which undermine the overall integrity of the machined surface. Removal of these microcracks is in practice very difficult, particularly in very hard materials such as hardened tool steels. Not all studies, however, indicate an adverse impact of the presence of the HAZ in tool steels. However, a small group of researchers, which include Dewangan and Sidhom [11,12], mention benefits of the presence of the HAZ over a certain period of technical life of the tool produced by die-sinking EDM technology. This is in particular the period of the first phase of the life-cycle of the tool, in which the integrity of the WL is not compromised. During this period, the tools produced by die-sinking EDM show increased durability compared to the tools made by traditional technologies. On the contrary, when WL is damaged [13], a higher wear occurs on the functional surfaces of the tool produced by die-sinking EDM when compared to the tool produced e.g., by milling. The overview shows that the heat affecting extent presents a significant impact on performance and life-cycle of tools made by progressive technology of die-sinking EDM. Suitable control of the heat affecting extent in sub-surface layers of eroded material, in terms of microhardness variation, should result in an increase of the durability and the working life of tools produced by this technology. Proper setting of significant technological and process parameters of EDM is essential to achieve a surface with predefined specific properties. The aim of the experimental research, therefore, was to contribute to the existing knowledge database by clearly articulating the particular laws in relation to the processes that occur directly below the eroded surface. Relying on the results of experimental measurements of the quality of machined surface reached in terms of microhardness variation, and the total HAZ depth of chrome-molybdenum-vanadium alloyed tool steel EN X32CrMoV12-28 (W.-Nr. 1.2365), mathematical models were designed using statistical methods. The purpose of the models is efficient computer simulation and prediction of the final quality of the machined surface in terms of the observed qualitative indicators on the basis of an appropriate combination of significant technological and process parameters.

2. Materials and Methods

As was already mentioned, the resulting structure of the subsurface layer of the eroded surface and its properties have a significant influence on the durability and service life of tools (e.g., parts of molds, shearing tools, etc.), produced by die-sinking EDM technology. Nevertheless, the quality of the machined surface after die-sinking EDM is often evaluated solely on the basis of selected parameters of roughness. These quality parameters are defined in detail in the valid standard EN ISO 4287 [14]. However, particular parameters of roughness of machined surfaces fail to include important parameters related to the microstructure changes of the sub-surface layers of the machined surfaces. In general, the microstructure change of metal materials is accompanied by a change of microhardness. This change occurs as a result of the influence of the secondary hardening caused by extreme heat effects of electrical discharge between the tool and the workpiece, with the consequent rapid cooling by dielectric fluid. On the basis of the facts mentioned above, it can be considered that microhardness measurement is the appropriate method for evaluating changes in the microstructure of the sub-surface layers after die-sinking EDM. A detailed procedure of Vickers's microhardness test is defined in standard EN ISO 6507 [15]. In addition to the characteristic change of the material microhardness in HAZ after die-sinking EDM, the total HAZ depth is also a decisive parameter. The change of microhardness and total HAZ depth is primarily dependent on the physical, mechanical, and chemical properties of the machined material. It is also dependent on the setting range of the main technological and process parameters. Table 1 shows the overview of the main technological parameters for die-sinking EDM that in general have an essential influence on microhardness change, and also have a direct impact on the total HAZ depth.

Table 1. Setting range of the main technological parameters for die-sinking electrical discharge machining (EDM), and their anticipated impact on the change of microhardness and the total depth of the heat affected zone (HAZ).

Technological Parameters	Operation	Setting Range	Influence of Technological Parameter on Microhardness	Influence of Technological Parameter on HAZ
Peak current I (A)	roughing semifinishing finishing	40.0–60.0 10.0–40.0 2.0–10.0	With an increase of value of parameter I microhardness grows.	With an increase of value of parameter I total HAZ depth grows markedly.
Pulse on-time duration t_{on} (µs)	roughing semifinishing finishing	150.0–300.0 50.0–150.0 5.0–50.0	With an increase of value of parameter t_{on} microhardness grows markedly.	With an increase of value of parameter t_{on} total HAZ depth grows markedly.
Pulse off-time duration t_{off} (µs)	roughing semifinishing finishing	75.0–120.0 35.0–75.0 5.0–35.0	With an increase of value of parameter t_{off} microhardness slightly declines.	With an increase of value of parameter t_{off} total HAZ depth slightly declines.
Voltage of discharge U (V)	roughing semifinishing finishing	70–90 70–95 75–95	With a change of value of parameter U microhardness varies only slightly.	With a change of value of parameter U total HAZ depth varies only slightly.

Table 1 shows the expected impact of the main technological parameters, i.e., peak current I, pulse on-time duration t_{on}, related pause for recovery of a discharge channel–pulse off-time duration t_{off}, and voltage of discharge U, during die-sinking EDM, on microhardness change and total HAZ depth.

As mentioned above, microhardness change and total HAZ depth is also dependent on process parameters. The overview of the main process parameters for die-sinking EDM, which substantially affect the change of microhardness and have a direct impact on the total depth of the HAZ, is given in Table 2.

Table 2. Setting range of process parameters at die-sinking EDM with a SF-Cu electrode of φ20 mm diameter, and their expected impact on microhardness change and total HAZ depth.

Processing Parameters	Operation	Setting Range	Influence of Processing Parameter on Microhardness	Influence of Processing Parameter on HAZ
Feed rate of the tool electrode v_f (mm·min^{-1})	roughing semifinishing finishing	0.2–0.4 0.1–0.2 0.05–0.1	With an increase of value of parameter v_f microhardness grows only slightly.	With an increase of value of parameter v_f total HAZ depth grows marginally.
Intensity of the volumetric material removal MRR (mm^3·min^{-1})	roughing semifinishing finishing	60.0–120.0 30.0–60.0 15.0–30.0	With an increase of value of parameter MRR microhardness grows substantially.	With an increase of value of parameter MRR total HAZ depth grows markedly.

Table 2 shows the setting range of process parameters, i.e., feed rate of the tool electrode v_f, and intensity of the volumetric material removal (MRR) during die-sinking EDM with a SF-Cu electrode of φ20 mm diameter. Table 2 also shows the expected impact of the given process parameters on the microhardness change and total HAZ depth.

2.1. Equipment and Tools Used in Experiments

The experimental samples were made using a CNC electrical discharge machining facility (Figure 1) Sodick AG60L Sinker EDM (Sodick Europe Ltd., Rowley Drive, Baginton, UK). The tool applied in machining was a cylindrical SF-Cu electrode with dimension $\phi20 \times 60$ mm. Die-sinking EDM was performed in dielectric liquid on the basis of non-ionized water with electric conductivity less than 10 μS·cm^{-1}.

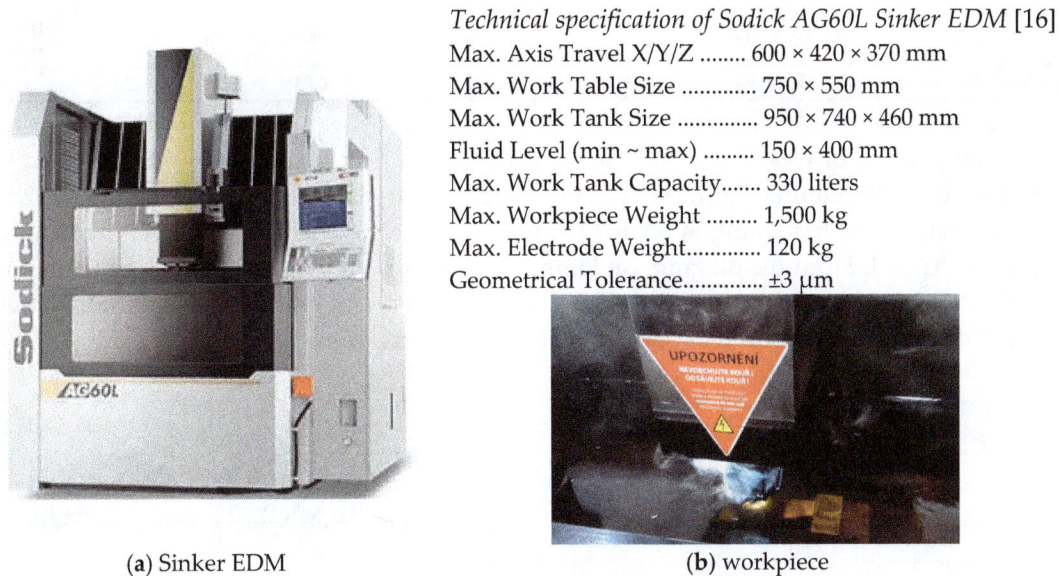

Technical specification of Sodick AG60L Sinker EDM [16]
Max. Axis Travel X/Y/Z .... 600 × 420 × 370 mm
Max. Work Table Size .... 750 × 550 mm
Max. Work Tank Size .... 950 × 740 × 460 mm
Fluid Level (min ~ max) .... 150 × 400 mm
Max. Work Tank Capacity....... 330 liters
Max. Workpiece Weight .... 1,500 kg
Max. Electrode Weight............. 120 kg
Geometrical Tolerance............. ±3 μm

(**a**) Sinker EDM　　　　　　　　　　　　　(**b**) workpiece

Figure 1. CNC Sodick AG60L Sinker EDM.

Table 3 shows basic mechanical properties and chemical composition of SF-Cu material which was used for the die-sinking EDM electrode production.

Table 3. Basic mechanical properties and chemical composition of SF-Cu electrode material [17].

Electrode Labeling	Chemical Composition in %					Mechanical Properties	
	Cu	Zn	Al	Bi	Pb	After Annealing HB max.	Tensile Strength MPa min.
SF-Cu (W.-Nr. 2.0090)	99.8	0.057	0.15	0.0011	0.0008	65	210

2.2. Production of Experimental Samples

In the experiments, samples made from steel block of dimensions $70 \times 40 \times 10$ mm were used. The experimental block was made from medium-alloyed tool steel EN X32CrMoV12-28 (W.-Nr. 1.2365, STN 19 541). Tensile strength limit (TS) after heat treatment (martensitic hardening + tempering for removal of internal stresses) of the material EN X32CrMoV12-28 ranges from 1250 MPa, at basic material (BM) hardness approx. 40 HRC, to 1850 MPa at BM hardness approx. 52 HRC. The material is medium alloyed chrome-molybdenum-vanadium tool steel with the following alloying element content: 0.28%–0.35% C, 2.7%–3.2% Cr, and 2.6%–3.0% Mo. The material has at 20 °C low heat conductivity (30 Wm^{-1}·K^{-1}), and high electrical conductivity (2.70 Siemens·m·mm^{-2}) so it is suitable for machining by die-sinking EDM technology. The material is mostly used for production of highly stressed tools for increased temperature applications, e.g., for processing of alloys containing heavy metals. These are the components such as pressing brackets and matrixes, die forging inserts, highly stressed parts of molds for pressure die casting, shearing tools, etc. The material is characterized by high temperature strength and tempering resistance, good resistance to burn-off, and good thermal conductivity. At the same time the material exhibits very good resistance to formation of heat-caused microcracks, and low sensitivity to sudden changes of temperature. Table 4 describes the basic chemical composition of tool steel EN X32CrMoV12-28 that was used for production of the experimental samples.

Table 4. Basic chemical composition of tool steel EN X32CrMoV12-28 [18].

| Steel Labeling | Chemical Composition in % | | | | | | | | Hardness in State | |
| | | | | | | | | | Soft Annealed | Refined |
	C	Si	Mn	Cr	Mo	V	P max.	S max.	HB max.	HRC min.
EN X32CrMoV12-28	0.28–0.35	0.1–0.4	0.15–0.45	2.7–3.2	2.6–3.0	0.4–0.7	0.03	0.03	230	52

The metal block of tool steel EN X32CrMoV12-28 was heat-treated before the experimental die-sinking EDM. The required hardness of 52 HRC and the material tensile strength of 1850 MPa were reached by martensitic hardening in oil at a temperature of about 1040 °C, followed by tempering at about 450 °C [18]. From the diagram on Figure 2 it can be seen that low tempering temperature (up to about 500 °C) has almost no effect on reducing the hardness of the basic material. Therefore, this material is suitable for tools that operate at elevated temperatures.

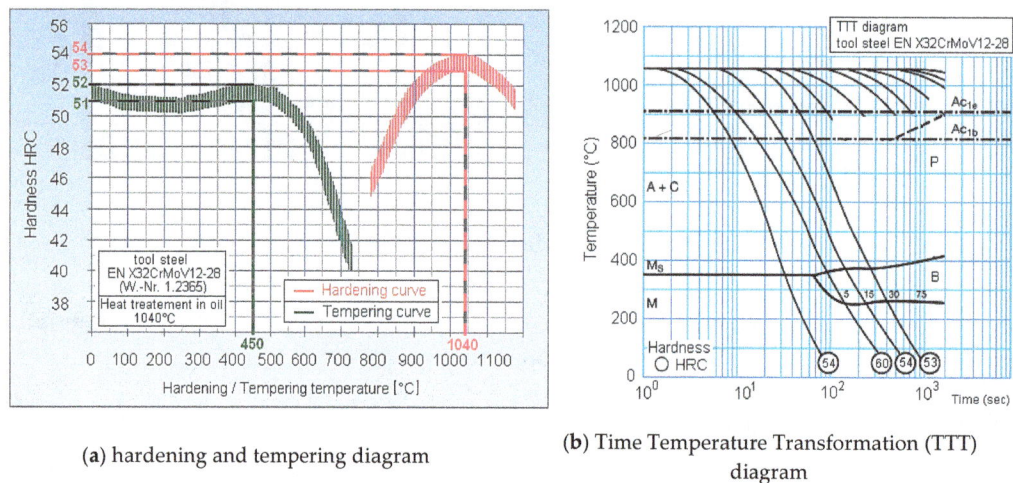

(a) hardening and tempering diagram

(b) Time Temperature Transformation (TTT) diagram

Figure 2. Hardening, tempering and TTT diagram of tool steel EN X32CrMoV12-28 [18].

Standard tempering temperature for the tool steels is around 200–250 °C. Since the tool steel EN X32CrMoV12-28 is stable at elevated temperatures, stress relief tempering had to be done at 450 °C. This temperature ensures a significant reduction of material internal stress after martensitic hardening, while the decrease of BM hardness is minimal.

Figure 3 shows the eroded areas of the experimental samples of tool steel EN X32CrMoV12-28 using a SF-Cu electrode after roughing, semifinishing, and finishing operations.

Ra > 3.6 µm   3.6 µm > Ra > 2.5 µm   Ra < 2.5 µm
Rz > 17.0 µm   17.0 µm > Rz > 8.0 µm   Rz < 8.0 µm

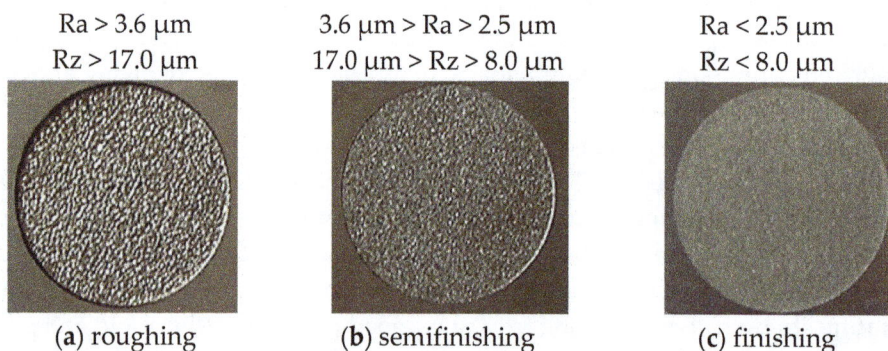

(a) roughing   (b) semifinishing   (c) finishing

Figure 3. Eroded areas of experimental samples from tool steel EN X32CrMoV12-28 after die-sinking EDM roughing, semifinishing, and finishing with SF-Cu electrode.

Photographs in Figure 3 illustrate the essential differences of the character of machined surfaces after die-sinking EDM with a SF-Cu electrode. The machined surface after the roughing operation (Figure 3a) shows an extremely varied topography and coarse structure, on the contrary, the surface after the finishing operation (Figure 3c) has a very fine surface structure and is smooth. For higher resolution, the surface of the experimental samples of tool steel EN X32CrMoV12-28 after die-sinking EDM was observed using a digital microscope Keyence VHX-5000 (Keyence International, Mechelen, Belgium) with 500× magnification. On the eroded surface, a characteristic change of relief and microstructure compared to the basic material (Figure 4) was observed. Surface integrity remained preserved. At the same time no significant microcracks or other surface discontinuities were observed. This can be attributed to the proper combination of the main technological and process parameters, in combination with the appropriate heat treatment of the experimental samples. Figure 4b,c show, besides microstructure change, also the presence of small metal particles out-melted from the SF-Cu electrode.

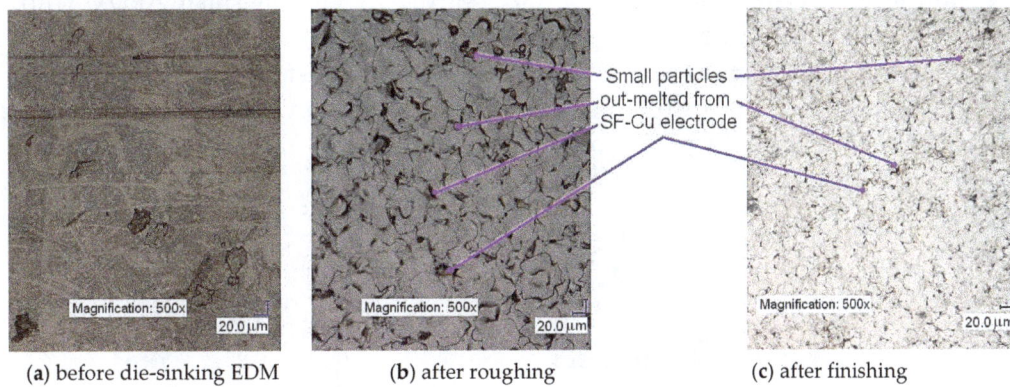

(a) before die-sinking EDM   (b) after roughing   (c) after finishing

Figure 4. Characteristic change of relief and microstructure of the surface of the material EN X32CrMoV12-28 after die-sinking EDM, roughing, and finishing in comparison to the source material, magnification 500×.

2.3. Measurements of Microhardness and Total HAZ Depth

The regular HAZ of the material EN X32CrMoV12-28 after die-sinking EDM, after roughing and finishing is created by several separate layers. Immediately after die-sinking EDM roughing, it is possible to observe a so called black layer (BL) visually—without any instruments—on the untreated eroded surface. This layer was named by a characteristic black burn-erosion which occurs on the surface due to the burning of the material. Its thickness at die-sinking EDM roughing ranges from 5 to 10 μm. After die-sinking EDM finishing, BL was not observed to be present. From the metallographic view, BL is created by the metal remnants from the tool SF-Cu electrode, and by the transformed BM.

Another layer of the HAZ of material EN X32CrMoV12-28 after die-sinking EDM with SF-Cu electrode is called the white layer (WL). The name was given by the characteristic white color that results from exposure to high temperature and subsequent rapid cooling of the workpiece surface. This is a highly carburized solidified melt on the surface of a material which is very hard, but brittle. Concerning metallography, WL is formed by nanocrystalline martensite. According to the authors Krastev and Lei [19,20], WL also has good corrosion resistance. WL is formed in die-sinking EDM roughing, as well as in finishing. The difference is in its thickness which ranges from 10 to 40 μm. The higher values apply to die-sinking EDM roughing, while lower values for the finishing. If the WL thickness exceeds 20 μm, there is a risk of occurrence of microcracks. These can penetrate up to the BM and thus reduce the service life of e.g., shearing tools and mold parts produced by this advanced technology. Figure 5 shows metallographic images of HAZ of steel EN X32CrMoV12-28 after die-sinking EDM roughing and finishing. The images were recorded by an electron microscope JEOL 5900 LV (JEOL Ltd., Tokyo, Japan) with magnification 1000×.

(**a**) roughing   (**b**) finishing

Figure 5. Total thickness of white layer (WL) in HAZ of steel EN X32CrMoV12-28 after die-sinking EDM roughing and finishing with SF-Cu electrode, recorded by electron microscope with 1000× magnification.

The third layer in HAZ of material EN X32CrMoV12-28 after die-sinking EDM with SF-Cu electrode is represented by a so-called transition layer (TL). Its name was given due to the gradual change of microstructure and microhardness. Its thickness in the tool steel EN X32CrMoV12-28 ranges from 40 to 250 μm (Figure 6). Microstructure and microhardness in TL gradually approaches the parameters of BM. Metallography of TL mostly represents tempered martensite.

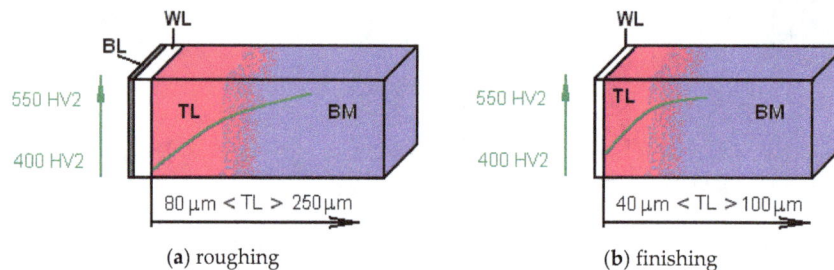

(**a**) roughing   (**b**) finishing

Figure 6. Total transition layer (TL) depth in HAZ of material EN X32CrMoV12-28 after die-sinking EDM roughing and finishing.

As was already mentioned, crystallographic structure and microhardness of TL and WL differs significantly from the structure and microhardness of BM. In both these layers, predominantly a dendritic structure can be observed, which results from crystallization processes. The extent of the crystallization processes largely depends on the physical properties of the machined material and its chemical composition. The range of the crystallization processes is also dependent on the settings of the main technological and process parameters, as well as on the conditions of dielectric liquid cooling. The presence of two different structures in the HAZ refers to the two-phase solidification process of the molten material during the die-sinking EDM. In the first phase, heat convection occurs with the surface material, due to the impact of thermal energy, which is released by the electrical discharge between the electrode and the machined material. In the second phase, the conductive removal of heat occurs from the material surface by the dielectric fluid and through the subsurface layer to the BM. Even though the structures of TL and WL are not homogeneous, the thickness of WL in the entire cross-section of the eroded surface is approximately constant. On the contrary, the thickness of TL is not constant in the surface cross-section. Its size is largely dependent on the intensity of the discharge energy during die-sinking EDM, and cooling intensity.

In addition to the problem of partial inhomogeneity of the total HAZ depth through the cross-section, there is also the problem of the choice of a suitable measurement methodology. Because the total depth of the HAZ after the die-sinking EDM with SF-Cu electrode is in the order of tens of microns (μm) graduation, it presents a certain problem with measurement methodology. Many researchers, from which we can mention Bátora [21], recommend for the detection of microhardness and total HAZ depth variation in thin sub-surface layers application of the beveled cross-sections method. However, this method is not suitable in this case, because as the test area is small, it is necessary to use a large angle of the cross-section cut. Such an angle does not guarantee sufficiently relevant results. A more appropriate method therefore seems to be the gradual removal of thin surface layers of material in the thickness range from 5 to 20 μm. This is a rather difficult and laborious method because it is necessary to make a total of several tens of metallographic sections.

The total numbers of cross-section samples n_{vc} for particular sub-surface layers (BL, WL, and TL), assuming (theoretical) thickness h_{HAZ}, were determined by following calculation:

$$n_{vc} = n_{vBL} + n_{vWL} + n_{vTL} \tag{1}$$

$$n_{vc} = \frac{h_{HAZ_t} - (n_{vBL} \cdot h_{vBL} + n_{vWL} \cdot h_{vWL})}{h_{vTL}} + (n_{vBL} + n_{vWL}) \tag{2}$$

where:

h_{HAZt}—assumed (theoretical) total HAZ thickness,
h_{vBL}—facet thickness in BL (for roughing approx. 5 μm),
h_{vWL}—facet thickness in WL (for roughing approx. 10 μm; for finishing approx. 5 μm),
h_{vTL}—facet thickness in TL (for roughing approx. 20 μm; for finishing approx. 10 μm),
n_{vBL}—number of facets in BL (for roughing 2×),
n_{vWL}—number of facets in WL (for roughing and finishing 4×),
n_{vTL}—number of facets in TL (for roughing and finishing as needed).

In the experimental assessment of changes of microhardness and total HAZ thickness of samples from tool steel EN X32CrMoV12 after die-sinking EDM roughing, with assumed total HAZ depth within the range 105–300 μm, it was necessary—according to the formula (2)—to carry out from 10 to 19 metallographic sections on every sample. Concerning samples made by die-sinking EDM finishing with assumed total HAZ depth within the range 50–120 μm, it was necessary to carry out from 8 to 14 metallographic sections on every sample. Figure 7 shows the composition of particular metallographic sections of the samples made by die-sinking EDM roughing and finishing.

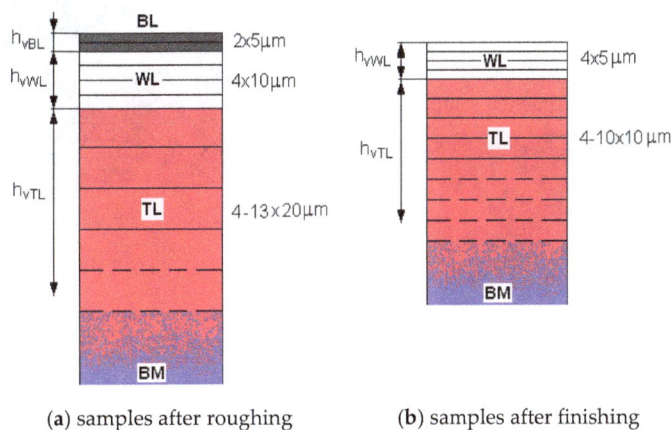

(a) samples after roughing   (b) samples after finishing

Figure 7. Composition of experimental metallographic sections for assessment of microhardness change in particular layers, and for assessment of total HAZ thickness of tool steel EN X32CrMoV12-28 after die-sinking EDM roughing and finishing with a SF-Cu electrode.

For an exact assessment of the sub-surface layers microhardness of experimental samples after die-sinking EDM with a SF-Cu electrode, metallographic sections according to Figure 7 were carried out gradually. For samples made by roughing, at first, two metallographic sections with thickness of 5 μm were prepared for assessment of BL. Subsequently, another four sections of thickness 10 μm were carried out for WL assessment. For assessment of WL in samples made by finishing, four sections of thickness 5 μm were carried out. For assessment of TL in samples made by roughing, sections of 20 μm thickness were carried out, and in samples made by finishing sections of 10 μm thickness were carried out. In both methods of machining (roughing and finishing) the metallographic sections were continually prepared for TL assessment until the measured microhardness values HV2 matched the hardness value of BM, i.e., approx. 540 HV2.

In the experimental assessment of total depth and extent of microhardness change in HAZ of tool steel EN X32CrMoV12 after die-sinking EDM with a SF-Cu electrode, it was necessary to take into account the possible dispersion of the recorded values due to the structural composition of the material. This fact is noted by several authors in their research, from which Banker [22] may be mentioned. He states that the course of microhardness and total depth of the HAZ is not constant across all cross-sections of the surface after die-sinking EDM. Because there is a reasonable presumption that the total thickness of the HAZ is not constant across all cross-sections, a pair of indentations was always carried out. The first of the pair of indentations was made on a circle with radius $r_1 = 5$ mm, the other on a circle with radius $r_2 = 9$ mm. To avoid re-injection of tetrahedral pyramids in the same place at the Vickers microhardness test, each sample was rotated by an angle α of about 20° after every pair of indentations, according to Figure 8. The experimental microhardness measurements were completed when three successive measured values of the microhardness HV2 of the both indentations stabilized at a constant value, i.e., at the hardness value of approx. 540 HV2—the value of BM. The total thickness of the HAZ was then determined as the difference between the original height and the height after carrying out of all metallographic sections, values were read on the Digimatic Test Indicator ID-CX Mitutoyo (Mitutoyo, Kawasaki, Japan).

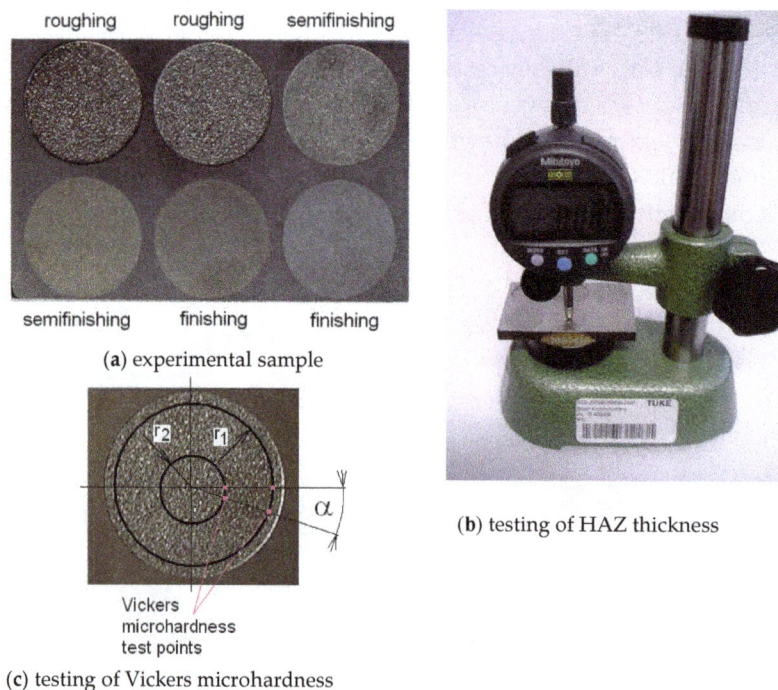

(a) experimental sample

(c) testing of Vickers microhardness

(b) testing of HAZ thickness

Figure 8. Testing of Vickers microhardness and measurement of total HAZ thickness on metallographic section of experimental sample from tool steel EN X32CrMoV12-28 after die-sinking EDM with a SF-Cu electrode.

Measurement of microhardness in HAZ on metallographic sections from experimentally prepared samples made from tool steel EN X32CrMoV12-28 after die-sinking EDM with a SF-Cu electrode was carried out using a device Zwick ZHV30 Vickers Hardness Tester (Zwick GmbH & Co. KG, Ulm,

Germany). The measuring range of the given device is HV 0.2 to HV 30. Because the hardness of the BM of tool steel EN X32CrMoV12-28 after heat treatment reaches up to 540 HV (52 HRC), it was not possible to apply Vickers microhardness test according to EN ISO 6507-1 for the measurements [15]. The load in this test is from 0.098 to 1.961 N, which is insufficient for the given BM hardness. With this load, the microhardness can be measured up to the maximum value 464 HV 0.2 which corresponds to a Rockwell hardness of approx. 48 HRC. Therefore, the Vickers hardness test at low load HV2 was applied, for which the prescribed load of the tetrahedral pyramid is 19.61 N, according to the standard EN ISO 6507.

3. Results and Discussions

The thicknesses of the individual layers of the HAZ (BL, WL, and TL) of tool steel EN X32CrMoV12-28 after die-sinking EDM with a SF-Cu electrode was at first identified on the basis of the microhardness changes. These results were then compared with the results obtained by identification of layer thicknesses based on micro-structural changes. Figure 9 shows the metallographic image of HAZ of tool steel EN X32CrMoV12-28 recorded by the electron microscope JEOL 5900 LV (JEOL Ltd., Tokyo, Japan). Figure 9 shows also the diagram of microhardness HV 2 course after die-sinking EDM roughing at maximum and minimum recorded values of total HAZ depth. The values of microhardness HV 2 in the particular HAZ layers were recorded with a Zwick ZHV30 Vickers Hardness Tester (Zwick GmbH & Co. KG, Ulm, Germany).

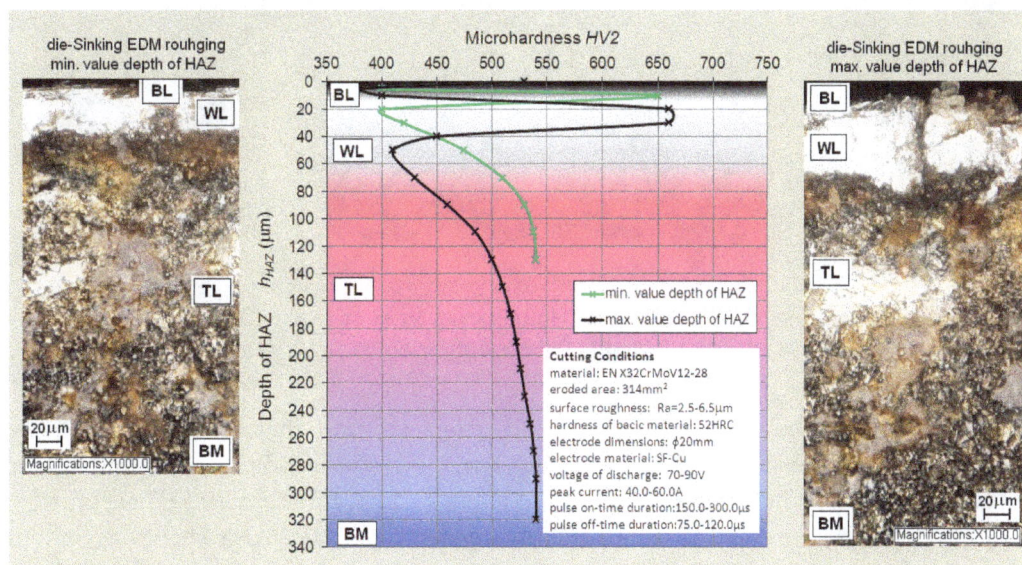

Figure 9. Boundary values of microhardness HV 2 recorded in particular HAZ layers (BL, WL, and TL) of tool steel EN X32CrMoV12-28 after die-sinking EDM roughing with a SF-Cu electrode.

In Figure 9 it can be observed that there are visible microstructural changes in the HAZ of tool steel EN X32CrMoV12-28 after die-sinking EDM roughing with a SF-Cu electrode. The microstructure of these layers differs significantly from that of the BM, as well as from each other. On the image, there are visible small microcracks, i.e., discontinuities which occur predominantly in the WL at the maximum value of the HAZ depth. However, these do not interfere with the TL or BM. The experiment showed the presence of BL directly on the surface of the eroded area. It consisted primarily of transformed BM and metal remnants melted out from the SF-Cu electrode. Visible are also traces left after material burning due to electric discharge, which causes the characteristic black color surface. From the recorded course of microhardness, fluctuations of HV 2 values in particular HAZ layers can be observed, compared to the hardness of BM. Microhardness of BL ranges from 400 to 500 HV 2. Thickness of this layer after roughing operations is between 5 and 10 μm. Another clearly visible layer in HAZ is the so-called white layer (WL). It is located just below BL. In terms of microstructure, WL is formed by nanocrystalline martensite as a result of high discharge energy and subsequent rapid quenching by dielectric fluid [23,24]. The resulting martensitic structure is involved in extreme growth of microhardness in WL, up to the level of 600 HV 2. Thickness of this layer at roughing operations ranges from 20 to 40 µm. Under the well-defined WL which is characterized by its white color, there is a so-called transition layer (TL). This layer is in terms of microstructure mainly composed of tempered martensite, which causes considerable reduction of microhardness [25]. Microhardness of TL stands at about 400 HV 2 and with increasing depth progressively approaches the microhardness of BM. The thickness of this layer after roughing operations is between 80 and 250 µm. Total HAZ thickness at die-sinking EDM roughing with SF-Cu electrode of 20 mm diameter ranges from about 120 to 280 µm.

Figure 10 shows the course of microhardness HV 2, and metallographic images of HAZ after die-sinking EDM finishing, at maximum and minimum recorded value of total HAZ depth.

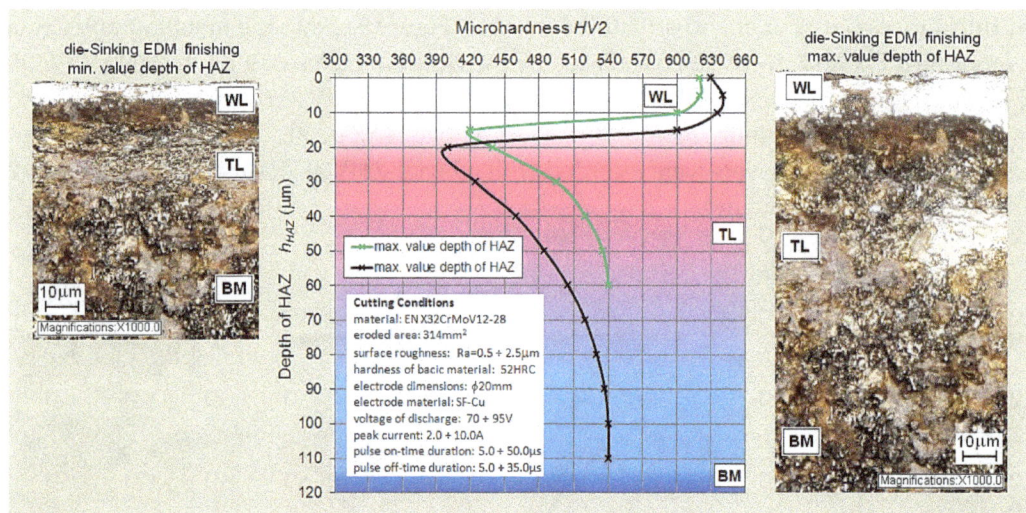

Figure 10. Microhardness HV 2 values recorded in particular HAZ layers (BL, WL, and TL) of tool steel EN X32CrMoV12-28 after die-sinking EDM roughing with SF-Cu electrode.

In Figure 10 it can be seen, as well as in the previous case, that in HAZ of tool steel EN X32CrMoV12-28 after die-sinking EDM finishing with a SF-Cu electrode visible microstructural changes occur. They are, however, to a lesser extent than for roughing operations. There are no visible microcracks or other discontinuities. This is mainly due to a lower discharge energy in combination with the proper heat treatment of BM [26]. The experiment did not show occurrence of BL directly on the eroded surface. Directly on the surface there was the so-called white layer (WL) with its typical character of microhardness increase. The microhardness value of WL ranges from 620 to 640 HV 2 and reaches a depth of about 20 µm. Below WL there was a so called transition layer (TL) in which microhardness values showed a decrease to levels from 390 to 420 HV2, and gradually from this value rose until they reached the hardness of the base material (BM) 540 HV2. The thickness of this layer after finishing operations ranges from 40 to 100 µm. Total HAZ thickness at die-sinking EDM finishing with a SF-Cu electrode of 20 mm diameter stands in the range from 60 to 100 µm.

Performed experimental measurements have shown significant effects of the type of die-sinking EDM on the microstructural changes, i.e., whether the roughing or the finishing is observed. Differences were recorded in the BL, WL, and TL. The microstructural changes are of the same nature in roughing, and in finishing [27,28]. The essential difference, however, is the extent of the changes. They manifested as microhardness variation, and achieved total HAZ thickness. Significantly lower values of total HAZ thickness were recorded in finishing operations. Differences in achieved microhardness change were recorded also [29]. The extent of the microhardness change was slightly less for finishing operations than for roughing operations. In the same time it maintained a steeper characteristic. In particular with HAZ layers, a significant difference between roughing and finishing operations can be observed. It is the absence of BL in finishing operations. In WL after die-sinking EDM roughing, the recorded microhardness values in experimental measurements were on average 40 HV 2 higher in comparison with finishing operations. On the contrary, in TL, recorded microhardness values were on average 20 HV 2 lower in roughing operations than in finishing operations. The differences were also in total TL thickness. The thickness of TL was significantly less in finishing operations compared with roughing operations.

Based on preliminary studies, relying on the results of other researchers, an assumption arose that there is a difference of total HAZ thickness between intermediate and boundary areas of the eroded surface [30]. In experimental investigations of the eroded surface of tool steel EN X32CrMoV12-28 after die-sinking EDM using a SF-Cu electrode this assumption was confirmed. Figure 11 in graphic form illustrates the differences of total HAZ thickness of a machined surface of 20 mm diameter from tool steel EN X32CrMoV12-28 after die-sinking EDM roughing and finishing with SF-Cu electrode.

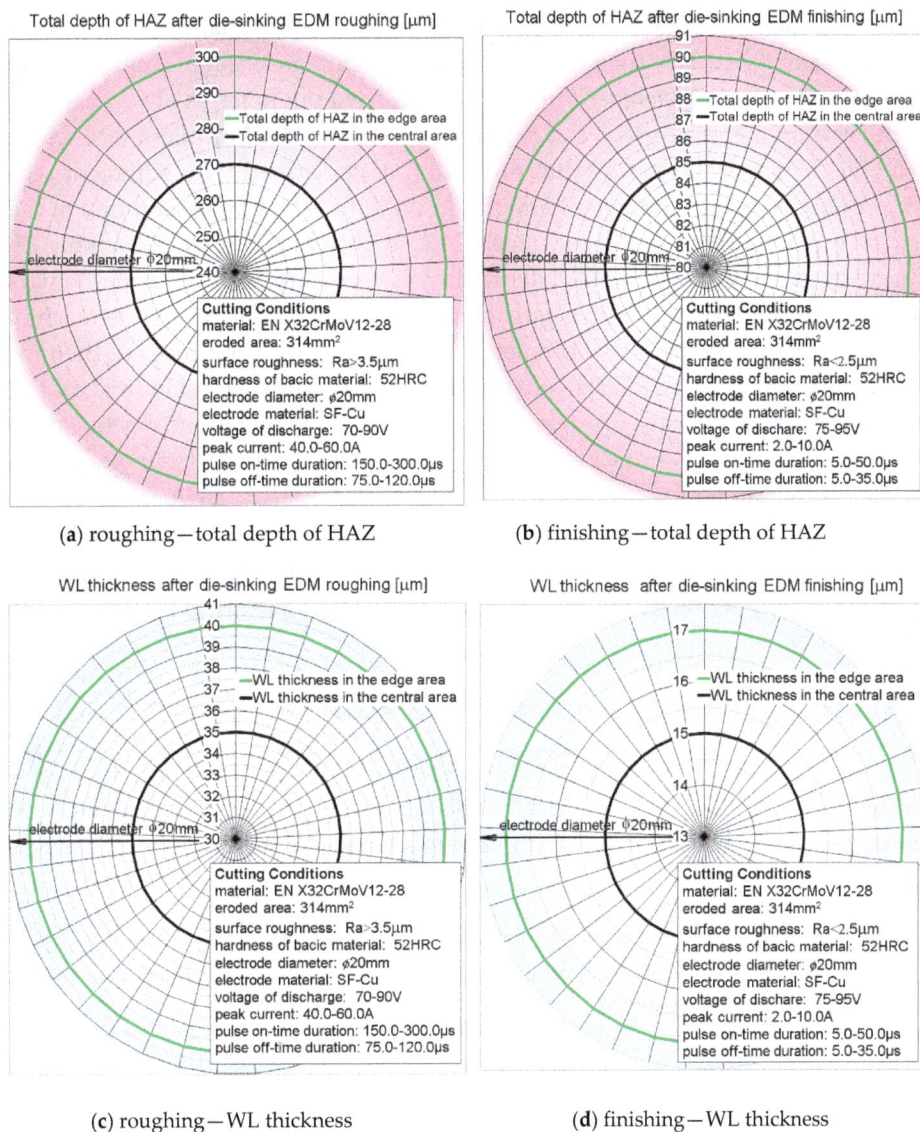

(a) roughing — total depth of HAZ

(b) finishing — total depth of HAZ

(c) roughing — WL thickness

(d) finishing — WL thickness

Figure 11. Recorded differences of total thickness of WL and HAZ of a machined surface of 20 mm diameter from tool steel EN X32CrMoV12-28 after die-sinking EDM roughing and finishing with a SF-Cu electrode.

From the plots in Figure 11, small variations of WL thickness and total thickness of the HAZ can be observed between intermediate and boundary areas of the machined surface from tool steel EN X32CrMoV12-28 after die-sinking EDM roughing, and also finishing with a SF-Cu electrode. These variations are obviously caused by excessive overheating of the base material in the middle part, due to the lower cooling intensity of the material by dielectric liquid. This effect is somewhat less pronounced for finishing operations compared to die-sinking EDM roughing because of the lower intensity of electrical discharges. Deviation of the measured values of the WL thickness and the total thickness of the HAZ in the middle and boundary area of the machined surface with 20 mm diameter was around 10% for roughing and 5% finishing operations. Overall lower values were obtained in finishing operations. However, there is a realistic assumption that with increasing eroded surface, such differences will grow even more. This is due to the lower efficiency of the washing out of residues from the point of electro-erosion by dielectric fluid. Therefore, with electrodes of larger dimensions, it is required to set up an additional internal flushing with dielectric fluid.

3.1. Prediction of HAZ Size of Tool Steel EN X32CrMoV12-28 after Die-Sinking EDM with SF-Cu Electrode

In practice it is quite an advantage if we can predict the resulting quality of the machined surface of a specific material on the basis of combination of major technological and procedural parameters [31]. Prediction of the quality of the machined surface can be carried out in terms of various indicators [32]. Surface roughness parameters, as well as the total depth of the HAZ or microhardness change can be predicted [33]. However, to ensure that prediction is performed with sufficient accuracy, it is necessary to propose an appropriate mathematical model. For the implementation of the resulting machined surface quality prediction of tool steel EN X32CrMoV12-28 after die-sinking EDM-with a SF Cu electrode in terms of the total HAZ depth, it was necessary to create a mathematical model. Its task is to predict the real relationship between the roughness parameters of the machined surface and the main technology and process parameter settings [34,35]. Since most of the modern CNC electrical discharge machining facilities do not allow selective adjustment of individual technological parameters, so for practical reasons, the roughness parameters relating to the machined surface were included in the model. These parameters reflect the settings of the main technological and process parameters, and at the same time they are coupled to the total HAZ thickness and microhardness change. The mathematical model was constructed using the theory of statistics. The best method for the purpose appeared to be the least squares method (LSM). This method properly approximates the n-tuple of the measured parameter values of the total HAZ thickness $[h_{HAZ1}, h_{HAZ2}, ..., h_{HAZm}, y]$ by the function of m variables in the form (3):

$$y = f(h_{HAZ1}, ..., h_{HAZm}) \tag{3}$$

Based on preliminary analysis of the character and distribution of experimentally measured values of total HAZ depth of tool steel EN X32CrMoV12-28 after die-sinking EDM with a SF-Cu electrode, an exponential function with the base of any natural number was chosen for the mathematical model in the formula (4):

$$y = a_{00} \cdot a_{10}^{h_{HAZ1}} \cdot a_{01}^{h_{HAZ2}} \cdot a_{11}^{h_{HAZ1}h_{HAZ2}} \tag{4}$$

while an important condition was that function $S(A)$, which expresses the sum of squares of the differences of calculated and measured values of the total HAZ thickness, in all cases reached its minimum in accordance with formula (5):

$$S(A) = \sum_{i=1}^{r} [y_i - f(h_{HAZ1}, ..., h_{HAZm}, A)]^2 \tag{5}$$

Subsequently, the individual values $f(h_{HAZ1}, ..., h_{HAZm}, A)$ were replaced by the selected function. Since this is a function of several variables, namely the unknown matrix A, taking into account a necessary condition of the existence of the extreme of such a function, the first partial derivatives $S(A)$ must equal zero. Then we get the formula (6) for calculating the unknown coefficients:

$$\frac{\partial S(a_{00}, \ldots, a_{rr})}{\partial a_{ij}} = 0 \text{ for } i, j = 0, \ldots, r. \tag{6}$$

From adjusted partial derivatives, we obtain a set of linear equations, the solution of which are the sought coefficients. The quality of replacement of the experimentally measured values of the particular parameters, provided that its course is described by regression model, expresses the index of correlation, which can be determined by the formula (7):

$$IK = \sqrt{1 - \frac{\sum\limits_{i=1}^{n} (y'_i - y_i)^2}{\sum\limits_{i=1}^{n} (\overline{y}_i - y_i)^2}} \tag{7}$$

where y'_i represents the calculated values according the selected function for $i = 1, \ldots, n$, $\overline{y}_i$ is the arithmetic mean of the measured values, and y_i are measured values.

In the diagram in Figure 12, the significance of the influence of major technological parameters on the total HAZ thickness in die-sinking EDM can be observed. Assessment of the influence significance was done by factor analysis. The values of the parameters A, B, C, and D point to the significance of the effect. The more the values are distant from the mean value of the total HAZ thickness, the higher the significance of the influence of the technological parameter.

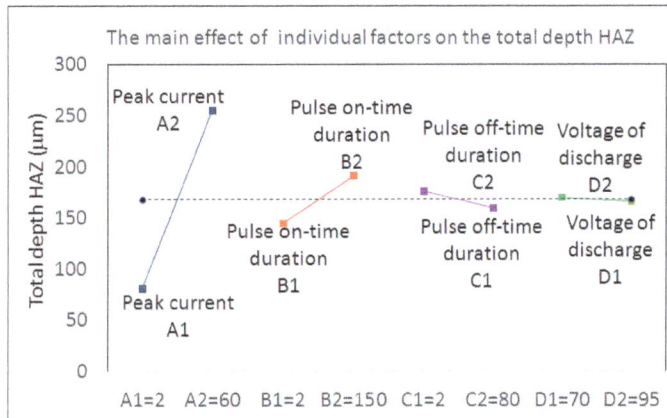

Figure 12. Significance of influence of main technological parameters on the total HAZ thickness at die-sinking EDM.

It can be observed from the diagram on Figure 12, that the greatest influence on the total HAZ thickness at die-sinking EDM is the peak current. On the contrary, the technological parameter, voltage of discharge, proved to have negligible influence and therefore was excluded from further consideration.

Based on factor analysis of the significance of the influence of major technological parameters on the total HAZ thickness at die-sinking EDM of tool steel EN X32CrMoV12-28, for the building of the mathematical model the following was further taken into consideration:

-    peak current I ranging from 2–60 A;
-    pulse on-time duration t_{on} ranging from 5–300 μs;
-    pulse off-time duration t_{off} ranging from 5–120 μs.

Parameter Voltage of discharge was not recognized in the factor analysis as a parameter that significantly contributes to the nature and size of the total HAZ thickness, therefore, it was not considered for further mathematical modeling.

Prediction of the total HAZ thickness only by mutual combination of the main technological parameters at die-sinking EDM would be very impractical. This is because we need to know the optimal combinations of the main technological parameters to maintain stability and efficiency of the electro-erosion process at die-sinking EDM. As was mentioned above, the machined surface roughness parameters defined in a valid standard EN ISO 4287 after die-sinking EDM are directly related to the main technological parameters. Therefore, the building of the mathematical models proceeded in two steps. First, the mathematical models were established to predict the qualitative parameters of machined surfaces Ra and Rz, based on approximation of the recorded values of technological parameters I, t_{on}, and t_{off}, as sub-models for predicting of the total HAZ thickness. In the second step, a mathematical model was made for predicting the total HAZ thickness on the basis of the qualitative parameters of the machined surfaces Ra and Rz as a function of seven variables.

I. The mathematical model describing the relationship between main technological parameters and machined surface roughness parameters Ra and Rz, is defined by formula (8):

$$Ra \; or \; Rz = a_{00}.a_{10}{}^{I}.a_{20}{}^{I^2}.a_{30}{}^{I^3}.a_{01}{}^{t_d}.a_{02}{}^{t_d^2}.a_{03}{}^{t_d^3} \qquad (8)$$

while parameter t_d is defined as a ratio of idling according to the formula (9):

$$t_d = \frac{t_{on}}{t_{on} + t_{off}} \qquad (9)$$

Subsequently the function of seven variables a_{ij} approximates the set of recorded values $[I_i, t_{di}, Ra \; resp. \; Rz]$ by the function (10):

$$Ra \; resp. \; Rz = f(I, t_d, A) = f(I, t_d, a_{00}, ..., a_{ij}) \qquad (10)$$

where the unknown parameters a_{ij}, $i, j = 0, \dots, r$ are calculated so that the area $S(A)$ best approximates the measured operational values. In this case, we can transform formula (4) into the formula (11):

$$S(A) = \sum_{i=1}^{n} [Ra \; resp. \; Rz - f(I_i, t_{di}, A)]^2 \qquad (11)$$

provided that the function reaches its minimum. The unknown in this case, is a matrix of variables a_{ij}.

By application of this procedure, there have been proposed mathematical models that describe the array of values of roughness parameters Ra and Rz of machined surface from tool steel EN X32CrMoV12-28 after die-sinking EDM with a SF-Cu electrode in dependence of significant technological parameters in the formulas (12) and (13):

$$Ra = 1.6327 \cdot 1.0984^{I} \cdot 0.9985^{I^2} \cdot 1.00001^{I^3} \cdot 0.8350^{t_d} \cdot 5,6739^{t_d^2} \cdot 0.0374^{t_d^3} \; [\mu m] \qquad (12)$$

correlation index is $IK^2 = 0.9810$

$$Rz = 7.4051 \cdot 1.0695^{I} \cdot 0.9991^{I^2} \cdot 1.000004^{I^3} \cdot 7.5201^{t_d} \cdot 0.0029^{t_d^2} \cdot 16.3326^{t_d^3} \; [\mu m] \qquad (13)$$

correlation index is $IK^2 = 0.9642$
where:

Ra, Rz—are machined surface roughness parameters [μm],
I—peak current [A],
t_d—ratio of idling, defined by formula (9).

The accuracy of the proposed mathematical model for prediction of machined surface roughness parameters of tool steel EN X32CrMoV12-28 is represented by the correlation index IK^2. The index has the value 0.9810 for roughness parameter Ra, and for roughness parameter Rz it has the value 0.9642. In both cases, this means a deviation of the predicted values from the actually measured values at the level of about 0.1 to 0.2%.

Based on the implementation of mathematical models (12) and (13) that were prepared in the first step, the 3D diagrams shown in Figure 13a,b were created in simulation program Graphis. These dependencies graphically illustrate the prediction of quality parameters of eroded surface roughness Ra (Figure 13a), and Rz (Figure 13b) of tool steel EN X32CrMoV12-28 after die-sinking EDM with SF-Cu electrode, on the basis of the mutual combination of the significant technological parameters I, t_{on} and t_{off}.

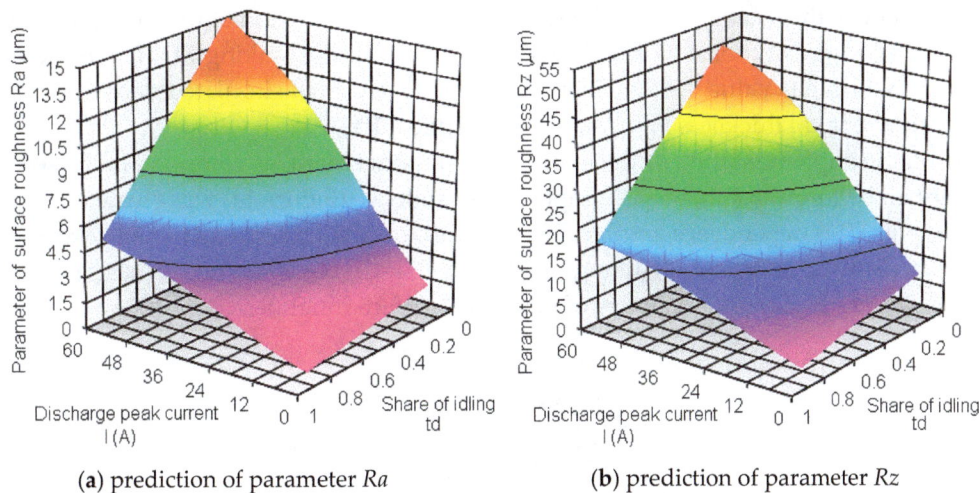

(**a**) prediction of parameter Ra (**b**) prediction of parameter Rz

Figure 13. Graphical representation of prediction of the surface roughness parameters Ra and Rz of tool steel EN X32CrMoV12-28 in dependence of significant technological parameters I, t_{on} and t_{off} at die-sinking EDM with a SF-Cu electrode.

It can be observed from the graphical dependencies that with an increase of the value of technological parameters I, t_{on} and t_{off}, there is a substantial growth of the machined surface roughness parameters Ra and Rz of tool steel EN X32CrMoV12-28 after die-sinking EDM.

II. The mathematical model describing relation between total HAZ thickness and machined surface roughness parameters, is defined by formula (14):

$$h_{HAZ} = a_{00}.a_{10}{}^{Ra}.a_{20}{}^{Ra^2}.a_{30}{}^{Ra^3}.a_{01}{}^{Rz}.a_{02}{}^{Rz^2}.a_{03}{}^{Rz^3} \tag{14}$$

which approximates the set of experimentally measured values $\left[Ra_i, Rz_i, h_{HAZ_i}\right]$ by the function according to the formula (15):

$$h_{HAZ} = f(Ra, Rz, \boldsymbol{A}) = f(I, t, a_{00}, ..., a_{ij}) \tag{15}$$

where unknown parameters a_{ij}, $i, j = 0, \ldots, r$ are calculated with account of minimization of area $S(A)$. In the given case, we can transform formula (4) into formula (16):

$$S(A) = \sum_{i=1}^{n}[h_{HAZi} - f(Ra_i, Rz_i, A)]^2 \tag{16}$$

on the condition that the given function $S(A)$ reaches its minimum.

Analogously, based on the mentioned procedure, mathematical models were compiled. The models describe an array of parameter values of total thickness h_{HAZ} of eroded surface of tool steel EN X32CrMoV12-28 after die-sinking EDM with a SF-Cu electrode in dependence of surface roughness parameters Ra and Rz in formula (17):

$$h_{HAZ} = 51.7876 \cdot 1.6772^{Ra} \cdot 0.9699^{Ra^2} \cdot 1.0005^{Ra^3} \cdot 0.9536^{Rz} \cdot 1.00029^{Rz^2} \cdot 1.0000054^{Rz^3} \; [\mu m] \qquad (17)$$

correlation index is $IK^2 = 0.9943$
where:

h_{HAZ}—is total thickness h_{HAZ} [μm],
Ra, Rz—are machined surface roughness parameters [μm].

The accuracy of the established mathematical model for prediction of the total thickness of the machined surface h_{HAZ} of tool steel EN X32CrMoV12-28 after die-sinking EDM, is represented by parameter IK^2 which has value 0.9943. This presents a deviation of simulated values from actually measured values of total thickness h_{HAZ} on the level up to 0.1%.

Based on the implementation of the mathematical model that was compiled in the second step and is described by formula (17), the 3D graphical dependence (Figure 14) was created in simulation program Graphis. This dependence graphically illustrates the prediction of total thickness h_{HAZ} of tool steel EN X32CrMoV12-28 after die-sinking EDM with a SF-Cu electrode on the basis of the mutual combination of the significant parameters of machined surface roughness Ra and Rz.

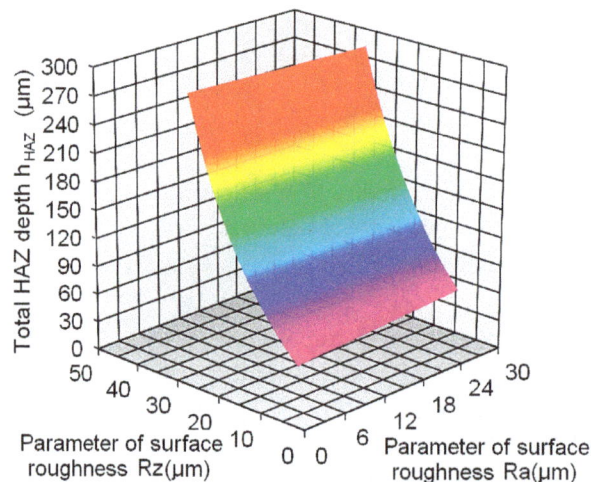

Figure 14. Graphical presentation of prediction of total thickness h_{HAZ} of tool steel EN X32CrMoV12-28 in dependence of mutual combination of significant machined surface roughness parameters Ra and Rz at die-sinking EDM with a SF-Cu electrode.

From the graphical dependence it can be observed that with increased values of machined surface roughness parameters Ra and Rz of tool steel EN X32CrMoV12-28 after die-sinking EDM, the total thickness h_{HAZ} grows substantially. At maximum values of roughness parameters Ra (25 μm), and Rz (50 μm) it reaches the level of 280 μm.

Based on preliminary studies, there is a realistic assumption that with an increase of electrode surface, there is also an increase of total thickness h_{HAZ} after die-sinking EDM. Therefore, the last step performed was the prediction of the total thickness h_{HAZ} of tool steel EN X32CrMoV12-28 after die-sinking EDM in dependence on the size of the area of a SF-Cu electrode of a round cross-section of diameter ranging 5 to 60 mm, and roughness parameters Ra ranging 0.5 to 2.5 μm, and Rz ranging 2.5 to 50 μm. The dependence is represented by the 3D diagrams on Figure 15.

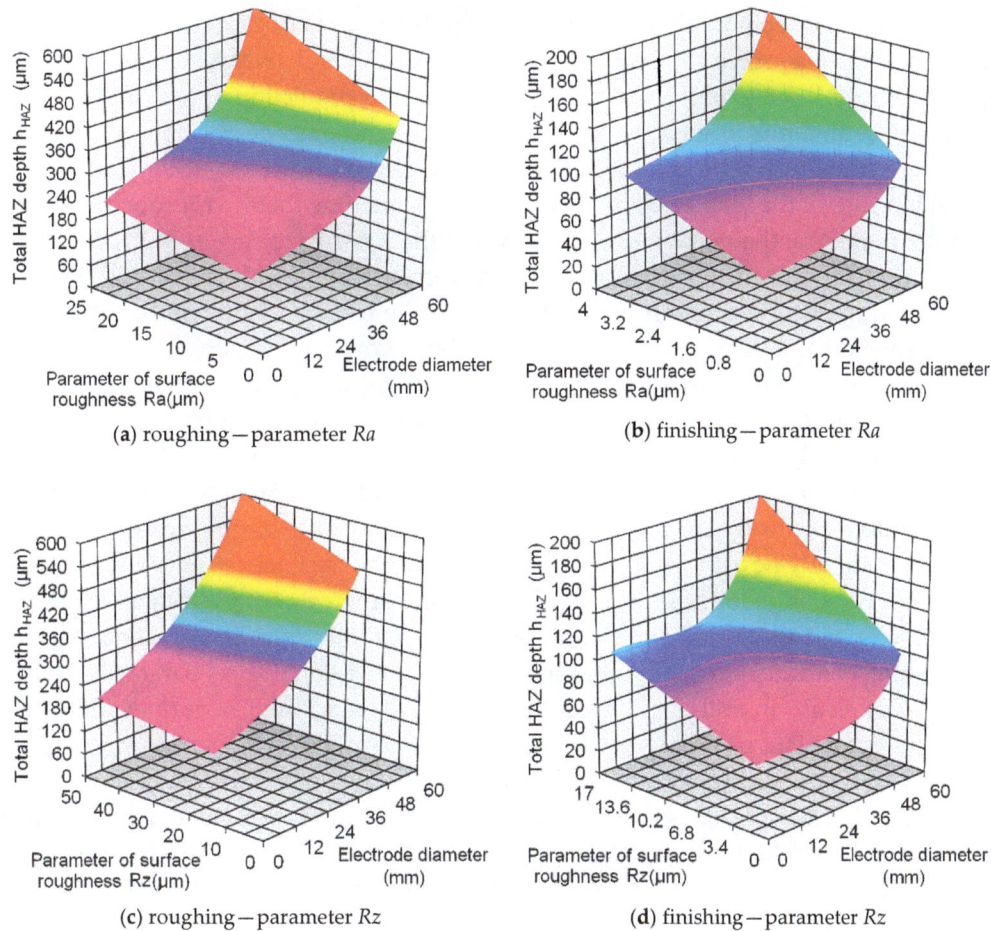

(a) roughing—parameter Ra

(b) finishing—parameter Ra

(c) roughing—parameter Rz

(d) finishing—parameter Rz

Figure 15. Graphical presentation of prediction of total thickness h_{HAZ} of tool steel EN X32CrMoV12-28 after die-sinking EDM with a SF-Cu electrode in dependence of machined surface roughness parameters Ra and Rz for an eroded surface of a round cross-section with diameter ranging from 5 to 60 mm.

The prediction of total thickness h_{HAZ} of tool steel EN X32CrMoV12-28 after die-sinking EDM with a SF-Cu electrode was performed for an eroded surface of a round cross-section with diameter ranging 5 to 60 mm. In the eroded surfaces made by a SF-Cu electrode of the given diameter range, the total thickness h_{HAZ} ranges from 50 to 600 µm in dependence of significant technological parameters settings that are coupled to the achieved qualitative roughness parameters Ra and Rz of the machined surfaces. Lower values of total thickness h_{HAZ} are generally achieved with smaller diameters of SF-Cu electrodes and finishing operations. On the contrary, higher values occur in roughing operations, extreme growth of the total thickness h_{HAZ} is present when SF-Cu electrodes with a diameter above 40 mm are used. Surfaces produced by an electrode with diameter less than 5 mm are very difficult to verify in terms of total thickness h_{HAZ}, so these were excluded from the prediction. Electrodes with diameter larger than 60 mm are in practice applied mostly with internal flushing, which completely changes the character of the machined surface in terms of variations of microstructure, microhardness, and total thickness h_{HAZ}. Surfaces eroded with a SF-Cu electrode with diameter larger than 60 mm with internal flushing require a specific approach. This approach goes beyond the given experimental research, and therefore was omitted from the prediction.

3.2. Practical Recommendation for Optimization of Total HAZ Thickness

Relying on the detailed experimental assessment of changes in microstructure, microhardness, and total HAZ thickness of the tool steel EN X32CrMoV12-28 after die-sinking EDM with a SF-Cu electrode, the significance of the influence of the selected technological parameters [36–40] can be confirmed. Technological parameters peak current I, pulse on-time duration t_{on}, and pulse off-time duration t_{off} in factor analysis showed that they have a direct influence on the change of microhardness and also on the overall range of thickness h_{HAZ}. The realized prediction of the resulting quality of the eroded surface in terms of total thickness h_{HAZ} using mathematical modeling highlighted the negative impact of the high setting values of the peak current I and the low idling ratio values t_d on the essential increase of the total HAZ thickness. At the same time their inappropriate combination has a negative impact on the microhardness change in the sub-surface layers WL, and TL. Suitable combination of settings of the main technological parameters can produce machined surfaces not only with the required surface roughness, but also with the required value of the total depth h_{HAZ}. This will improve the quality indicators in all directions. Low total thickness h_{HAZ} has a positive influence on the extent of the structural changes and changes of the microhardness in WL and TL. It is necessary to note, however, that the setting of the main technological parameters must strictly respect the stability of the die-sinking EDM process. In die-sinking EDM, it is also important to take into account its efficiency and productivity [41]. It is therefore convenient to increase the quality of the machined surface with regard to the economic efficiency of the electro-erosion process, while maintaining its stability. One of the solutions of how to minimize the total extent of HAZ in die-sinking EDM of tool steel EN X32CrMoV12-28, is an increase of the setting value of the technological parameter t_{off}. However, to avoid a substantial decrease of the idling ratio t_d and the subsequent substantial decrease of cutting performance, it is appropriate to increase the technological parameter t_{off} by 25% as a maximum. This results in a slightly longer time to cool the machined surface by dielectric liquid between successive discharges. This leads to a substantially more homogenous microstructure in the HAZ and also to a decrease of the total HAZ thickness. Another problem at die-sinking EDM of tool steel EN X32CrMoV12-28 is an increase of the difference of surface quality with increasing area of the SF-Cu electrode. In the middle of the eroded surfaces the recorded HAZ thickness was approximately 10% larger compared to the boundaries of the surface. This increase is quite significant for SF-Cu electrodes with diameters larger than 40 mm. Above this threshold, a greater decrease of microhardness in TL occurs. There is a drop of microhardness in comparison with the boundaries of over 20 HV 2. This is mainly related to the lower dielectric liquid cooling effect and therefore the overheating of the material at greater depth [42,43]. A possible solution of how to avoid this undesirable phenomenon in die-sinking EDM with a SF-Cu electrode, is the application of an internal flushing system in the SF-Cu electrodes with diameter larger than 40 mm.

4. Conclusions

The research was focused on the experimental assessment of particular sub-surface layers of machined surface of mildly-alloyed chrome-molybdenum-vanadium tool steel EN X32CrMoV12-28 (W.-Nr. 1.2365) after die-sinking EDM with a SF-Cu electrode. The subject of the research was the influence of the main technological parameters on the selected qualitative parameters relating to the individual layers of HAZ. The heat affected zone (HAZ) was investigated in terms of the microstructural changes, changes of microhardness, and thickness of each HAZ layer (BL, WL, and TL). The results of the experiment were oriented on the practical application to achieve a higher quality of machined surface after die-sinking EDM. In practice, great emphasis is often given only on qualitative indicators in terms of machined surface roughness. However, the microstructural changes in the sub-surface layers have a primary effect on the operational functionality, durability, and service life of surfaces machined by the progressive technology of die-sinking EDM. For a more effective applicability of the obtained experimental results in practice, the prediction of the achieved quality was made in terms of the parameters mentioned above. The prediction was made on the basis of mathematical modeling. Implementation of the established mathematical models into the simulation program may serve to predict the total HAZ depth on the basis of the selection of the main technological parameters in die-sinking EDM with different diameters of a SF-Cu electrode. It also allows the determination of an appropriate combination of significant technological parameters to achieve the required thickness h_{HAZ}.

Summary of experimental research results:

- based on the results of experimental research of sub-surface layers of mildly-alloyed chrome-molybdenum-vanadium tool steel EN X32CrMoV12-28 (W.-Nr. 1.2365) after die-sinking EDM with SF-Cu electrode, the main technological parameters (I, t_{on} and t_{off}) that significantly affect the quality of machined surface in terms of microstructural changes were selected;

- in the heat affected zone (HAZ) three layers (BL, WL and TL) were identified as having a specific microstructure, mechanical, physical, and chemical properties;

- for each layer of HAZ of tool steel EN X32CrMoV12-28 the range of thicknesses, and the microhardness change depending on the setting of important technological parameters in die-sinking EDM with a SF-Cu electrode were determined;

- total thickness h_{HAZ} was determined for the roughing (120 to 280 μm), and finishing operation (60 to 100 μm) with a SF-Cu electrode of 20 mm diameter;

- based on mathematical modeling the total HAZ thickness was predicted of the tool steel EN X32CrMoV12-28 at die-sinking EDM for SF-Cu electrodes with diameters ranging from 5 to 60 mm;

- there is the possibility of application of established mathematical models for the optimal selection of values of significant technological parameters in die-sinking EDM of tool steel EN X32CrMoV12-28 with SF-Cu electrode of diameter in the range from 5 to 60 mm, based on the required maximum thickness h_{HAZ};

- on the basis of the defined causes, the measures for elimination of the increase of the total thickness h_{HAZ} were proposed;

- on the basis of the defined causes of HAZ inhomogeneity deviations in the middle of the eroded area compared to the area boundaries, reaching a level of 10%, the measures for elimination of the deviations were proposed;

- experimental research of HAZ of the machined surface of tool steel EN X32CrMoV12-28 after die-sinking EDM with a SF-Cu electrode was oriented on the practical application of the results in theory, as well as in technical practice;

- the achieved results of the experimental measurements represent partial results of an extensive set of experimental measurements focused on the creation of a complex database containing data that describe the influence of process parameters on surface roughness and total HAZ depth at die-sinking EDM of tool steels [44];

- the achieved results and proposed solutions concerning prediction make it possible to meet a much closer specification of the requirements, imposed on quality of the machined surface at die-sinking EDM with a SF-Cu electrode.

Acknowledgments: This research work was supported by the Slovak Research and Development Agency under the contract No. APVV-15-0602 and also by the Project of the Structural Funds of the EU, ITMS code 26220220103.

Author Contributions: Ľuboslav Straka conceived and designed the experiments; Ivan Čorný and Ján Piteľ performed the experiments; Ľuboslav Straka and Slavomíra Hašová analyzed the data; Ľuboslav Straka wrote the paper.

Conflicts of Interest: The authors declare no conflicts of interests.

References

1. Kompella, S.; Moylan, S.; Chandrasekar, S. Mechanical properties of thin surface layers affected by material removal processes. *Surf. Coat. Technol.* **2001**, *146*, 384–390. [CrossRef]

2. Ťavodová, M. Research state heat affected zone of the material after wire EDM. *Acta Fac. Tech.* **2014**, *19*, 145–152.

3.   Choudhary, R.; Kumar, H.; Gark, R.K. Analysis and evaluation of heat affected zones in electric discharge machining of EN-31 die steel. *Indian J. Eng. Mater. Sci.* **2010**, *2*, 91–98.

4.   Shrestha, T.; Alsagabi, S.F.; Charit, I.; Potirniche, G.P.; Glazoff, M.V. Effect of heat treatment on microstructure and hardness of grade 91 steel. *Metals* **2015**, *5*, 131–149. [CrossRef]

5.   Čada, R.; Zlámalík, J. Materials comparison of cutting tools functional parts for cutting of electrical engineering sheets. *Trans. VŠB–Ostrava Mech. Ser.* **2012**, 33–41. Available online: http://transactions. fs.vsb.cz/2012--1/1892.pdf (accessed on 18 April 2016).

6.   Švecová, V.; Madaj, M. Surface characteristics evaluation of the VANADIS 23 high speed steel punch after Wire Electrical Discharge Machining. *Stroj. Technol.* **2012**, 40–41. Available online: http://casopis. strojirenskatechnologie.cz/templates/obalky_casopis/XVII_1,2_2012.pdf (accessed on 24 January 2017).

7.   Abu Zeid, O.A. On the effect of electro-discharge machining parameters on the fatigue life of AISI D6 tool steel. *J. Mater. Process. Technol.* **1997**, *68*, 27–32. [CrossRef]

8.   Marafona, J. Black layer characterisation and electrode wear ratio in electrical discharge machining (EDM). *J. Mater. Process. Technol.* **2007**, *184*, 27–31. [CrossRef]

9.   Ekmekci, B. Residual stresses and white layer in electric discharge machining (EDM). *Appl. Surf. Sci.* **2007**, *253*, 9234–9240. [CrossRef]

10. Zang, Y.; Liu, Y.; Ji, R.; Cai, B. Study of the recast layer of a surface machined by sinking electrical discharge machining using water-in-oil emulsion as dielectric. *Appl. Surf. Sci.* **2011**, *257*, 5989–5997. [CrossRef]

11. Dewangan, S.; Gangopadhyay, S.; Biswas, C.K. Study of surface integrity and dimensional accuracy in EDM using Fuzzy TOPSIS and sensitivity analysis. *Measurement* **2015**, *63*, 364–376. [CrossRef]

12. Sidhom, H. Effect of electro discharge machining (EDM) on the AISI316L SS white layer microstructure and corrosion resistance. *Int. J. Adv. Manuf. Technol.* **2012**, *1–4*, 141–153. [CrossRef]

13. Puri, A.B.; Bhattacharyya, B. Modeling and analysis of white layer depth in a wire-cut EDM process through response surface methodology. *Int. J. Adv. Manuf. Technol.* **2005**, *3–4*, 301–307. [CrossRef]

14. STN ISO 4287 Surface Roughness Testing. Available online: http://www.gagesite.com/documents/Training/ Mitutoyo%20Surface%20Analysis_April%202%202014%20at%20PQI.pdf (accessed on 17 August 2016).

15. STN ISO 6507 Vickers Hardness Testing. Available online: http://infostore.saiglobal.com/store/PreviewDoc. aspx?saleItemID=40729 (accessed on 17 August 2016).

16. Sodick Technical Specifications. Available online: http://www.sodick.com/products/sinkeredm/ag60l.htm (accessed on 17 August 2016).

17. Material Card of Copper SF-Cu (W.-Nr. 2.0090). Available online: https://www.google.sk/#q=Material+ Card+of+Copper+SF-Cu+(W.-Nr.2.0090) (accessed on 18 April 2016).

18. Material Card of Steel EN X210Cr12. Available online: http://www.usbcosteels.com/pdffile/OCR12.pdf (accessed on 18 April 2016).

19. Krastev, D. Improvement of Corrosion Resistance of Steels by Surface Modification.   Available online: http://cdn.intechopen.com/pdfs/34491/InTech-Improvement_of_corrosion_resistance_of_steels_ by_surface_modification.pdf (accessed on 24 January 2017).

20. Lei, M.K.; Zang, Z.L. Microstructure and Corrosion Resistance of Plasma Source Ion Nitrided Austenitic Stainless Steel. *J. Vac. Sci. Technol.* **1997**, *2*, 421–427. [CrossRef]

21. Bátora, B.; Vasilko, K. *Obrobené Povrchy (Machined Surfaces)*; University of Trenčín: Trenčín, Slovakia, 2000; p. 183.

22. Banker, K.S.; Parmar, S.P.; Parekh, B.C. Review to Performance Improvement of Die Sinking EDM Using Powder Mixed Dielectric Fluid. *Int. J. Res. Modern Eng. Emerg. Technol.* **2013**, *1*, 57–62.

23. Cusanelli, G.; Hessler-Wyser, A.; Bobard, F.; Demellayer, R.; Perez, R.; Flükiger, R. Microstructure at submicron scale of the white layer produced by EDM technique. *J. Mater. Process. Technol.* **2004**, *1*, 289–295. [CrossRef]

24. Zhang, Y.; Cheng, X.; Zhong, H.; Xu, Z.; Li, L.; Gong, Y.; Miao, X.; Song, C.; Zhai, Q. Comparative Study on the Grain Refinement of Al-Si Alloy Solidified under the Impact of Pulsed Electric Current and Travelling Magnetic Field. *Metals* **2016**, *6*. [CrossRef]

25. Ekmekci, B. White layer composition, heat treatment and crack formation in electric discharge machining process. *Metall. Mater. Trans. B: Process Metall. Mater. Processing Sci.* **2009**, *40B*, 70–81. [CrossRef]

26. Straka, Ľ.; Čorný, I. Heat treating of chrome tool steel before electroerosion cutting with brass electrode. *Acta Metall. Slovaca* **2009**, *15*, 180–186.

27. Mathew, S.; Varma, P.R.D.; Kurian, P.S. Study on the Influence of process parameters on surface roughness and MRR of AISI 420 stainless steel machined by EDM. *Int. J. Eng. Trends Technol.* **2014**, *2*, 54–58. [CrossRef]

28. Straka, Ľ.; Čorný, I.; Piteľ, J. Properties evaluation of thin microhardened surface layer of tool steel after wire EDM. *Metals* **2016**, *6*. [CrossRef]

29. Kim, J.M.; Ha, T.H.; Park, J.S.; Kim, H.G. Effect of laser surface treatment on the corrosion behavior of FeCrAl-coated TZM alloy. *Metals* **2016**, *6*. [CrossRef]

30. Saha, S.K.; Chaudhary, S.K. Experimental investigation and empirical modeling of the dry electrical discharge machining process. *Int. J. Mach. Tool. Manuf.* **2009**, *49*, 297–308. [CrossRef]

31. Das, S.; Klotz, M.; Klocke, F. EDM simulation finite element-based calculation of deformation, microstructure and residual stresses. *J. Mater. Process. Technol.* **2003**, *142*, 434–451. [CrossRef]

32. Kiyak, M.; Cakir, O. Examination of machining parameters on surface roughness in EDM of tool steel. *J. Mater. Process. Technol.* **2007**, *1–3*, 41–144. [CrossRef]

33. Salonitis, K.; Stournaras, A.; Stavropoulos, P.; Chryssolouris, G. Thermal modeling of the material removal rate and surface roughness for die-sinking EDM. *Int. J. Adv. Manuf. Technol.* **2009**, *40*, 316–323. [CrossRef]

34. Ho, K.H.; Newman, S.T. State of the art electrical discharge machining (EDM). *Int. J. Mach. Tools Manuf.* **2003**, *13*, 1287–1300. [CrossRef]

35. Mičietová, A.; Neslušan, M.; Čilliková, M. Influence of surface geometry and structure after non-conventional methods of parting on the following milling operations. *Manuf. Technol.* **2013**, *13*, 199–204.

36. Amorim, F.L.; Weingaertner, W.L. The behavior of graphite and copper electrodes on the finish die-Sinking electrical discharge machining (EDM) of AISI P20 tool Steel. *J. Braz. Soc. Mech. Sci. Eng.* **2007**, *29*, 366–371. [CrossRef]

37. Marafona, J.; Wykes, C. A new method of optimizing material removal rate using EDM with copper–tungsten electrodes. *Int. J. Mach. Tools Manuf.* **2000**, *40*, 153–164. [CrossRef]

38. Kopac, J. High precision machining on high speed machines. *J. Achiev. Mater. Manuf. Eng.* **2007**, *24*, 405–412.

39. Che Haron, C.H. Investigation on the influence of machining parameters when machining tool steel using EDM. *J. Mater. Process. Technol.* **2001**, *1*, 84–87. [CrossRef]

40. Han, X.L.; Wu, D.Y.; Min, X.L.; Wang, X.; Liao, B.; Xiao, F.R. Influence of Post-Weld Heat Treatment on the Microstructure, Microhardness, and Toughness of a Weld Metal for Hot Bend. *Metals* **2016**, *6*. [CrossRef]

41. Straka, Ľ.; Hašová, S. Study of tool electrode wear in EDM process. *Key Eng. Mater.* **2016**, *669*, 302–310. [CrossRef]

42. Govindan, P.; Joshi, S.S. Analysis of micro-cracks on machined surfaces in dry electrical discharge Machining. *J. Manufacturing Process.* **2012**, *14*, 277–288. [CrossRef]

43. Panda, A.; Duplák, J.; Kormoš, M.; Ružbarský, J. Comprehensive durability identification of ceramic cutting materials in machining process of steel 80MoCrV4016. *Key Eng. Mater.* **2016**, *663*, 286–293. [CrossRef]

44. Straka, Ľ.; Hašová, S. Prediction of the heat-affected zone of tool steel EN X37CrMoV5–1 after die-sinking electrical discharge machining. *Proc Inst. Mech. B: J. Eng. Manuf.* **2016**, *9*, 1–12. [CrossRef]

9

Urban Mining and Electrochemistry: Cyclic Voltammetry Study of Acidic Solutions from Electronic Wastes (Printed Circuit Boards) for Recovery of Cu, Zn, and Ni

Ma. Isabel Reyes-Valderrama, Eleazar Salinas-Rodríguez *, J. Fabian Montiel-Hernández, Isauro Rivera-Landero †, Eduardo Cerecedo-Sáenz, Juan Hernández-Ávila and Alberto Arenas-Flores

Área Académica de Ciencias de la Tierra y Materiales, Universidad Autónoma del Estado de Hidalgo, Carretera Pachuca—Tulancingo km. 4.5, C.P. 42184, Mineral de la Reforma, Hidalgo, México; isareyv@hotmail.com (M.I.R.-V.); jfmomt17@hotmail.com (J.F.M.-H.); isaurorivera@yahoo.es (I.R.-L.); mardenjazz@yahoo.com.mx (E.C.-S.); herjuan@uaeh.edu.mx (J.H.-Á.); arenasa@uaeh.edu.mx (A.A.-F.)
* Correspondence: salinasr@uaeh.edu.mx
† This author died last September, that is why the sign.

Academic Editors: Jae-chun Lee, Bong-Gyoo Cho and Kyoungkeun Yoo

Abstract: We report potentiodynamic studies to characterize copper, nickel and zinc leaching solutions from electronic waste. The metals were leached using oxygen and sulfuric acid (pH = 1.5). As is known, reduction potentials are determined using thermodynamics laws, and metal recovery strategies from electronic waste are usually considered according these thermodynamic values. Pourbaix-type diagrams are not appropriate to plan strategies in electrochemical processing. Therefore, knowledge of electrode potentials for the metal deposit/dissolution process is the basis for the selective recovery planning. For this reason, potentiodynamic studies, specifically cyclic voltammetry, are revealed as a good way to decide the best conditions for the process of electrochemical recovery of metals from electronic waste, which is also cost-efficient and has no interference from strange ions, such as lead, in this case.

Keywords: urban mining; leaching solutions; printed circuit boards; electronic waste; electrochemical recovery; cyclic voltammetry

1. Introduction

The implementation of urban mining in recent years has been due to a concern for the environment and the scarcity of natural resources. This term has no specific definition, because if some authors refer to resources that can be obtained only from landfills, others also include the recycling of construction debris, plastics, glass and scrap metal (electronic waste, in many cases nowadays). The term denotes the systematic reuse of anthropogenic materials, from urban areas [1] and this represents a potential source of material resources that would be available for reuse at the end of the product lifetime.

As new generations of electronic gadgets are developed, the capabilities of these tend to change, and their useful lifetimes get shorter. The above began in the 1980s when the development and production of electronic equipment were promoted to increase its consumption among the population. For some materials, a recycling system cannot be spontaneously implemented with high effectiveness, so is an issue that concerns technology, planning, and economic and environmental regulation. For this reason, at present, due to the increase in the stringency of recycling policies, countries such as China, Japan, Taiwan, and South Korea, including the European Union and some states of the US, have established norms for the management and regulation of electronic waste (e-waste) [2,3]. According to this, strategies for the implementation of an appropriated recycling system should be based on three points: (i) technical viability; (ii) economic sustainability of the process; and (iii) a high and real level of social support for the program [4]. An important aspect that has led to the production of "e-waste" is the steady rise of the computing industry, so it is estimated that one billion computers have been disposed [5]. The e-waste is mostly discharged with domestic waste without receiving any special treatment; on the other hand, 80% of the waste collected is sent to several poor countries [6]. These electronic wastes are received in the form of "donations" by some organizations in rich countries; however, it is simply a way to get rid of electronic equipment considered obsolete, without breaking the Basel Convention, which regulates these kinds of exportations. Seventy percent of "e-waste" is shipped to China [7], but other countries that also receive significant amounts are India, Pakistan, Vietnam, the Philippines, Malaysia, Nigeria and Ghana; finally, on the American continent, the greatest amount discarded is received by Brazil and Mexico. However, it is difficult to quantify the amount of "e-waste" that is being exported due to the semi-clandestine nature of these operations [8,9].

The recovery of materials from electronic scrap involves the disassembly and destruction of the equipment. According to the estimate by Robinson [9], 20 million tons of "e-waste" are generated annually, and this was confirmed according to the data collected by Morf et al. [8]. Thus, it is possible to calculate the annual amount of certain recoverable elements in the electronic scrap (Table 1).

Table 1. Elements present in the environment from the disposal of e-waste [10]

Element	Typical E-Waste Concentration (mg/kg)	Annual Global Emission in E-Waste (Tons)
Cadmium (Cd)	180	3600
Chromium (Cr)	9900	198,000
Copper (Cu)	41,000	820,000
Lead (Pb)	2900	58,000
Mercury (Hg)	0.68	13.6
Nickel (Ni)	10,300	206,000
Tin (Sn)	2400	48,000
Zinc (Zn)	5100	102,000

As is shown in Table 1, we must highlight that 820,000 tons of copper, 206,000 tons of nickel and 102,000 tons of zinc are involved during the generation of the "e-waste", which contributes to pollution and damages to health; at the same time, this production is becoming an important economic opportunity for metals recovery. Furthermore, gold and silver are also present in the "e-waste", increasing its importance from an economic point of view.

Gold, silver, copper, nickel and zinc are examples of elements in the "e-waste" [10]. These are basically present in computer components, printed circuits boards (PCBs), cell phones, etc. Specifically, circuit boards from computers have a design consisting of a gold layer deposited on a substrate of a non-noble metal such as nickel, zinc and/or copper, which is also inserted into a polymer base. Obviously, there is a huge potential source of gold in the mentioned metals, which nowadays are partially recovered or disposed in the "e-waste" [2,11]. There are previous studies on the leaching of the "e-waste" that report the dissolution of copper, nickel and zinc, and the subsequent recovery of metallic gold with a purity close to 99% [12,13]. Because of the highly heterogeneous mixture of organic materials with metals and ceramics, the PCBs are particularly hard to recycle [14]. Viewed from the different perspectives of the recycling of PCBs, scientists have developed many and varied methods involving physical and mechanical technology for separation, bio-technology, supercritical fluid technology, microwave treatment and processes of extractive metallurgy (hydro and pyrometallurgy). Despite the existence of these routes, the processing of "mechanical crushing + hydrometallurgy" is the more predominant technology and the most commonly used, because of the advantages reached in the recycling of PCBs [10,15,16].

In an optimal process, PCBs should be processed in an environmentally sustainable way with the purpose of taking advantage of the treatment of the "e-waste" [14,17]. On the other hand, these kinds of residues could, if possible, be naturally leached and some metal contents can migrate to the water and contaminate important aquifers, which increases the degree of contamination not only by e-waste, but also by the more dangerous contamination of metallic ions. Although some works have been developed to detect metals ions, particularly Cu (II), and to remove them properly by using mesoporous materials, calorimetric detection, and adsorbent composites, among others, and with excellent results, it is necessary to initially remove the possible source of contamination of these types of ions [18–21]. There are certain reviews about progress in the recycling of PCBs made by some researchers [2,17,22–24]; however, some of these reviews deal with too many issues and the depth of these studies, related to the summary and discussion of the use of hydrometallurgy technology, tends to be poor. Every year, the amount of discarded computer equipment around the world increases [25], which triggers the exponential growth of the "e-waste", including PCBs, so it is necessary to accelerate the research to develop an optimal process for the recovery of the PCBs, which can be environmentally friendly, compatible with hydrometallurgical processes and more cost-efficient, and which involves the use of sulfuric acid and oxygen, in comparison with those implicating the use of nitric acid, hydrochloric acid, peroxides and even aqua regia, as they are potential sources of pollution and health risks [26]. An optimum mechanism for the recovery of metal products is to submit them to the process of acid leaching using oxygen and sulfuric acid (pH 1.5) to dissolve the copper, nickel and zinc, with later gold retrieval [11,12,27].

Year after year, the interest in the development of effective electrochemical methods for elimination and recovery of metallic ions from residual waters has increased because studies have demonstrated the possibility of recovering the metals in a pure state [28,29].

The concentrations of metal ions allowed in the effluents have been reduced dramatically since the government laws have been directed to protect the environment, so regulations on their discharge are becoming more stringent [29]. On the other hand, the conventional technique of hydroxide precipitation allows higher concentrations than those established. Accordingly, there is an increasing interest in taking into account the importance of developing ion-exchange techniques as an alternative, instead of the precipitation of hydroxides [30,31].

Previous works about the electrolytic recovery of copper and other metals describe the recovery of a few ppm of metallic ions with low energetic consumption, operating at a relative high electric current density [32]. Cathodic metal deposition is the principle method that leads to the removal of metal ions (Me^{z+}) from waste waters [33,34].

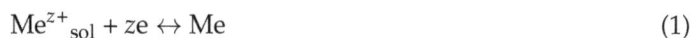

$$Me^{z+}_{sol} + ze \leftrightarrow Me \tag{1}$$

From a thermodynamic point of view, the Nernst equation predicts the individual electro-deposition potential of different metals and their displacement increasing the concentrations, as shown in the Pourbaix diagrams. The strategies to pursue the recovery of the leachate are not perfectly defined with these thermodynamic values. In this regard, it is necessary to perform potentiodynamic studies to characterize the composition of the leachates. The aim of this work is therefore the development of cyclic voltammetry studies with synthetic and real solutions to qualitatively characterize the acid leaching solutions. Through these strategic studies, we try to reach electrochemical recovery for non-noble metals.

2. Materials and Methods

The potentiodynamic studies have been carried out using a typical electrochemical cell of three-electrodes: reference, counter and working electrode, as shown in Figure 1. The silver/silver chloride electrode is used as reference, the platinum wire as counter and the Toray paper TGPH—120, (carbonaceous paper from Toray) acquired in Global Proventus, a company located in Monterrey,

Mexico, as working electrode, the electrodes configuration was dependent on the way the experiments were performed. The cyclic voltammetry were performed in 5 mM Cu + 1 mM Ni + 1 mM Zn at a pH 1.5 (sulfuric medium) synthetic solution and leaching solution with sulfuric acid (pH 1.5). In the same way, the cyclic voltammetry were performed in 1 mM Ni + 1 mM Zn at a pH 5 for synthetic solutions and leaching solutions with sulfuric acid (pH 5) without copper. The electrode potential was controlled using a PGSTAT30 AUTOLAB system, from the brand Ω Metrohm Autolab B.V., equipment located at the University of Alicante, Spain and used during a research stay at the University Institute of Electrochemistry. All the cyclic voltammetry were carried out at a scan rate 50 mV·s^{-1} and at room temperature. All the solutions were purged with nitrogen for 20 min before the experiments were conducted.

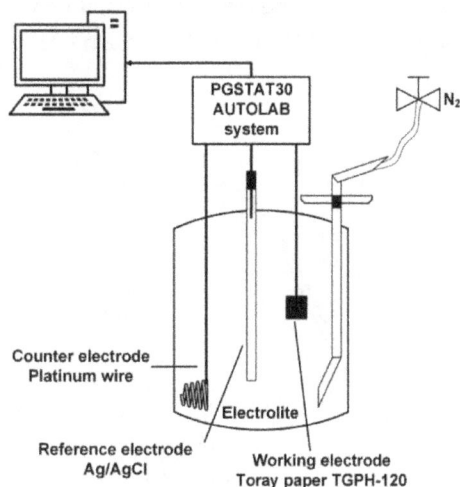

Figure 1. Scheme of a three-electrode electrochemical cell.

Furthermore, morphological and compositional analysis of the electrodeposited metal on working electrodes, were carried out using Scanning Electron Microscopy (SEM, Hitachi S—3000 N at 20 kV, acquired in the company Hitachi in Spain, Madrid, Spain) and Energy Dispersive Spectrometry of X-rays (EDS, Bruker XFlash 3001, detector acquired joint with SEM Hitachi S-3000 N in the company Hitachi in Spain, Madrid, Spain). The metals were electrodeposited on paper Toray, fixing the working electrode potential at -1.6 V vs. Ag/AgCl during 600 s for each one of the solutions described before, i.e., chronoamperometry at -1.6 V vs. Ag/AgCl during 600 s.

3. Results

In previous works, the chemical species and the equilibrium potential in the leached solution have been established through thermodynamic studies using Pourbaix-type diagrams [35–37]. For the results concerning Cu, the predominant chemical species is $CuSO_4$, found by carrying out the following reaction:

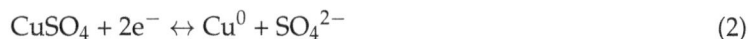

$$CuSO_4 + 2e^- \leftrightarrow Cu^0 + SO_4^{2-} \tag{2}$$

For Ni, the predominant chemical species is $NiSO_4$, found by carrying out the reaction:

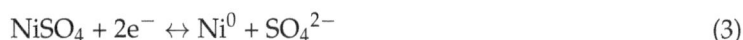

$$NiSO_4 + 2e^- \leftrightarrow Ni^0 + SO_4^{2-} \tag{3}$$

In addition, for Zn, the predominant chemical species is $[Zn(SO_4)_2 \cdot (H_2O)_4]$, found by carrying out the reaction:

$$[Zn(SO_4)_2 \cdot (H_2O)_4] + 4e^- \leftrightarrow Zn^0 + 2SO_4^{2-} + 4H_2O \tag{4}$$

In the same way, it was possible to observe the influence of the concentration of each metal with respect to the reduction potential. In general, a displacement of the reduction potential for each one was observed. Obviously, thermodynamic studies are fundamental, but the viability of the chemical processes is obtained by kinetic studies. In this case, cyclic voltammetry in both synthetic and leaching solutions was carried out.

Firstly, cyclic voltammetry was carried out in synthetic solutions of 5 mM Cu, 1 mM Ni and 1 mM Zn, at pH 1.5. In Figure 2a, the cyclic voltammetry for the Cu shows a wave at -0.36 V vs. Ag/AgCl where the deposit of copper starts, and a stripping zone at 0.25 V vs. Ag/AgCl where the electrodeposited Cu is dissolved. In Figure 2c, the cyclic voltammetry for Zn shows a little stripping zone at -1.0 V vs. Ag/AgCl. Finally, in Figure 2b, the cyclic voltammetry for Ni does not show a relevant electrochemical process, because the metallic Ni at a pH of 1.5 is not electrodeposited. It is important to notice that all electrodeposited metals are characterized by the existence of stripping zones, and although lead ions were present in the real solutions, they were present in marginal amounts and did not influence the recovery of Cu (II) or Zn (II and Zn (II) in their respective experimental recovery conditions.

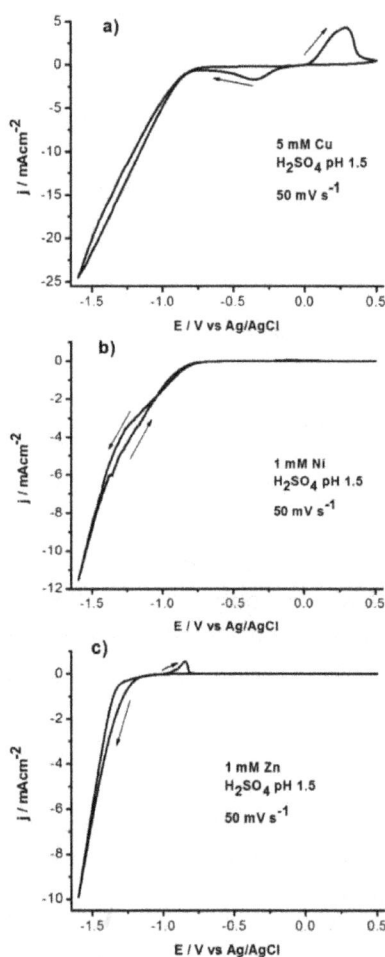

Figure 2. Cyclic voltammetries for: (**a**) 5 mM of Cu; (**b**) 1 mM of Ni and; (**c**) 1 mM of Zn, in H_2SO_4, pH 1.5 solution. Scan rate of 50 mV·s^{-1}.

After studying the behavior of each metal, we carried out cyclic voltammetry in a solution with all the elements together and at the same pH, as shown in Figure 3a. In this case, only the Cu profile is shown. When all the elements are present in the synthetic solution, only the copper electrodeposition appears because the Ni and Zn metals were not electrodeposited in this medium. Similarly, cyclic voltammetry in the leaching solution has the same behavior, as is illustrated in Figure 3b. Both figures are identical, which affirms that we only obtain metallic copper during the electrochemical recovery of solutions of copper, nickel and zinc at pH 1.5, because Zn and Ni are not electrodeposited under these conditions, without the influence of lead ions. This affirmation is corroborated by SEM and EDS analysis, done on the metal electrodeposited on Toray paper. This electrodeposition was carried out with the fixing of the working electrode potential at -1.6 V vs. Ag/AgCl during 600 s (as shown in Figure 4) on the solution described previously. As is shown, the cyclic voltammetry and the potentiodynamic studies, were a good choice for the qualitative characterization of lixiviate solutions from electronic waste, and the interpretation of the curves obtained is a good way to plan a correct strategy for the electrochemical recovery of the metals involved in these residues.

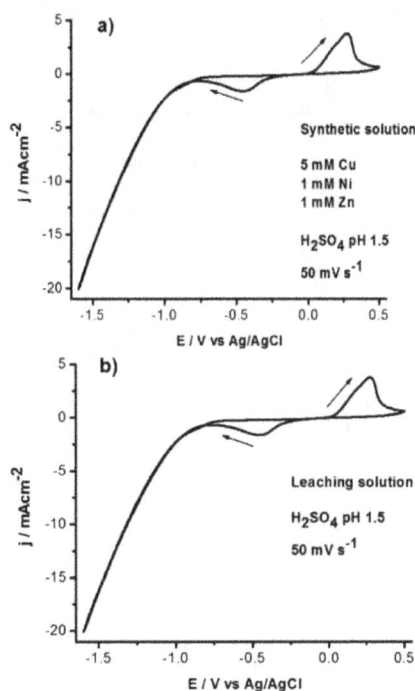

Figure 3. Cyclic voltammetries in: (**a**) H_2SO_4, pH 1.5 synthetic solution; and (**b**) H_2SO_4, pH 1.5 leaching solution. Scan rate of 50 mV·s^{-1}.

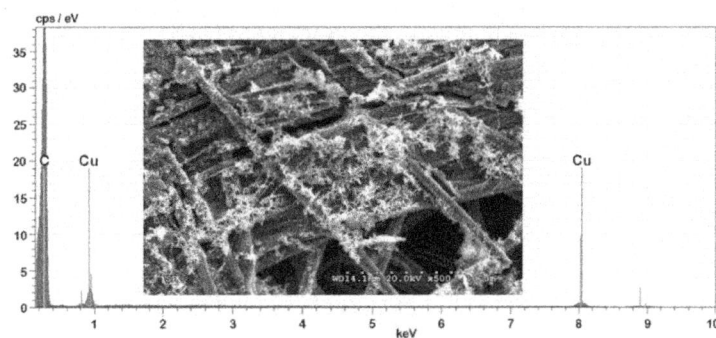

Figure 4. SEM micrograph and EDS analysis of electrodeposition at pH 1.5 on Toray paper carried out with a fixed working electrode potential at -1.6 V vs. Ag/AgCl during 600 s in synthetic and leaching solutions.

Secondly (with the same idea to characterize solutions involved in the electrochemical recovery of non-precious metals), the pH of the synthetic solutions without copper was raised to pH 1.5 as a possible option to recover Ni and Zn. Likewise, the analysis of the solutions at pH 1.5 in Figure 5a, can reach a Zn deposition at -1.4 V vs. Ag/AgCl, and a stripping zone at -1.0 V vs. Ag/AgCl, which was carried out by cyclic voltammetry in Zn and Ni solutions at pH 5, respectively. Ag/AgCl can be seen clearly, while the processes of electrodeposited/stripping for Ni are not observed in Figure 5b.

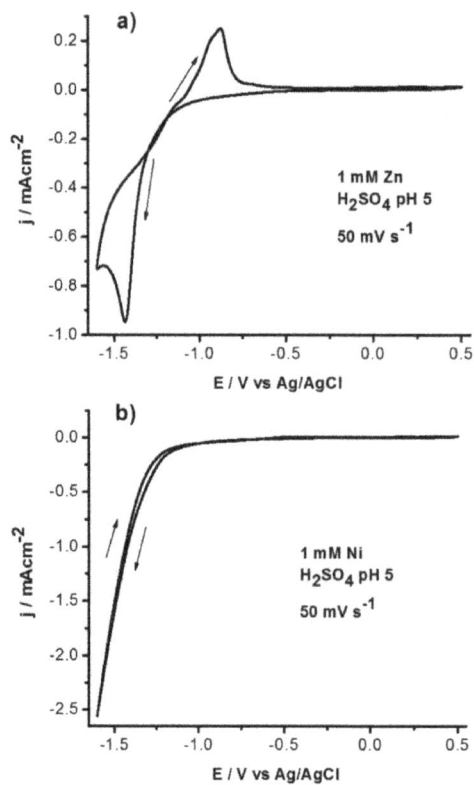

Figure 5. Cyclic voltammetries in: (**a**) 1 mM Zn and (**b**) 1 mM Ni, in H_2SO_4 solutions. Scan rate 50 mV·s^{-1}.

When the cyclic voltammetry with the two elements together was carried out, stripping processes were not observed due to the corrosion resistance of the Ni-Zn deposit [38], as shown in Figure 6. In this case, SEM micrographs and EDS analysis of the electrodeposition on Toray paper were carried out with fixing the working electrode potential at −1.6 V vs. Ag/AgCl in Zn and Ni solution during 600 s, as is observed in Figure 7. Indeed, the composition of the electrodeposited solids showed only the presence of Ni and Zn, as is shown by the EDS analysis. Clearly, the use of a potentiometer and a simple electrochemical cell could be enough for a qualitative analysis of these metals involved in the leaching process from electronic waste.

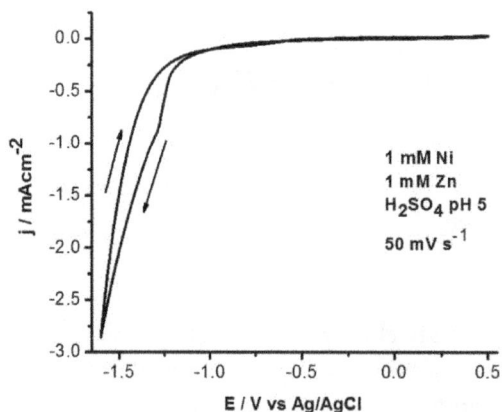

Figure 6. Cyclic voltammetry of 1 mM Ni and 1 mM Zn in H_2SO_4, pH 5 solution. Scan rate: 50 mV·s^{-1}.

Figure 7. SEM micrograph and EDS analysis of electrodeposition at pH 5 on Toray paper carried out with a fixed working electrode potential at −1.6 V vs. Ag/AgCl during 600 s.

4. Conclusions

Generally, thermodynamic studies based on Pourbaix-type diagrams determine the individual electrodeposition potentials of different metals and their displacement, with an increasing concentration in the leaching processes. These values are not appropriate to plan strategies for the electrochemical recovery of metals. In this work, potentiodynamic studies were conducted using cyclic voltammetry studies for synthetic and real leaching solutions from e-waste printed circuit boards. In this sense, the interpretation of the cyclic voltammetry showed the viability of the design of an electrochemical selective recovery of Cu, Zn and Ni for different pHs: the Cu was obtained by electrodeposition at pH 1.5 and the recovery of Ni and Zn was reached when the pH was increased to 5. Furthermore, the previous leaching solution (pH 1.5) can be re-utilized with the adjustment of pH and this, joined with the use of the electrolytic recovery, can make the overall process of leaching and electro-recovery, selective and profitable.

Acknowledgments: The authors want to thank the PRODEP-SEP of the Mexico Government for their financial support. Thanks also go to Universidad Autonoma del Estado de Hidalgo, especially to the Materials and Metallurgy Research Center.

Author Contributions: I. Rivera and M.I. Reyes conceived and designed the experiments; J.F. Montiel, E. Salinas and J. Hernandez performed and reviewed the experiments; E. Cerecedo and A. Arenas analyzed and helped in characterization of the data. I. Rivera, E. Salinas and J. Hernandez contributed with reagents, materials and tools. Finally, E. Salinas, M.I. Reyes and J. Montiel wrote the paper.

Conflicts of Interest: The authors declare no conflict of interest.

References

1. Brunner, P.H. Urban mining—A contribution to reindustrializing the city. *J. Ind. Ecol.* **2011**, *15*, 339–341. [CrossRef]
2. Cui, J.; Zhang, L. Metallurgical recovery of metals from electronic waste: A review. *J. Hazard. Mater.* **2008**, *158*, 228–256. [CrossRef] [PubMed]
3. Rocchetti, L.; Vegliò, F.; Kopacek, B.; Beolchini, F. Environmental impact assessment of hydrometallurgical processes for metal recovery from residues using a portable prototype plant. *Environ. Sci. Technol.* **2013**, *47*, 1581–1588. [CrossRef] [PubMed]
4. Kang, H.Y.; Schoenung, J.M. Electronic waste recycling: A review of US infrastructure and technology options. *Resour. Conserv. Recycl.* **2005**, *45*, 368–400. [CrossRef]
5. LaDou, J.; Lovegrove, S. Export of electronics waste. *Int. J. Occup. Environ. Health* **2008**, *14*, 1–10. [CrossRef] [PubMed]

6.   Schmidt, C.W. Unfair trade: E-waste in Africa. *Environ. Health Perspect.* **2006**, *114*, A232–A235. [CrossRef] [PubMed]

7.   Liu, X.; Tanaka, M.; Matsui, Y. Generation amount prediction and material flow analysis of economic waste: A case study in Beijing, China. *Waste Manag. Res.* **2006**, *24*, 434–445. [CrossRef] [PubMed]

8.   Morf, L.S.; Tremp, J.; Gloor, R.; Schuppisser, F.; Stengele, M.; Taverna, R. Metals, non-metals and PCB in electrical and electronic waste—Actual levels in Switzerland. *Waste Manag.* **2007**, *27*, 1306–1316. [CrossRef] [PubMed]

9.   Robinson, B.H. E-waste: An assessment of global production and environmental impacts. *Sci. Total Environ.* **2009**, *408*, 183–191. [CrossRef] [PubMed]

10.   Xiu, F.R.; Qi, Y.; Zhang, F.S. Co-treatment of waste printed circuit boards and polyvinyl chloride by subcritical water oxidation: Removal of brominated flame retardants and recovery of Cu and Pb. *Chem. Eng. J.* **2014**, *237*, 242–249. [CrossRef]

11.   Reyes, M.I.; Rivera, I.; Patiño, F.; Flores, M.U.; Reyes, M. Total recovery of gold contained in computer circuit boards. Leaching kinetics of Cu, Zn and Ni. *J. Mex. Chem. Soc.* **2012**, *56*, 144–148.

12.   Montiel, J.F.; Reyes, M.I.; Rivera, I.; Patiño, F.; Hernández, J. Caracterización de circuitos impresos vía SEM-EDS y su lixiviación en el sistema O_2-H_2SO_4. *Bol. Soc. Quim. Mex.* **2012**, *6*, 21–23.

13.   Montiel, J.F.; Reyes, M.I.; Rivera, I.; Patiño, F.; Hernández, J. Recuperación de Au, Cu, Ni y Zn contenidos en desechos electrónicos. *Bol. Soc. Quim. Mex.* **2013**, *7*, 120.

14.   Hall, W.J.; Williams, P.T. Separation and recovery of materials from scrap printed circuit boards. *Resour. Conserv. Recycl.* **2007**, *51*, 691–709. [CrossRef]

15.   Li, N.; Lu, X.; Zhang, S. A novel reuse method for waste printed circuit boards as catalyst for wastewater bearing pyridine degradation. *Chem. Eng. J.* **2014**, *257*, 253–261. [CrossRef]

16.   Zhang, Y.; Liu, S.; Xie, H.; Zeng, X.; Li, J. Current status on leaching precious metals from waste printed circuit boards. *Proced. Environ. Sci.* **2012**, *16*, 560–568. [CrossRef]

17.   Xing, M.; Zhang, F.S. Degradation of brominated epoxy resin and metal recovery from waste printed circuits board through batch sub/supercritical water treatments. *Chem. Eng. J.* **2013**, *219*, 131–136. [CrossRef]

18.   Awal, M.R. New type mesoporous conjugate material for selective optical copper (II) ions monitoring & removal from polluted waters. *Chem. Eng. J.* **2017**, *307*, 85–94.

19.   Awal, M.R.; Hasan, M.M. Colorimetric detection and removal of copper (II) ions from wastewater samples using tailor-made composites adsorbent. *Sens. Actuators B Chem.* **2015**, *2016*, 692–700. [CrossRef]

20.   Awual, M.R. A novel facial composite adsorbent for enhanced copper (II) detection and removal from wastewater. *Chem. Eng. J.* **2015**, *266*, 368–375. [CrossRef]

21.   Awal, M.R.; Hasan, M.M.; Khaleque, M.A.; Sheikh, M.C. Treatment of copper (II) containing wastewater by a newly developed ligand based facial conjugate materials. *Chem. Eng. J.* **2016**, *288*, 368–376. [CrossRef]

22.   Liu, W.Q.; Shang, T.M.; Lei, W.N.; Zhou, Q.F. Progress in the study of recycling and innocuous treatment of waste printed circuit boards. *Huanjing Kexue yu Jishu.* **2011**, *34*, 48–54.

23.   Ningtao, Y.; Zhanxu, T.; Fahui, W. The Methods for Recycling of Waste Printed Circuits Boards. *China Resour. Compr. Util.* **2011**, *7*, 024.

24.   Yao, Y.Q.; Xu, X.P.; Liu, Y.; Qiu, X.Y. Review on the comprehensive recovery technologies of waste printed circuits boards. *Mater. Res. Appl.* **2011**, *5*, 17–20.

25.   Hadi, P.; Ning, C.; Ouyang, W.; Lin, C.S.K.; Hui, C.W.; McKay, G. Conversion of an aluminosilicate-based waste material to high-value efficient adsorbent. *Chem. Eng. J.* **2014**, *256*, 415–420. [CrossRef]

26.   Habbache, N.; Alane, N.; Djerad, S.; Tifouti, L. Leaching of copper oxide with different acid solutions. *Chem. Eng. J.* **2009**, *152*, 503–508. [CrossRef]

27.   Kim, E.Y.; Kim, M.S.; Lee, J.C.; Pandey, B.D. Selective recovery of gold from waste mobile phone PCBs by hydrometallurgical process. *J. Hazard. Mater.* **2011**, *198*, 206–215. [CrossRef] [PubMed]

28.   Kaminari, N.; Schultz, D.; Ponte, M.; Ponte, H.; Marino, C.; Neto, A. Heavy metals recovery from industrial wastewater using Taguchi method. *Chem. Eng. J.* **2007**, *126*, 139–146. [CrossRef]

29.   Lapicque, F.; Storck, A.; Wragg, A.A. *Electrochemical Engineering and Energy*; Plenum Press: New York, NY, USA, 2008; Volume 1.

30.   Kammel, R.; Lieber, H.W. Possibilities of treatment electroplating effluents without sludge formation. Pt. 3. metal recovery instead of sludge disposal. *Galvanotechnik* **1977**, *68*, 413–418.

31. Kreysa, G. Festbettelektrolyse-ein Verfahren zur Reinigung metaahaltiger Abwässer. *Chem. Ing. Tech.* **1978**, *50*, 332–337. [CrossRef]

32. Scott, K. Metal recovery using a moving-bed electrode. *J. Appl. Electrochem.* **1981**, *11*, 339–346. [CrossRef]

33. Gao, J.; Liu, F.; Ling, P.; Lei, J.; Li, L.; Li, C.; Li, A. High efficient removal of Cu(II) by a chelating resin from strong acidic solutions: Complex formation and DFT certification. *Chem. Eng. J.* **2013**, *222*, 240–247. [CrossRef]

34. Wong, E.T.; Chan, K.H.; Idris, A. Kinetic and equilibrium investigation of Cu (II) removal by Co(II)-doped iron oxide nanoparticle-immobilized in PVA-alginate recyclable adsorbent under dark and photo condition. *Chem. Eng. J.* **2015**, *268*, 311–324. [CrossRef]

35. Espinoza, E.; Escudero, R.; Tavera, F.J. Waste water treatment by precipitating copper, lead and nickel species. *Res. J. Recent Sci.* **2012**, *1*, 1–6.

36. Granados, M.N.; Huizar, L.H.M.; Rios-Reyes, C.H. Electrochemical study about zinc electrodeposition onto GCE and HOPG substrates. *Quím. Nova* **2011**, *34*, 439–443. [CrossRef]

37. Montiel, J.H.; Reyes, M.I.; Rivera, I.; Rios-Reyes, C.; Rodríguez, V.; Patiño, F.; Reyes-Cruz, V. Thermodynamic study of leached metals (Cu, Zn and Ni) from waste printed circuits by electrochemical method. *Adv. Mat. Res.* **2014**, *1*, 86–89.

38. Hadi, P.; Barford, J.; McKay, G. Synergistic effect in the simultaneous removal of binary cobalt-nickel heavy metals from effluents by a novel e-waste-derived material. *Chem. Eng. J.* **2013**, *228*, 140–146. [CrossRef]

10

The Effect of Diffusion Welding Parameters on the Mechanical Properties of Titanium Alloy and Aluminum Couples

Enes Akca [1,*] and Ali Gursel [1,2]

[1] Department of Mechanical Engineering, Faculty of Engineering and Natural Sciences, International University of Sarajevo, Hrasnička cesta 15, 71210 Sarajevo, Bosnia and Herzegovina; agursel@ius.edu.ba

[2] Department of Mechanical Engineering, Faculty of Engineering, Duzce University, 81620 Duzce, Turkey

* Correspondence: eakca@ius.edu.ba

Academic Editors: Halil Ibrahim Kurt, Necip Fazil Yilmaz and Adem Kurt

Abstract: Ti-6Al-4V alloy and commercially pure aluminum, which are commonly used in aerospace, medical, and automotive industries, are bonded by diffusion welding. Different welding parameters (560, 600, and 640 °C—0, 45, and 60 min—under argon shielding) are used in this process to make the materials more applicable in the industry. Here, the effects of parameters on the strength of joints were studied. The bonded samples were subjected to microhardness and tensile tests in order to determine their interfacial strength. The hardness values were found to decrease with increasing distance from the interface on the titanium side while it remained constant on the aluminum side. Maximum tensile strength was taken from the maximum bonding temperatures of 600 and 640 °C. A morphology examination of the diffusion interfaces was carried out with scanning electron microscopy.

Keywords: Ti-6Al-4V alloy; diffusion welding; dissimilar metal bonding; solid state welding; SEM

1. Introduction

Aluminum and titanium alloys are considered to be the most ideal structure material for aerospace and aircraft vehicles due to their low density, high specific, and strength [1]. Titanium is a strong metal that is quite ductile, and it has low thermal conductivity such that less heat can transfer through boundaries. The relatively high melting point (1660 °C) makes it useful as a refractory metal. Furthermore, aluminum is also remarkable for its ability to resist corrosion due to the phenomenon of passivation. Structural components made from titanium and aluminum play a vital role in the aerospace and defense industries [2,3]. These materials are also important in other applications such as transportation, structural materials, automotive, medical prostheses, orthopedic implants, dental implants, sporting goods, jewelry, and mobile phones. The reduction of weight and costs by use of aluminum and the improvement of strength and corrosion resistance by use of titanium are the main reasons for the joining of these dissimilar materials.

Large differences in the physical properties between the aluminum and titanium alloy prevent the use of conventional welding methods such as fusion welding to join these dissimilar metals [3]. Vaidya et al. [4] have shown in the frame of a feasibility study that the laser beam welding of Ti-6Al-4V and AA 6056 can be performed without any formation of cracks and pores, respectively. Chemical components, crystal structure, and melting points can be given as examples. Nevertheless, diffusion welding is a recent, non-conventional joining process that has attracted the considerable interest of researchers in recent times [5], and it is one of the solid state welding (SSW) processes [6]. According to literature research, many dissimilar metals have been bonded by SSW as well [7,8].

Additionally, many other metals are joined by diffusion bonding [9–11]; however, joining commercially pure aluminum and Ti-6Al-4V alloy does not have its place in the reported literature. In diffusion bonding, the bond strength is achieved by the pressure, temperature, time of contact, and cleanness of the surfaces, and these combinations are called as diffusion parameters [12].

In this study, the diffusion parameters were determined to be as follows: the temperatures were 520, 560, 600, and 640 °C, and the process times were 30, 45, and 60 min, under argon gas shielding. After all necessary preparation of the bonded samples and the metallographic process was complete, processed samples were subjected to Vickers microhardness and tensile tests to observe the strength of the joints. Additionally, the morphologies of the diffusion interfaces were examined via scanning electron microscopy (SEM).

2. Materials and Methods

The chemical compositions of the two materials are given in Tables 1 and 2. Ti-6Al-4V and aluminum samples were prepared for SEM, microhardness measurement, and tensile tests as shown in Figure 1.

Table 1. Aluminum chemical composition.

Aluminum	Al	Si	Fe	Mn	Mg	Cr
wt. %	99.90	0.033	0.059	0.0006	0.0004	0.0004

Table 2. Ti-6Al-4V chemical composition.

Ti-6Al-4V	Ti	Al	V	N	H	Y
wt. %	Balance	6.75	4.5	0.5	0.0125	0.005

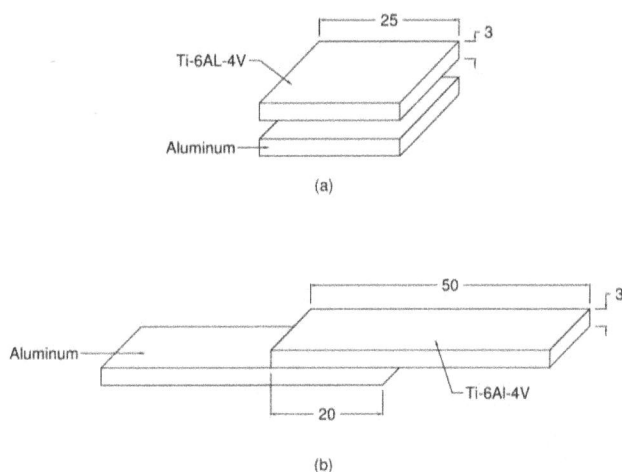

Figure 1. Dimensions of test samples for (**a**) SEM and microhardness (**b**) tensile test (All dimensions are in mm).

Surfaces samples were ground with SiC, paper grade 120–280. The cleaning process were carried out by either acetone or carbon tetra chlorine. Although cleaning with carbon tetra chlorine improves the joining strength 14% more than the acetone cleaning process [13], surface cleaning with linen achieved a successful result in diffusion bonding as well.

Properly controlled and monitored atmospheric furnace was used for the process. A pressure of 3 MPa was applied to the bonding surfaces to improve the interfacial diffusion. Firstly, the bonding furnace was completely filled with argon gas at a flow rate of 6 L/min. The furnace was programmed to be heated at a rate of 30 °C/min until process temperature was achieved. The samples were held in the furnace for specific times (30, 45, and 60 min). At the end of the process, the samples were allowed to cool down in the bonding furnace. The bonding processes were completed with different welding parameters as shown in Table 3.

Table 3. Diffusion welding parameters.

Sample No.	Tests and Examinations	Welding Temperature (°C)	Welding Time (min)
A1	SEM/Hardness	520	30
A2		520	45
A3		520	60
A4		560	30
A5		560	45
A6		560	60
A7		600	30
A8		600	45
A9		600	60
A10		640	30
A11		640	45
A12		640	60
T1	Tensile	560	30
T2		560	45
T3		560	60
T4		600	30
T5		600	45
T6		600	60
T7		640	30
T8		640	45
T9		640	60

The bonded samples were firstly cut perpendicular to the bonding surface. The cut samples were mounted as shown in Figure 2. The mounting operations were carried out at 9 min of heating, 3 min of cooling at a temperature of 180 °C, and a force of 40 kN. The grinding processes were done with SiC, paper grade 180, 500, 800, 1200, 2000, and 2500, respectively. The ground samples were subjected to a polishing operation with 1 μm alumina suspension. Both grinding and polishing processes were done with Struers LaboPol-5 at a velocity of 500 rev/min. The samples were etched with a chemical solution: 1% HF–1.5% HCl–2.5% HNO_3–95% H_2O [14].

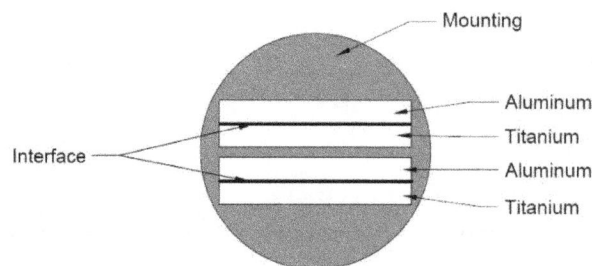

Figure 2. Samples mounted in bakelite.

Microstructure and morphologies of diffusion interfaces were examined by scanning electron microscopy (SEM, Philips XL30S FEG, Tustin, CA, USA). Changes in joint compositions across the joints were examined using energy dispersive spectroscopy (EDS). Mechanical properties were evaluated by tensile and microhardness tests. A universal Instron 5569 (Norwood, UK) was used for the tensile tests. The load was applied to the material gripped at two sides until fracture occurred with a velocity of 1 mm/min. Tensile tests were applied to 9 samples that were prepared for tensile tests with different parameters shown in Table 3. The tests were carried out with Instron 5569 tensile tester. Hardness tests were carried out using the Vickers (HV) method. Micro HV at 50 gram force (gf) was used. Hardness measurements were carried out on etched surfaces and mounted samples by using micro HV. Hardness measurements were taken from an Instron Wolpert Testor 2100 (Norwood, UK).

3. Results and Discussions

The samples have been prepared for SEM, microhardness, and tensile tests. Bonding did not occur in A1, A2, and A4 samples either because temperatures were too low or because there was not enough time. While 480 °C temperatures, even after 60 min, were too low to weld, 680 °C was too high. Because of the yielding of aluminum, 680 °C temperatures were not investigated in tests and analyses [15]. Bonding did not occur after 30 or 45 min at 520 °C, but successful bonding did occur after 60 min at 520 °C. In addition, bonding did not occur after 30 min at 560 °C; thus, in order for atoms to diffuse, appropriate temperatures and times are required.

3.1. Microhardness Tests

Microhardness measurements were performed on diffusion couples at different intervals, and hardness distribution profiles were determined on two sides of the bonded joints. All bonded samples with different parameters were subjected to microhardness tests, and the Micro-Vickers method was used. The microhardness measurement method and marks are shown in Figure 3. The distances between microhardness marks are 100 μm. A logical connection between the different measurement results was attempted with respect to different welding temperatures and times. Table 4 shows the microhardness results of all the bonded samples, and Figures 4 and 5 were prepared according to the hardness results. The results are grouped with respect to constant temperature and time, separately. It can be seen that the titanium sides have hardness values of 450 HV, while the aluminum sides have hardness values of about 33 HV. The microhardness profiles of diffusion couples that bonded with different welding parameters were examined. As expected, the hardness values of the aluminum sides were all lower than those of the titanium sides, and the hardness values in the transition zone are all higher than those of the aluminum sides, but lower than those of the titanium sides.

According to Table 4 and Figure 4, the microhardness values move wavily independent of temperature and time; there is no remarkable change with respect to temperature when compared. However, low temperatures may lead to the absence of higher hardness values on the titanium side. In the literature, it has been observed that the hardness values are higher on the titanium side, and the welding temperature values are higher as well [16].

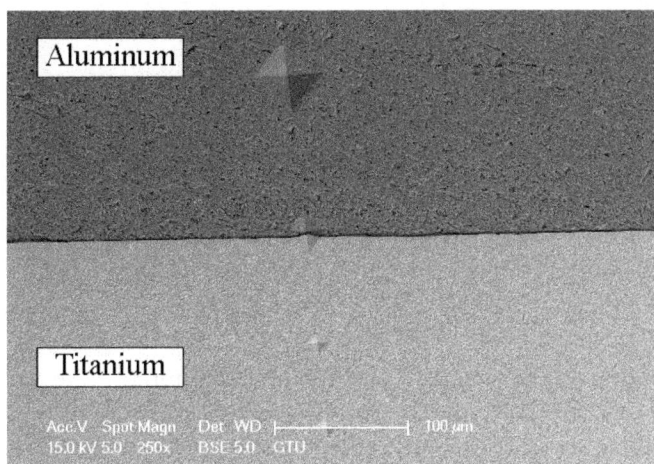

Figure 3. Microhardness measurement marks with SEM.

Table 4. Microhardness test results for all samples with different points.

Sample No.	Point 1 (−400 μm)	Point 2 (−300 μm)	Point 3 (−200 μm)	Point 4 (−100 μm)	Interface	Point 5 (100 μm)	Point 6 (200 μm)	Point 7 (300 μm)	Point 8 (400 μm)
A12	33	33	32	32	86	445	406	406	422
A11	33	33	34	33	104	372	422	422	411
A10	35	36	35	35	92	350	346	350	330
A9	31	32	32	34	131	354	354	342	342
A8	32	32	32	33	102	363	330	342	363
A7	34	35	32	32	141	417	417	406	434
A6	33	33	33	33	120	354	359	386	326
A5	36	33	34	35	92	381	372	464	439
A3	32	33	35	23	106	372	450	422	350

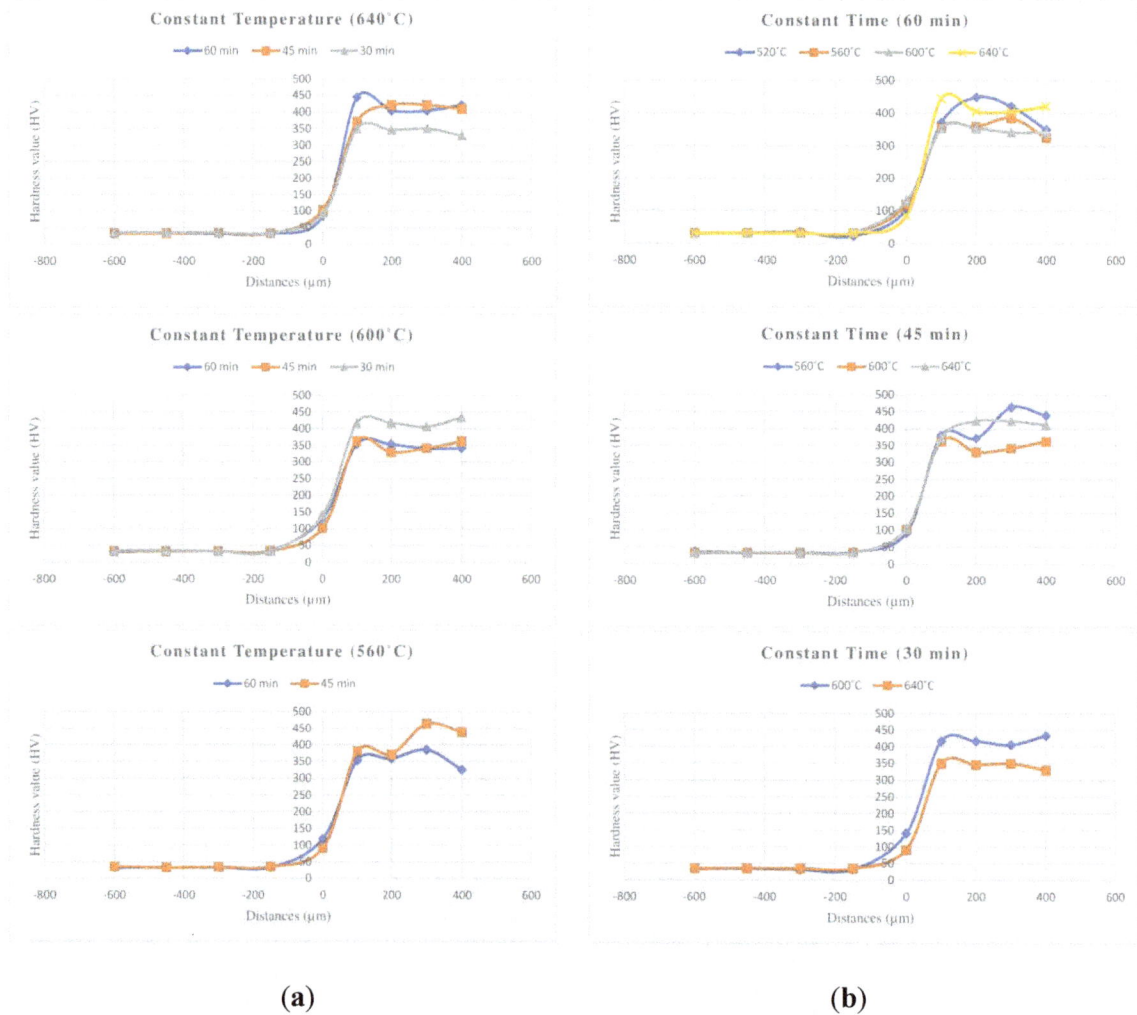

Figure 4. Hardness profiles of samples bonded based on (**a**) temperature and (**b**) time.

3.2. Tensile Tests

According to the tensile test results, T1, T2, T3, T4, and T5 samples were fractured on the welding zone; however, T6, T7, T8, and T9 samples were fractured on the aluminum side (see Figure 5), and these results actually show the strength of the welding zone. Tensile stresses at the crack are shown in Table 5. Maximum loads and extensions at the crack are also presented. The load was applied to all samples with a velocity of 1 mm/min.

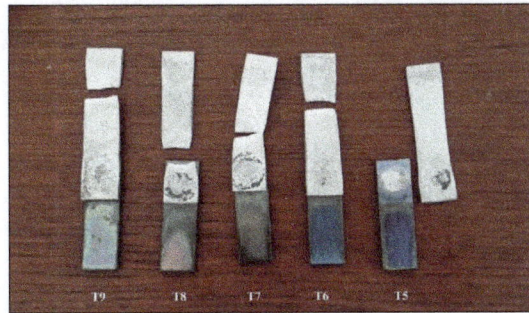

Figure 5. Successful tensile test parameters.

Table 5. Tensile test results.

Sample No.	Maximum Load (N)	Extension at Crack (mm)	Tensile Strain %	Stress at 0.2% Yield (MPa)
T1	819.92	0.43	0.0053336	-
T2	3221.60	4.07	0.0508468	46.399
T3	942.66	0.48	0.0059383	-
T4	2600.34	2.84	0.0355398	45.882
T5	3341.79	12.95	0.1618671	-
T6	3241.74	11.10	0.1388743	48.274
T7	2855.94	9.35	0.1168766	41.950
T8	3095.25	11.60	0.1453258	-
T9	3069.24	14.33	0.1791618	-

Figure 6 shows the tensile curves of the failure tensile test result; when maximum stresses are reached or become closed, fracture occurs in the graphs. On the other hand, after reaching maximum stresses, the samples continue to extend until the fracture occurs on the aluminum parts in Figure 7. Thus, those tensile test results present successful bonding. A successful weld between dissimilar metals is one that is as strong as the weaker of the two metals being bonded, i.e., possessing sufficient tensile strength and ductility so that the joint will not fail in the weld. When the parameters are compared, it is observed that extensions increase with increasing time and temperature. Maximum extension occurred in the sample welded at 640 °C for 60 min, and this is the highest value parameter according to welding temperature and time. More comparisons can be drawn from a detailed examination of Table 5. Nevertheless, while the sample welded at 600 °C for 60 min has less extension, the sample was fractured from the aluminum side as expected, and this result shows the quality of bonding with less extension. This may be the best sample according to the tensile tests.

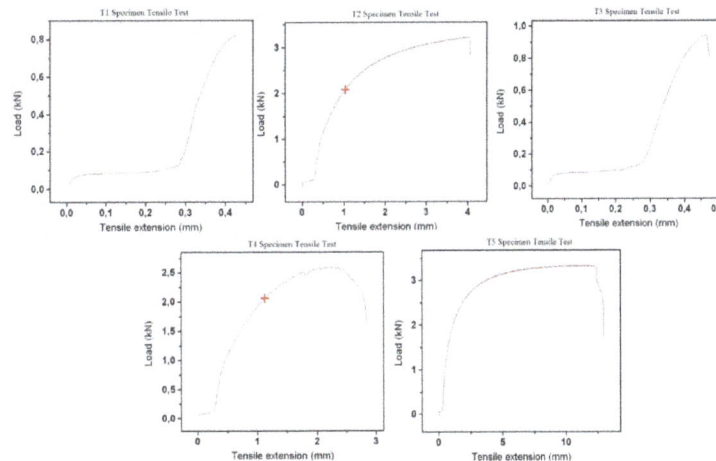

Figure 6. Force–extension curves of the failing results.

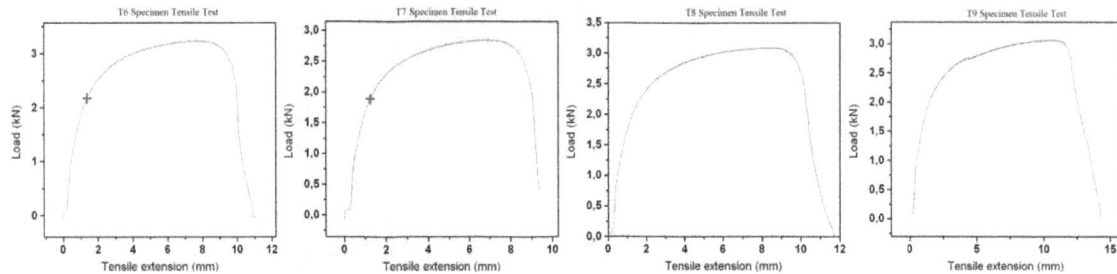

Figure 7. Force–extension curves of the successful results.

3.3. Morphology of the Welds

Morphological examination contains the study of the shape, the size, the phase distribution, and the concentration of the joints. SEM micrographs were selected to illustrate the principal defects or discontinuities that occur in the transition zone. The transition zone between the titanium alloy and aluminum, which does not have to be molten, is investigated morphologically. Thus, the transition zone is mostly affected by diffusion parameters. Luo and Acoff [17] applied a 680 °C temperature for 4 h, which is more than the melting point of aluminum; as a result, diffusion interfaces were extremely discontinuous. Aluminum atoms migrated from the Al to the Ti side during the bonding process, but the transition zone contained mostly aluminum, and this transformation decreased the strength of joints.

The amount of heat input during the welding process also plays an important role, processes such as oxyfuel welding use a high heat input that increases the size of the heat-affected zone (HAZ). A region in which the structure is affected by the applied heat is defined as the HAZ [18]. Processes such as laser beam welding and electron beam welding provide a highly concentrated, limited amount of heat, resulting in a small HAZ. Although the diffusion welding process does not cause HAZ, the bonded samples have enough strength according to the tensile test results.

In this study, if the welding temperature was further increased, aluminum parts would start to melt, potentially causing the HAZ, because more heat input would be applied, and it is known that heat and temperature are proportional [19].

Figure 8 shows SEM micrographs of the bonded samples with 560 °C for 45 min, 600 °C for 45 min, 600 °C for 60 min, 600 °C for 60 min, and 640 °C for 30 min, respectively. Discontinuities and continuities at the interface are shown in Figure 8. Sufficient diffusion was found for all parameters; however, it was observed that the diffusion interface became more discontinuous when the welding temperature increased. Applying a higher temperature results in more heat, but the irregularity in Figure 8e is still acceptable because the materials bonded without any gaps.

Figure 9 shows the EDS analysis of Sample A3. EDS analysis has been performed as line scanning from the left side to the right side. As a result, an element profile has been plotted. The results are as expected, because the purpose of diffusion welding is to weld dissimilar metals without any deformation—chemically, mechanically, or physically [20]. Thus, in the figure, it is possible to see a concentration of the elements in the transition zone. Figure 10 shows the sample welded at 520 °C for 60 min (Sample A3) and the selected areas on which EDS analysis was carried out; Figure 11 represents the results of the EDS analysis, and Area 1 shows the aluminum side. In fact, Spot 1 shows the diffusion interface, and it is obvious that the bonding occurred as a result of the elemental table in the figure. Areas 2 and 3 show that Ti-6Al-4V alloy kept its chemical origin; however, vanadium has different percentages such 2.71%, 4.16%, and 5.49% in Spot 1, Area 2, and Area 3, respectively. Vanadium has little neutron-adsorption ability and does not deform in creeping under high temperatures.

All results of the EDS analysis have been shown in the literature [21]; aluminum concentration increases, and titanium simultaneously decreases when the temperature increases in the transition zone. Bonding between titanium alloy and aluminum were also attempted at 480 °C for 60 min, but it was found that insufficient diffusion bonding takes place due to the low temperature.

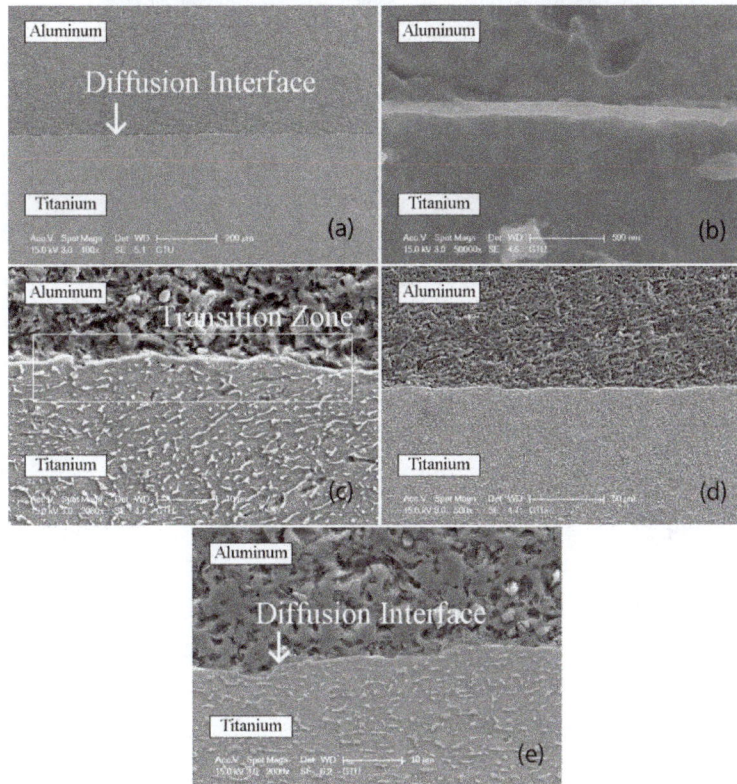

Figure 8. SEM images of different parameters: (**a**) 560 °C for 45 min; (**b**) 600 °C for 45 min; (**c,d**) 600 °C for 60 min; (**e**) 640 °C for 30 min.

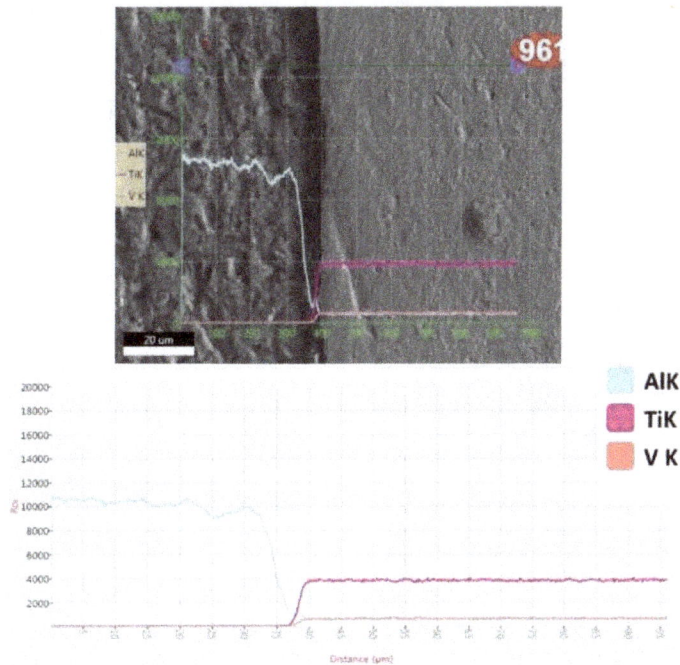

Figure 9. Linear EDS direction and line scanning result: element profile plot of Sample A3.

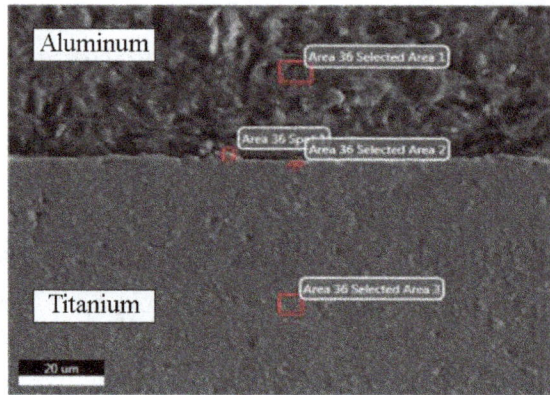

Figure 10. EDS analysis areas on the sample welded at 520 °C for 60 min.

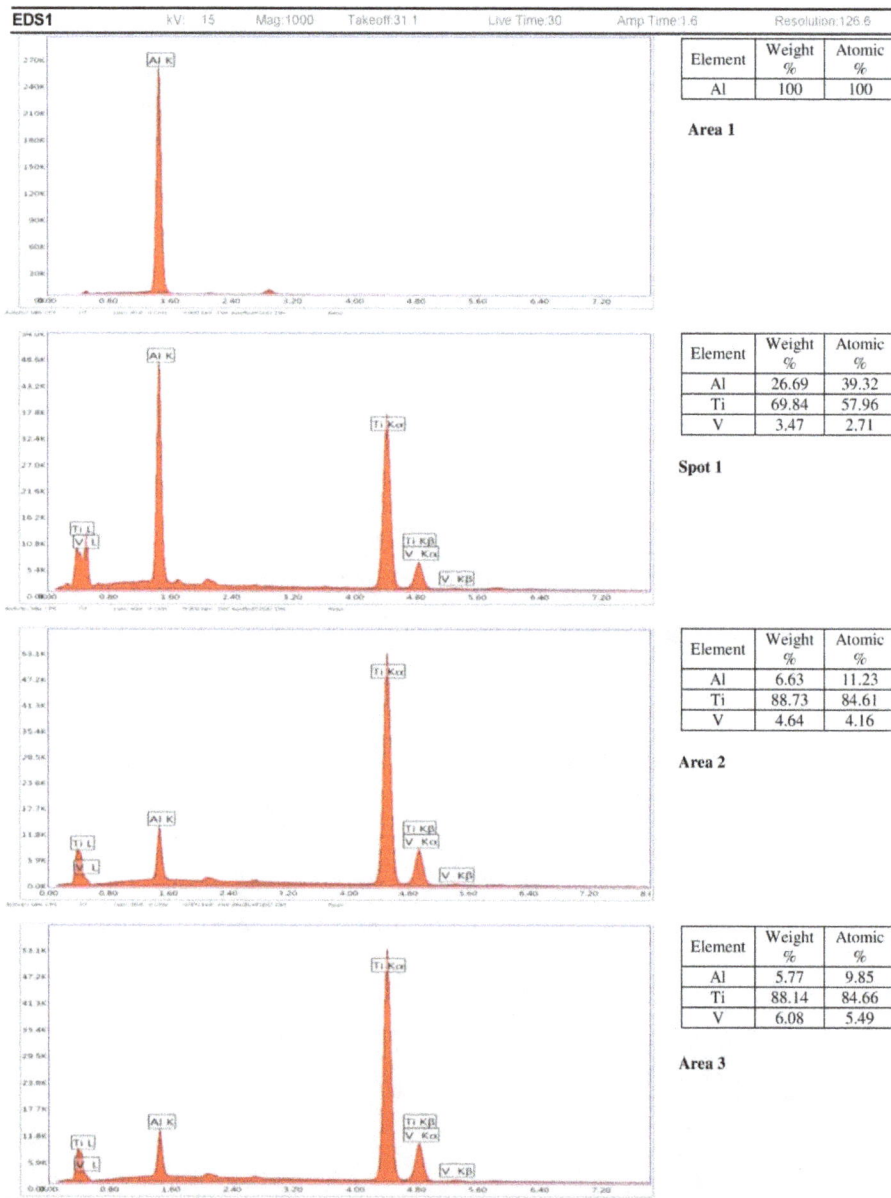

Figure 11. EDS analysis graphs of selected areas of Sample A3.

4. Conclusions

Titanium alloy and aluminum material couples were bonded by diffusion welding using different diffusion parameters. In this study, the results can be summarized as follows:

1. In the experiments, the samples were exposed to heat at temperatures from 480 °C to 680 °C, but it has been observed that 480 °C is too low to join and 680 °C is too high due to the melting point of aluminum. Additionally, welding parameters were determined according to the observation, and it shows the importance of the diffusion parameters as well. In fact, activation energy is inherent in a diffusion-controlled process, which cannot be altered by changing process parameters (temperature and time). Rather, a longer time and a higher temperature become necessary for a slow diffusion-limiting process.

2. In the microhardness results, hardness measurements are increasing from aluminum to the diffusion interface, towards the titanium side, as expected. The highest hardness value of 450 HV was obtained on the titanium side. On the aluminum side of the joints, the hardness value was found to be 35 HV, which remained constant as the distance from the interface increased.

3. When the welding temperature increased, hardness values increased as well, but with very small changes; furthermore, the β-phase of the titanium started to take place in the structure.

4. Among the parameters used in diffusion welding, maximum strain in the tensile tests occurred in the sample welded at 640 °C for 60 min; thus, this result shows the integrity of the diffusion interface.

5. According to the tensile test results, the bonded samples fractured on the aluminum side, and these results satisfy the strength of the welding zone.

6. Sufficient diffusion was found for all the parameters; however, it was observed that the diffusion interface became more discontinuous when the welding temperature increased.

Acknowledgments: This study was supported by the IUS-RDC (International University of Sarajevo—Research and Development Center) and Gebze Technical University (GTU), and Technological Research Laboratory and Scanning Electron Microscope Laboratory (SEM) of GTU.

Author Contributions: This study is a part of E. Akca's master thesis, A. Gursel was the supervisor of the thesis. Gursel's main contribution of this study is giving the research idea and supervising. Not only giving the topic, he guided to choose materials, bounding method and parameters etc. During the study, Gursel supervised the bounding and test steps and adjusts the parameters regarding the test results. The both authors prepared the paper and Gursel participated to ICWET'16 conference, and he presented the paper himself.

Conflicts of Interest: The authors declare no conflict of interest.

References

1. Williams, J.C.; Starke, E.A., Jr. Progress in structural materials for aerospace systems. *Acta Mater.* **2003**, *51*, 5775–5799. [CrossRef]

2. Leyens, C.; Peters, M. *Titanium and Titanium Alloys; Fundamentals and Applications*; WILEY-VCH Verlag GmbH & Co. KGaA: Weinheim, Germany, 2003.

3. Kahraman, N.; Gulenc, B.; Findik, F. Corrosion and mechanical-microstructural aspects of dissimilar joints of Ti-6Al-4V and Al plates. *Int. J. Impact Eng.* **2007**, *34*, 1423–1432. [CrossRef]

4. Vaidya, W.V.; Horstmann, M.; Ventzke, V.; Petrovski, B.; Koçak, M.; Kocik, R.; Tempus, G. Improving interfacial properties of a laser beam welded joint of aluminum AA6056 and titanium Ti6Al4V for aeronautical applications. *J. Mater. Sci.* **2010**, *45*, 6242–6254. [CrossRef]

5. Kicukov, E.; Gursel, A. Ultrasonic welding of dissimilar materials: A review. *Period. Eng. Nat. Sci.* **2015**, *3*, 28–36. [CrossRef]

6. Akca, E.; Gursel, A. Solid state welding and application in aeronautical industry. *Period. Eng. Nat. Sci.* **2016**, *4*, 1–8. [CrossRef]

7. Messler, R.W. *Principles of Welding: Processes, Physics, Chemistry, and Metallurgy*; John Wiley & Sons: New York, NY, USA, 1999.

8.   Avery, R.E. *Pay Attention to Dissimilar Metal Welds*; Nickel Development Institute: Toronto, ON, Canada, 1991; pp. 1–7.

9.   Aonuma, M.; Nakata, L. Dissimilar metal joining of 2024 and 7075 aluminium alloys to titanium alloys by friction stir welding. *Mater. Trans.* **2011**, *52*, 948–952. [CrossRef]

10. Hoppin, G.S.; Berry, T.F. Activated diffusion bonding. *Weld. J.* **1970**, *49*, 505–509.

11. Akca, E.; Gursel, A. The importance of interlayers in diffusion welding. *Period. Eng. Nat. Sci.* **2015**, *3*, 12–16. [CrossRef]

12. Rusnaldy, R. Diffusion Bonding: An advanced of material process. *Rotasi* **2001**, *3*, 23–27.

13. Kazakov, V.N. *Diffusion Bonding of Materials*; Pergamon Press: Moscow, Russia, 1985; pp. 157–170.

14. Aydın, K.; Kaya, Y.; Kahraman, N. Experimental study of diffusion welding/bonding of titanium to copper. *Mater. Des.* **2012**, *37*, 356–368. [CrossRef]

15. Itharaju, R.R. Friction Stir Processing of Aluminum Alloys. Master's Thesis, University of Kentucky, Lexington, KY, USA, 2004.

16. Barış, B. Diffusion Bonding of Ti-6Al-4V/304L Steels Couple using Copper Interlayer. Master's Thesis, Fırat University, Elazığ, Turkey, 2007.

17. Luo, J.; Acoff, V. Interfacial reactions of titanium and aluminum during diffusion welding. *Weld. J.* **2000**, *79*, 239–243.

18. Balkan, O.; Demirer, H.; Yildirim, H. Morphological and mechanical properties of hot gas welded PE, PP and PVC sheets. *J. Achiev. Mater. Manuf. Eng.* **2008**, *31*, 60–71.

19. Bapat, B. *Heat and Temperature*; Eklavya Publication: Bhopal, India, 2013.

20. Lundin, C.D. Dissimilar metal welds—Transition joints literature review. *Weld. J.* **1982**, *62*, 58–63.

21. Akca, E. Ti-6Al-4V and Aluminum Bonding by Diffusion Welding. Master's Thesis, International University of Sarajevo, Sarajevo, Bosnia and Herzegovina, September 2015.

11

Properties of Mechanically Alloyed W-Ti Materials with Dual Phase Particle Dispersion

František Lukáč [1,*], Monika Vilémová [1], Barbara Nevrlá [1], Jakub Klečka [1], Tomáš Chráska [1] and Orsolya Molnárová [2]

[1]   Institute of Plasma Physics, Czech Academy of Science, Za Slovankou 3, 18200 Prague, Czech Republic; vilemova@ipp.cas.cz (M.V.); nevrla@ipp.cas.cz (B.N.); klecka@ipp.cas.cz (J.K.); tchraska@ipp.cas.cz (T.C.)

[2]   Mathematics and Physics Faculty, Charles University, Ke Karlovu 3, 12116 Prague, Czech Republic; mopersze@gmail.com

*   Correspondence: lukac@ipp.cas.cz

Academic Editor: Chun-Liang Chen

Abstract: W alloys are currently widely studied materials for their potential application in future fusion reactors. In the presented study, we report on the preparation and properties of mechanically alloyed W-Ti powders compacted by pulsed electric current sintering. Four different powder compositions of W-(3%–7%)Ti with Hf or HfC were prepared. The alloys' structure contains only high-melting-point phases, namely the W-Ti matrix, complex carbide (Ti,W,Hf)C and HfO_2 particle dispersion; Ti in the form of a separate phase is not present. The bending strength of the alloys depends on the amount of Ti added. The addition of 3 wt. % Ti led to an increase whereas 7 wt. % Ti led to a major decrease in strength when compared to unalloyed tungsten sintered at similar conditions. The addition of Ti significantly lowered the room-temperature thermal conductivity of all prepared materials. However, unlike pure tungsten, the conductivity of the prepared alloys increased with the temperature. Thus, the thermal conductivity of the alloys at 1300 °C approached the value of the unalloyed tungsten.

Keywords: tungsten-titanium alloys; mechanical alloying; particle dispersion; pulsed electric current sintering; thermal conductivity; bending strength

---

1. Introduction

With the progress of nuclear fusion research, the need for new, advanced materials is becoming more urgent. For the International Thermonuclear Experimental Reactor (ITER), the choice of materials has been made. Thus, the area of the ITER's first wall will be covered by armor produced from beryllium and the exhaust components will be covered by tungsten. However, materials for the next step of reactors will have to satisfy strict requirements for the lifetime and safety levels. Thus, nontoxic, highly durable and functional materials would be the prime choice for tokamak such as DEMO tokamak.

Until recently, pure tungsten was considered the most suitable plasma-facing material for the future reactor's first wall. Its superiority over other materials was granted by the following group of properties: high resistance to sputtering, high melting point, good thermal conductivity, low thermal expansion and low tritium retention. Nevertheless, tungsten also has certain disadvantages, with thermally induced grain growth (depending on its thermomechanical history, starting at temperatures as low as 1000 °C) among the most serious. Thus, in the conditions of fusion plasma and plasma disruptions, excessive grain growth leads to the degradation of mechanical properties, which subsequently causes premature failure of the plasma-facing component during heat cycling. A further problem arises during exposure of the tungsten to the irradiated particles, such as those of helium and deuterium present in the fusion plasma. The particles penetrate into the material bulk and under specific conditions, such as high particle fluencies, the crystalline lattice becomes supersaturated which leads to blister or helium fuzz formation [1] and, therefore, the further degradation of the mechanical properties. Another concern of tungsten's behavior is related to the formation of tungsten oxides in the presence of oxygen. Tungsten trioxide represents a serious risk in the case of reactor accidents under which oxygen or oxygen-containing compounds (such as coolant water) enter the reactor chamber. The formation of volatile tungsten trioxide could lead to radiation escaping into the surrounding environment. Thus, efforts to develop a smart tungsten material that can suppress thermal and radiation degradation as well as the formation of volatile oxides have emerged recently.

Most recent attempts to improve tungsten properties have been conducted through the modification of tungsten's microstructure and the addition of minor alloying elements. Many of the developed materials show promising results. For example, some studies point out that ultrafine-grained tungsten not only has better mechanical properties but is also significantly more resistant to irradiation from ions [2]. In order to stabilize the grain size at higher temperatures, tungsten with a particle dispersion has been developed. In the case of tungsten, usually small amounts of oxides, e.g., Y_2O_3, La_2O_3 (ODS—oxide dispersion strengthening), or carbides of transition metals, e.g., TiC (CDS, carbide dispersion strengthening), are added [3,4]. The suppression of the formation of volatile tungsten oxide has been suggested by the formation of tungsten self-passivating alloys [5] consisting of W with the addition of Cr and Ti or Si. In the oxidation atmosphere, complex Cr-W oxide layers are formed, encapsulating the tungsten oxides at the surface of the armor. However, the effect of the various alloying elements is still being studied.

Since the melting temperature of tungsten is the highest among the chemical elements, the powder metallurgy accompanied by mechanical alloying is the obvious choice of alloying method. The spark plasma sintering method provides the exceptional advantage of fast heating/cooling rates when compared to conventional sintering methods. Therefore, a high sintering temperature is achieved simultaneously with the suppression of unwanted grain growth. In the present study, we attempted to prepare W-Ti alloys with the addition of HfC or Hf by means of mechanical alloying. According to our experience, HfC has a tendency to oxidize either during mechanical alloying or during sintering to form HfO_2. In this way, the mitigation of Ti oxidation was approached. The alloys were analyzed for their phase composition during each individual step of the preparation. Thus, the alloying process can be better understood and used for further tailoring of the alloys. The phases were correlated with the microstructural information. Basic mechanical properties of the prepared alloys were analyzed and the effect of the addition of Ti was discussed. The effect of the alloying elements on room- and high-temperature thermal conductivity was studied as it remains overlooked in the majority of published studies due to the low amounts of alloying elements.

2. Materials and Methods

The powder batches of composition W-3 wt. % Ti-2 wt. % HfC, W-7 wt. % Ti-2 wt. % HfC, W-3 wt. % Ti-2 wt. % Hf and W-7 wt. % Ti-2 wt. % Hf (see Table 1) were prepared in a planetary ball mill Pulverisette 5 (Fritsch, Germany). The starting powders were W (99.9% purity, 1.2 μm average powder size), Ti (99.4% purity, 5 μm average powder diameter size), Hf (bimodal powder diameter size distribution, 15 μm and 45 μm) and HfC (bimodal powder diameter size distribution, 5 μm and 20 μm). For the ball milling process, the powders were loaded in tungsten carbide bowls with tungsten carbide grinding balls in the ball to powder ratio (BPR) 11:1. High purity argon was used as protective atmosphere in order to prevent oxidation during the milling process. The summary of the milling parameters can be found in Table 1. The powders were consolidated by pulsed electric current sintering machine SPS 10-4 (Thermal Technology, Santa Rosa, CA, USA) under similar sintering conditions, i.e., sintering temperature 1750 °C, pressure of 70 MPa, vacuum of 10 Pa and sintering time of 3 min. Sintering was performed in graphite molds and graphite foils or graphite foils covered by hexagonal boron nitride (BN). All milling and sintering parameters are summarized in Table 1.

Table 1. Summary of the materials and production conditions.

Sample Designation *	Starting Powders (Average Size, μm)	Ball Milling Conditions	Sintering Conditions
W-3Ti-2HfC	W (1.2), Ti (5), HfC (bimodal 5 and 20)	11:1 (BPR), 30 min mixing at 80 RPM 270 rpm (milling speed), 40 h (milling time), Argon atmosphere, WC-Co milling bowls and balls.	100 °C/min (Heating speed), 1750 °C sintering temperature, 3 min hold time at 1750 °C, powder surrounded by a graphite foil, vacuum, 120 ms/30 ms on/off pulses at started at sintering temperature.
W-7Ti-2HfC			
W-3Ti-2Hf	W (1.2), Ti (5), Hf (bimodal 15 and 45)		Identical to the above except that powder was surrounded by a graphite foil covered by BN (boron nitride) layer.
W-7Ti-2Hf			
W	W (1.2)	no ball milling	Identical to the above (W + Ti + Hf case)

* Numbers refers to weight percentage.

Phase compositions and lattice parameters were determined from X-ray diffraction (XRD) patterns obtained at room temperature by CuKα (divergent beam was used for mechanically alloyed powders and parallel beam of 1 mm diameter in the middle of the cross section cut was used for sintered samples) and 1D LynxEye detector (Bruker, Karlsruhe, Germany) (Ni β filter in front of the detector) mounted on Bruker D8 Discover (Bruker, Karlsruhe, Germany) and subsequent Rietveld refinement [6] was performed in TOPAS 5 (Bruker, Karlsruhe, Germany) [7].

The flexure strength was measured using universal tensile test machine Instron 1362 (Instron, High Wycombe, UK) with support diameter of 5 mm, support span of 14.55 mm and loading rate of 0.2 mm/min.

The hardness was evaluated on a universal hardness tester Nexus 4504 (Innovatest, Maastricht, The Netherland) using Vickers indenter, load equivalent to 1 kg and dwell time of 10 s.

The microstructure was evaluated on polished cross-sections using SEM EVO MA 15 (Carl Zeiss SMT, Oberkochen, Germany) in backscattered mode and equipped with EDS detector XFlash® 5010 (Bruker, Karlsruhe, Germany).

The thermal diffusivity (α) and specific heat capacity (C_p) of the samples were measured by a laser-flash method on an LFA 1000 apparatus (Linseis, Selb, Germany) in vacuum at RT, 100, 300, 500, 700, 900, 1100 and 1300 °C. The thermal conductivity (λ) was calculated utilizing the relationship $\lambda = \varrho \alpha C_p$, where ϱ is material density. The samples were cut into 10 mm × 10 mm size and thickness was approximately 2 mm. Data were averaged from at least four measurements at each temperature.

3. Results

3.1. Microctructural and Phase Analysis

Table 2 shows the results of the Rietveld refinement of XRD diffractograms in the mechanically alloyed powders used in this study. The lattice parameters were significantly larger than the lattice parameter of pure tungsten at room temperature $a = 0.3165$ nm [8]. The dissolution of Ti in W continuously increased the lattice constant of the bcc phase β-(Ti,W) up to the lattice constant of the bcc Ti phase, which was close to 0.3283 nm [9]. Since no other phases of pure Ti or Hf were observed, one can conclude that the alloying elements Ti and Hf were introduced to the matrix of bcc W and formed a solid solution. Significant refinement of the crystallites' size was found in all milled powders after the high-energy milling, and microstrain values suggested that a substantial degree of deformation energy was stored in the powders.

Table 2. Results of Rietveld refinement fit for tungsten matrix phase in mechanically alloyed powders for 40 h. Fit errors of the last digit are given in parentheses.

Composition	Lattice Parameter a (nm)	Crystallite Size (nm)	Microstrain e_0 (10^{-3})
W-3Ti-2HfC	0.317536 (8)	11.9 (1)	3.88 (3)
W-7Ti-2HfC	0.31785 (1)	11.2 (1)	3.78 (4)
W-3Ti-2Hf	0.317207 (7)	13.5 (1)	3.51 (3)
W-7Ti-2Hf	0.317665 (8)	12.4 (1)	3.45 (3)

Figure 1 shows the measured diffraction patterns with Rietveld refinement analysis for the sintered samples.

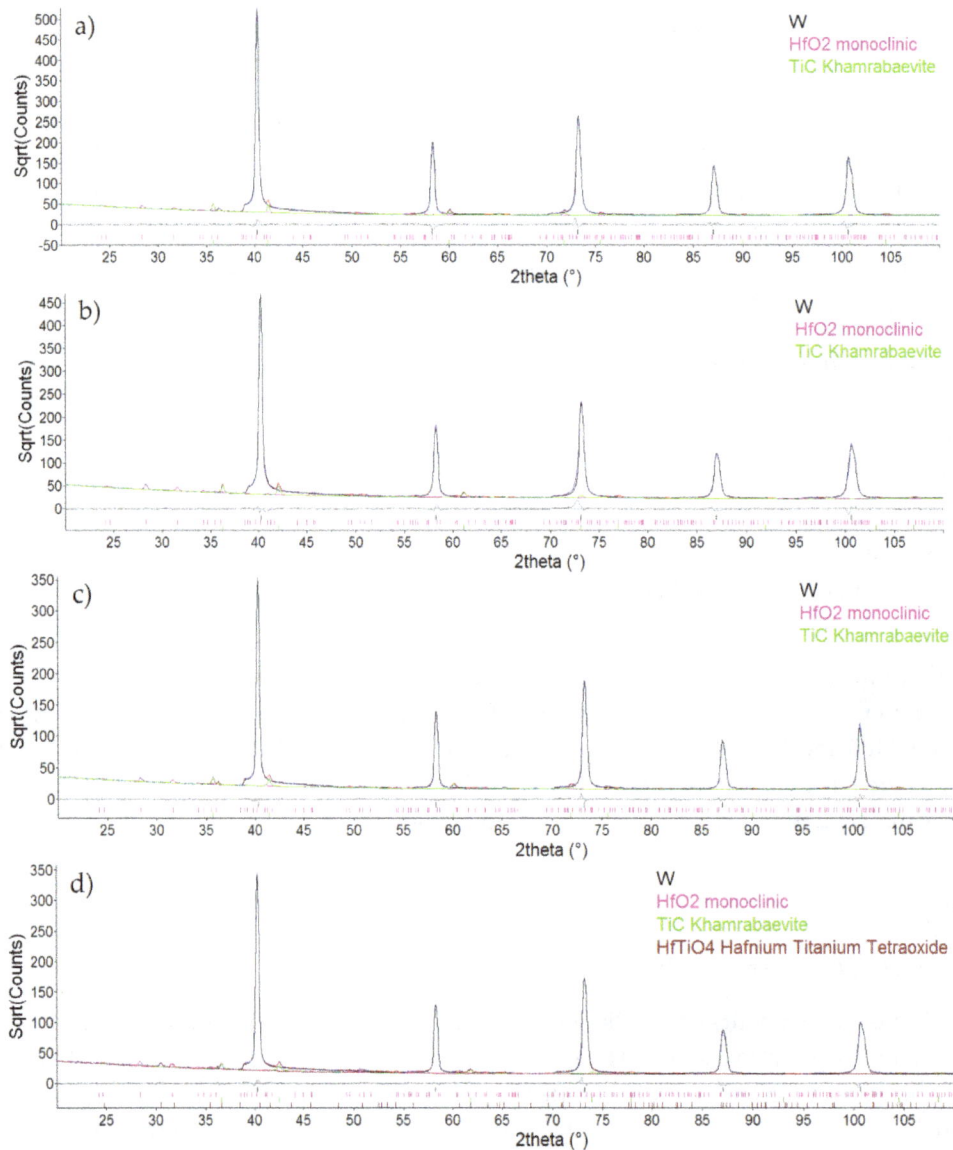

Figure 1.   X-ray diffraction patterns with Rietveld refinement analysis for sintered samples, (**a**) W-3Ti-2HfC; (**b**) W-7Ti-2HfC; (**c**) W-3Ti-2Hf; (**d**) W-7Ti-2Hf.

XRD quantitative Rietveld phase analysis of sintered samples, summarized in Table 3, revealed that hafnium oxide was present in the tungsten-based matrix in the form of a monoclinic HfO_2 phase and a cubic phase with the space group $Fm\bar{3}m$. Since W, Ti and Hf all form monocarbides with this space group and these isomorphous carbides were found miscible [10,11], a complex carbide, (Ti,W,Hf)C, was also found in all sintered samples in this work. In addition, sample W-7Ti-2Hf showed the presence of hafnium titanium tetraoxide (HfTiO$_4$ with the orthorhombic space group *Pbcn*). The lattice parameters of the tungsten matrix in the sintered samples (Table 4) were smaller than those of the powder samples (Table 2).

Table 3. Phase analysis of sintered samples by XRD quantitative Rietveld refinement noted in weight percent.

Identified Phases	W-3Ti-2HfC	W-7Ti-2HfC	W-3Ti-2Hf	W-7Ti-2Hf
W matrix	94.8 (6)	91.6 (4)	93.5 (8)	90.1 (8)
HfO$_2$ monoclinic	0.9 (3)	2.2 (2)	1.5 (4)	2.3 (3)
(Ti,W,Hf)C	4.3 (6)	6.2 (4)	5.0 (7)	6.1 (7)
HfTiO$_4$	-	-	-	1.6 (3)

Table 4. Lattice parameters of W-Ti alloy in sintered samples.

Phases	W-3Ti-2HfC	W-7Ti-2HfC	W-3Ti-2Hf	W-7Ti-2Hf
W matrix	3.16476 (7)	3.16631 (4)	3.16410 (7)	3.16464 (6)
(Ti,W,Hf)C	4.356 (1)	4.2866 (6)	4.353 (1)	4.245 (1)

The microstructure of compacted W-3Ti-2HfC and W-7Ti-2HfC shows features identical to W-3Ti-2Hf and W-7Ti-2Hf; therefore, only representative micrographs of chosen compositions will be presented here (Figure 2). Generally, the microstructure consists of regular polygonal equiaxed grains representing the W-Ti matrix and numerous smaller particles located mainly at the grain boundaries. Virtually no porosity was found in the sintered samples. According to the results of EDS (Figure 3), the brighter intergranular particles were rich in Hf and O and the darker particles were rich in Ti, an indication of Hf content being visible as well. A local increase in the C concentration was not observed due to the carbon contamination layer already present and/or deposited by the electron beam on the sample surface. Therefore, it can be expected that the lighter and darker particles are oxides (HfO$_2$, HfTiO$_4$) and a complex carbide (Ti,W,Hf)C as reported by XRD, respectively.

The size distribution of the W-Ti matrix grains is depicted in Figure 4. Local maxima within 1–5 µm^2 represent the most frequent area grain sizes. Lower size frequencies are not shown in the plot as they most likely represent HfO$_2$ particles and/or noise. The main difference between microstructures containing different amounts of Ti was that the frequency of larger carbide particles (i.e., 0.6 µm and larger) was higher for samples containing 7% Ti (Figure 5); a similar, though less obvious, trend shows the size distribution of HfO$_2$ (Figure 6). Fine spherical particles in the size range of tens of nanometers present within the matrix grains were more frequent in the samples containing 3% Ti (see Figure 2a).

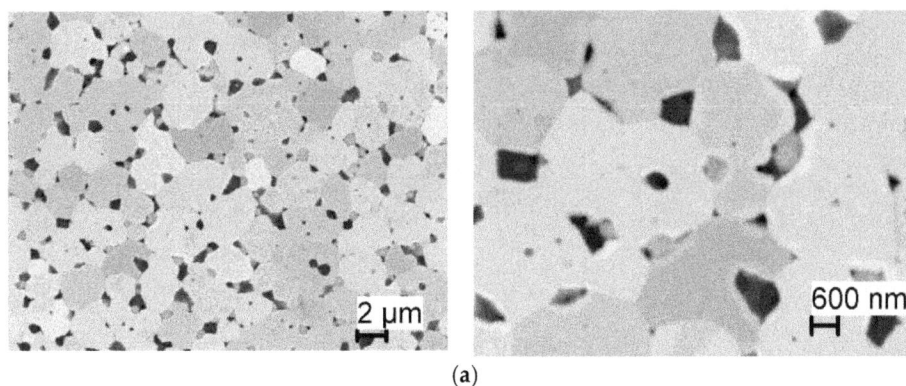

(a)

Figure 2. *Cont.*

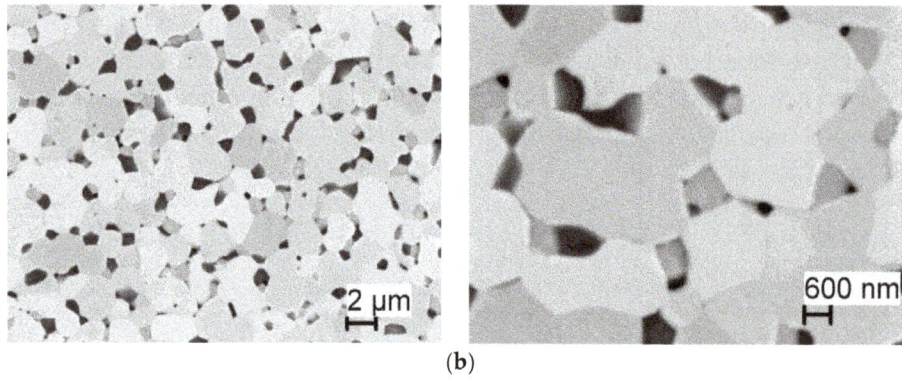

Figure 2. Microstructure of W-3Ti-2Hf and W-7Ti-2HfC consisting of W-Ti matrix (bright larger grains), (Ti,W,Hf)C (dark grains) and HfO_2 ($HfTiO_4$) particles (bright small grains). (**a**) W-3Ti-2Hf; (**b**) W-7Ti-2HfC.

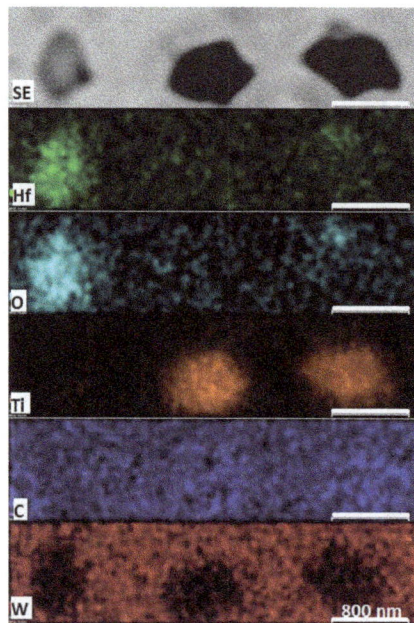

Figure 3. EDS map of area in W-3Ti-2HfC sample containing dispersed particles. All the scale bars represent the same size of 800 nm.

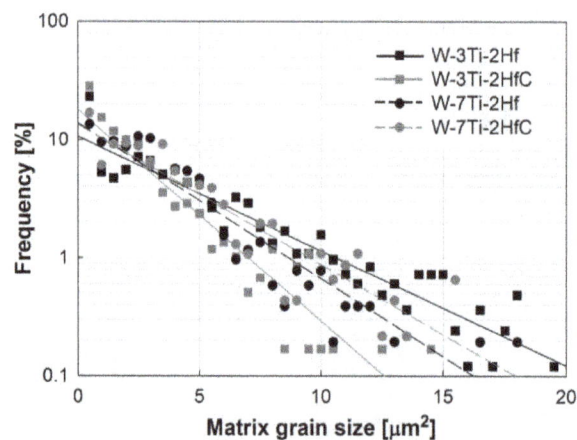

Figure 4. Frequency plot of matrix grain sizes.

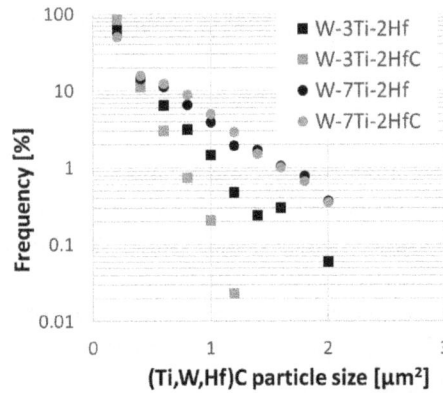

Figure 5. Frequency plot of (Ti,W,Hf)C carbide particle sizes.

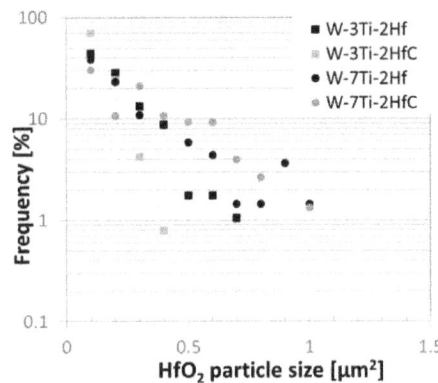

Figure 6. Frequency plot of HfO_2 ($HfTiO_4$) particle sizes.

3.2. Mechanical and Thermal Properties

Figure 7 summarizes results on the basic mechanical properties of the prepared alloys in comparison to pure tungsten sintered at similar conditions. A major influence of Ti content on the material strength was apparent. W-3Ti-2Hf and W-3Ti-2HfC showed an increase in the flexural strength when compared to pure tungsten whereas W-7Ti-2Hf and W-7Ti-2HfC showed a significant decrease in the strength. Slightly better values of flexural strength could be also observed for materials mechanically alloyed with the addition of HfC.

The hardness of the prepared alloys significantly increased when compared to pure tungsten. A larger increase was observed for materials containing 7% Ti. However, there was no apparent trend for the Hf/HfC content.

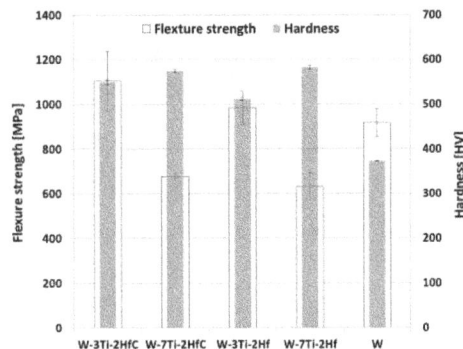

Figure 7. Flexural strength and hardness of the prepared samples in comparison with properties of pure tungsten.

The small addition of alloying elements had a significant effect on the thermal conductivity (Figure 8). In the most extreme case, i.e., for materials containing 7% Ti, the drop in the conductivity reached almost 80% at room temperature when compared to pure tungsten. The addition of HfC slightly improved the thermal conductivity at higher temperatures (above 300 °C) when compared to the materials with the addition of Hf. Generally, the thermal conductivity of metals inherently decreases with an increasing material temperature as in the case of pure tungsten. However, some alloys have an inverse dependency on the temperature, as is the case for the W-Ti-Hf(C) samples. Consequently, the difference between the prepared alloys and the pure sintered tungsten tended to decrease at higher temperatures. A minimal difference was reached for W-3Ti-2HfC above 900 °C, and for other alloys except W-7Ti-2Hf at 1300 °C. At this point, the conductivity of W-3Ti-2Hf, W-3Ti-2Hf and W-7Ti-2HfC was 20% lower than the thermal conductivity of pure tungsten.

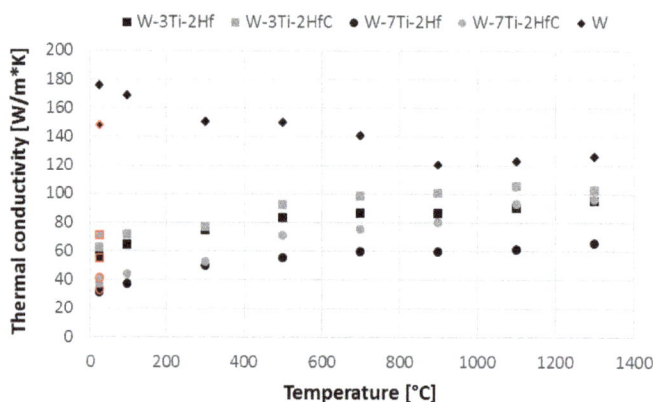

Figure 8. Thermal conductivity of the sintered samples in comparison with properties of sintered pure tungsten; points with red outlines depict the second room-temperature measurement after the high-temperature analysis.

In order to evaluate the stability of the thermal conductivity, the room-temperature value was measured twice. The first measurement was performed at 25 °C and the second measurement after the thermal cycle of the laser-flash method (appropriate points are outlined in red), i.e., after 14 h at gradually elevating temperatures from 25–1300 °C. For the W-Ti-Hf(C) samples the value of the first and second measurement is almost identical.

4. Discussion

After the sintering of the alloyed powders, new phases emerged, i.e., (Ti,W,Hf)C and HfO$_2$. Micrographs of the sintered materials proved the TiC phase was located at grain boundaries, where it can play an important role together with HfO$_2$ in grain stabilization at high temperatures by pinning grain boundaries. The dispersion of TiC particles supports the assumption of carbon surface diffusion through the powder materials and carbon grain boundary diffusion (in the later stage of sintering). The high affinity of carbon to titanium caused the depletion of excess Ti from the W-Ti matrix. Alloys with 7% Ti contained higher number of larger (Ti,W,Hf)C particles, as the higher amount of Ti led to particle coarsening. Besides the W-Ti solid solution, (Ti,W,Hf)C and HfO$_2$ (and HfTiO$_4$ in one case), the presence of additional phases was not confirmed. A number of research results reported on the formation of Ti-rich phases in W-Ti, e.g., Ti pools in W-2 wt. %Ti-0.5%Y$_2$O$_3$ and W-4 wt. %Ti-0.5%Y$_2$O$_3$ [12], in W-10 wt. %Ti [13] and in WCr12Ti2.5 [5]. Considering the potential high-temperature applications of the alloys, the formation of Ti pools or Ti-rich solid solution is undesirable due to the low melting point of such phases (for pure titanium it is 1668 °C). The materials studied in this work maintained a high melting point, as the only Ti-rich phase confirmed by XRD was complex (Ti,W,Hf)C with a melting point most likely around that of TiC (i.e., 3160 °C), and the melting point of WC is around 2830 °C.

The matrix of the prepared materials consists of fine grains, mostly in the size range of 1 μm–5 μm diameter. It seems there is only a slight effect of the addition of titanium on the matrix grain size distribution. Alloys with 7% Ti have a slightly higher frequency of fine grains, likely due to higher occurrence of larger carbide particles.

The effect of titanium on the bending strength of the prepared materials strongly depends on the added amount of Ti. It seems there is a threshold content of titanium in W, above which the bending strength starts to decrease and becomes smaller than that of pure sintered tungsten. Nevertheless, it also seems that the effect might be different when other secondary phases are added. Authors in [14] report an opposite trend in the bending strength for W-4Ti-0.5Y_2O_3 and W-2Ti-0.47-Y_2O_3, i.e., with an increasing content of Ti, the strength increases. However, the bending strength of alloys in the mentioned study still remained below the room-temperature strength of pure tungsten. The hardness of the prepared alloys increased with respect to the pure tungsten due to the increase in the dislocation density caused by ball milling, grain refinement and mainly due to the presence of hard particle dispersion. The hardness value was slightly lower for alloys containing 3% Ti as the frequency of carbide particles in each size category was lower.

The results of the thermal conductivity showed a major decrease for the prepared alloys when compared to pure tungsten. According to the rule of mixture and the Maxwell-Garnett model applied on W-3Ti-2HfC and W-7Ti-2HfC with respect to the volumetric representation of the XRD results, the room-temperature thermal conductivity due to the second-phase dispersion should not decrease by more than 26% (according to rule of mixture the values are 155 W/(K·m) and 144 W/(K·m), respectively; according to the Maxwell-Garnett model they are 146 W/(K·m) and 133 W/(K·m), respectively) [15]. The calculated numbers were significantly higher than the measured values which might imply a significant influence of Ti dissolved in the tungsten lattice. The thermal conductivity of the prepared materials grew with the temperature, which is a less common phenomenon. That can also be attributed to the Ti dissolved in the W matrix. However, from 900 °C the thermal conductivity of W-3Ti-HfC alloy was only 20 W/(K·m) smaller than that of pure tungsten. Although at higher temperatures the difference in the thermal conductivity was minimized, the authors believe that an additional improvement might be possible by lowering the titanium content or additional thermomechanical processing. For example, in [16] a dramatic improvement of thermal conductivity for W-TaC alloys was reached after hot rolling.

The lower rate in the thermal conductivity increase for W-7Ti-2Hf was probably caused by the presence of $HfTiO_4$.

5. Conclusions

Four types of tungsten materials were prepared in this study, i.e., W-3Ti-2Hf, W-3Ti-2HfC, W-7Ti-2Hf and W-7Ti-2HfC. Due to the sintering in the presence of graphite, a certain amount of Ti and Hf was transformed into the complex carbide phase. Thus, no Ti pools were formed, unlike in the results in a number of other studies. The microstructure consisted of a W-Ti matrix with fine (Ti,W,Hf)C and HfO_2 particles dispersed at the grain boundaries. Besides the mixed Hf-Ti oxide in the sample with the highest Ti content, no other phases within a detectable limit were present. It can thus be expected that the prepared alloys possess a high melting point and can be applied in high-temperature environments. Moreover, it can be expected that the particle dispersion improves the high-temperature stability.

The effect of the addition of titanium on the materials' strength was found to depend on the amount of Ti added. The bending strength of the alloys with 3% Ti increased with respect to pure tungsten, whereas 7% Ti led to a significant decrease in strength.

The addition of Ti into tungsten significantly lowers the thermal conductivity which is the major drawback of the prepared alloys. Nevertheless, it was proved that the conductivity has a tendency to increase with the temperature. Thus, in the temperature window predicted for future fusion reactors, i.e., between 700 °C and 1300 °C, the conductivity reached nearly the value of conductivity for pure tungsten.

Acknowledgments: This work was financially supported by the grant GAČR 15-15609S. One author, O.M., is grateful for financial support from the grant SVV-2016-260213.

Author Contributions: All the authors contributed equally to this research by conducting the experiments and by finalization of the manuscript. F.L., M.V. and T.C. contributed with preparation and finalization of manuscript, F.L. with XRD method results, M.V. with SEM method results , milling and sintering optimization, B.N. with laser-flash method, J.K. with mechanical properties testing, O.M. with powders and samples preparation.

Conflicts of Interest: The authors declare no conflict of interest.

References

1. Shu, W.M.; Kawasuso, A.; Yamanishi, T. Recent findings on blistering and deuterium retention in tungsten exposed to high-fluence deuterium plasma. *J. Nucl. Mater.* **2009**, *386–388*, 356–359. [CrossRef]
2. El-Atwani, O.; Gonderman, S.; Efe, M.; de Temmerman, G.; Morgan, T.; Bystrov, K.; Klenosky, D.; Qiu, T.; Allain, J.P. Ultrafine tungsten as a plasma-facing component in fusion devices: Effect of high flux, high fluence low energy helium irradiation. *Nucl. Fusion* **2014**, *54*, 83013. [CrossRef]
3. Rieth, M.; Boutard, J.L.; Dudarev, S.L.; Ahlgren, T.; Antusch, S.; Baluc, N.; Barthe, M.-F.; Becquart, C.S.; Ciupinski, L.; Correia, J.B.; et al. Review on the EFDA programme on tungsten materials technology and science. *J. Nucl. Mater.* **2011**, *417*, 463–467. [CrossRef]
4. Vilémová, M.; Pala, Z.; Jäger, A.; Matějíček, J.; Chernyshova, M.; Kowalska-Strzęciwilk, E.; Tonarová, D.; Gribkov, V.A. Evaluation of surface, microstructure and phase modifications on various tungsten grades induced by pulsed plasma loading. *Phys. Scr.* **2016**, *91*, 34003. [CrossRef]
5. López-Ruiz, P.; Ordás, N.; Iturriza, I.; Walter, M.; Gaganidze, E.; Lindig, S.; Koch, F.; García-Rosales, C. Powder metallurgical processing of self-passivating tungsten alloys for fusion first wall application. *J. Nucl. Mater.* **2013**, *442*, S219–S224. [CrossRef]
6. Rietveld, H.M. Line profiles of neutron powder-diffraction peaks for structure refinement. *Acta Crystallogr.* **1967**, *22*, 151–152. [CrossRef]
7. Coelho, A.A. *TOPAS*, version 5 (Computer Software); Coelho Software: Brisbane, Australia, 2016.
8. Waseda, Y.; Hirata, K.; Ohtani, M. High-temperature thermal expansion of platinum, tantalum, molybdenum, and tungsten measured by X-ray diffraction. *High Temp. High Press.* **1975**, *7*, 221–226.
9. Levinger, W.B. Lattice parameters of beta titanium at room temperature. *J. Met.* **1953**, *5*, 195.
10. Rudy, E. Constitution of ternary titanium-tungsten-carbon alloys. *J. Less Common Met.* **1973**, *33*, 245–273. [CrossRef]
11. Murray, P.; Weston, J.E. The 1700 °C isothermal section of the pseudoternary system TiC-ZrC-HfC. *J. Less Common Met.* **1981**, *81*, 173–179. [CrossRef]
12. Aguirre, M.V.; Martín, A.; Pastor, J.Y.; LLorca, J.; Monge, M.A.; Pareja, R. Mechanical properties of Y_2O_3-doped W-Ti alloys. *J. Nucl. Mater.* **2010**, *404*, 203–209. [CrossRef]
13. Dai, W.; Liang, S.; Luo, Y.; Yang, Q. Effect of W powders characteristics on the Ti-rich phase and properties of W–10 wt. % Ti alloy. *Int. J. Refract. Met. Hard Mater.* **2015**, *50*, 240–246. [CrossRef]
14. Aguirre, M.V.; Martín, A.; Pastor, J.Y.; LLorca, J.; Monge, M.A.; Pareja, R. Mechanical properties of tungsten alloys with Y_2O_3 and titanium additions. *J. Nucl. Mater.* **2011**, *417*, 516–519. [CrossRef]
15. Garnett, J.C.M. Colours in Metal Glasses, in Metallic Films, and in Metallic Solutions. II. *Philos. Trans. R. Soc. Math. Phys. Eng. Sci.* **1906**, *205*, 237–288. [CrossRef]
16. Miao, S.; Xie, Z.M.; Yang, X.D.; Liu, R.; Gao, R.; Zhang, T.; Wang, X.P.; Fang, Q.F.; Liu, C.S.; Luo, G.N.; et al. Effect of hot rolling and annealing on the mechanical properties and thermal conductivity of W-0.5 wt. % TaC alloys. *Int. J. Refract. Met. Hard Mater.* **2016**, *56*, 8–17. [CrossRef]

12

Microstructure and Oxidation Behavior of CrAl Laser-Coated Zircaloy-4 Alloy

Jeong-Min Kim [1,*], Tae-Hyung Ha [1], Il-Hyun Kim [2] and Hyun-Gil Kim [2]

[1]   Department of Advanced Materials Engineering, Hanbat National University, 125 Dongseo-daero, Yuseong-gu, Daejeon 34158, Korea; htman15@naver.com

[2]   Light Water Reactor Fuel Technology Division, Korea Atomic Energy Research Institute, 989-111 Daedek-daero, Yuseong-gu, Daejeon 34057, Korea; s-weat@hanmail.net (I.-H.K.); hgkim@kaeri.re.kr (H.-G.K.)

*   Correspondence: jmk7475@hanbat.ac.kr

Academic Editor: Hugo F. Lopez

Abstract: Laser coating of a CrAl layer on Zircaloy-4 alloy was carried out for the surface protection of the Zr substrate at high temperatures, and its microstructural and thermal stability were investigated. Significant mixing of CrAl coating metal with the Zr substrate occurred during the laser surface treatment, and a rapidly solidified microstructure was obtained. A considerable degree of diffusion of solute atoms and some intermetallic compounds were observed to occur when the coated specimen was heated at a high temperature. Oxidation appears to proceed more preferentially at Zr-rich region than Cr-rich region, and the incorporation of Zr into the CrAl coating layer deteriorates the oxidation resistance because of the formation of thermally unstable Zr oxides.

Keywords: laser coating; Zircaloy-4; CrAl; microstructure; oxidation

1. Introduction

Zirconium alloys have been widely used as nuclear materials because of their high chemical stability under the normal operating conditions of a pressurized or boiling water reactor, low absorption cross-section of thermal neutrons, and fairly good mechanical properties. However, Zr alloys are vulnerable to the high temperature oxidation that can occur in the case of accidents. Since explosive hydrogen is formed by the rapid zirconium oxidation, a reduced oxidation rate of Zr alloys at high temperatures is necessary to improve the accident tolerance [1–4]. As a short-term solution to the problem, protective coating through an efficient and economical method such as laser coating and thermal spray can be considered [2,4].

It has been previously reported that a laser coating of chromium on Zircaloy-4 cladding tube could enhance the high-temperature oxidation resistance significantly [2]. The diffusion of oxygen into the Zr substrate was observed to be effectively restricted by the Cr coating layer during the oxidation. When a coated alloy is exposed to a high temperature, a significant diffusion and microstructural variation may occur at the coating/substrate interface, even for a short time. In the case of laser-treated FeCrAl-coating on Mo alloy, an interfacial reaction was found to occur at high temperature [5].

Meanwhile, it was reported that Cr–Al composite coatings were very effective in enhancing the oxidation resistance of metals at high temperatures [6]. Since Al_2O_3 and Cr_2O_3 can form on the surface of Cr–Al coated alloys, the high temperature oxidation resistance is expected to increase even further. Although the superiority of Cr–Al laser coating has already been demonstrated, research on the stability of the microstructure at high temperature is not yet sufficient. In the case of Cr-coated Zr alloy, comparatively a few phases can be formed between Cr coating layer and Zr substrate according to the Zr–Cr binary phase diagram [7]. However, in the case of CrAl-coated Zr alloy, more complicated interfacial reactions would occur during the solidification and when it is exposed to high temperatures. Therefore, in the present research, CrAl laser-coated Zr alloys were exposed to high temperatures, and then their microstructural variation and oxidation behavior were investigated.

2. Materials and Methods

Zircaloy-4 alloy (Zr-1.38%Sn-0.2%Fe-0.1%Cr, wt. %) sheets were used as the substrate, and CrAl (30 wt. % Al) coating layer with the average thickness of about 300 μm was deposited on the surface of the Zr alloy through a laser coating process. A photograph of the laser equipment for coating is shown in Figure 1. The laser coating was carried out by using a continuous wave (CW) diode laser (wavelength of 1062 nm) with a maximum power of 300 W (PF-1500F model; HBL Co., Daejeon, Korea) and a power supply (Pwp14Y04K model; Yesystem Co., Daejeon, Korea). Coating process variables such as laser power, powder injection speed, specimen moving velocity, and gas flow speed were adjusted based on previous research results [8]. The applied power for the laser treatment ranged up to 300 W, and the scanning speed was 14 mm/s. To prevent any oxidation during the process, an inert gas (Ar) was continuously blowing into the melted surface of specimen. The mean size of the CrAl alloy powders as a raw material for coating was 90 μm.

As previously mentioned, when used for nuclear fuel claddings or the like, it can be accidently exposed to a high temperature for a long time, and it is presumed that the atmosphere is mainly water vapor. For ease of experiment, firstly, microstructural variations of the laser-coated specimens were investigated under argon or air atmosphere at 1100 °C for different holding times. Then, oxidation test of the CrAl and CrAlZr alloys was conducted in the steam atmosphere at 1200 °C for 1 h. Cr-30 wt. % Al and Cr-30 wt. % Al-20 wt. % Zr alloy specimens were prepared through vacuum arc remelting process to directly investigate the characteristics of the coating layers without Zr substrate. Microstructural analyses were performed using SEM (JEOL, Tokyo, Japan) equipped with an energy dispersive X-ray spectrometer (EDS, JEOL, Tokyo, Japan), and an X-ray diffractometer (XRD, Rigaku, Tokyo, Japan).

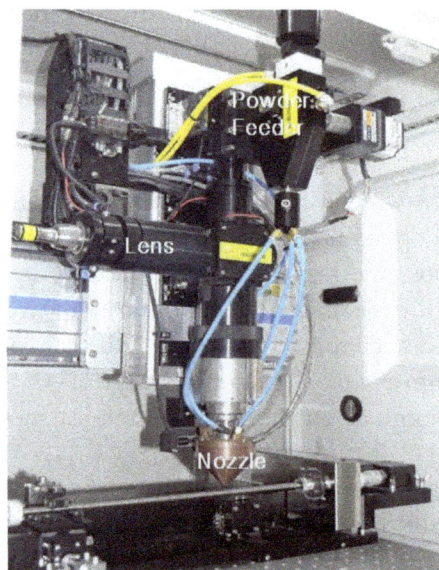

Figure 1. Appearance of the laser equipment for coating.

3. Results and Discussion

3.1. Micorstucture of CrAl Laser-Coated Zr Alloy

Figure 2 shows SEM micrographs of CrAl laser coating layers on Zircaloy-4 alloy. Some Zr content could be measured in the CrAl coating layer that is far away from the Zr substrate (sometimes even near the coating surface), implying that significant intermixing of CrAl coating and Zr substrate occurred during the laser coating. The light bottom area is Zr substrate, and the Zr content is increased as distance from the substrate is increased. A Zr-rich region appears between the Zr substrate and the CrAl coating layer. The composition of the Zr-rich part indicates the formation of a solid solution of Cr and Al in Zr. As indicated in Figure 2b, the majority of the Cr-rich coating layer is composed of two rapidly solidified regions: Cr-rich and AlZr(Cr) phases.

Element	at.%	Element	at.%
Zr	69.64	Zr	21.60
Cr	17.10	Cr	43.81
Al	13.26	Al	34.59

Element	at.%	Element	at.%
Zr	14.24	Zr	4.21
Cr	44.99	Cr	60.42
Al	40.77	Al	35.38

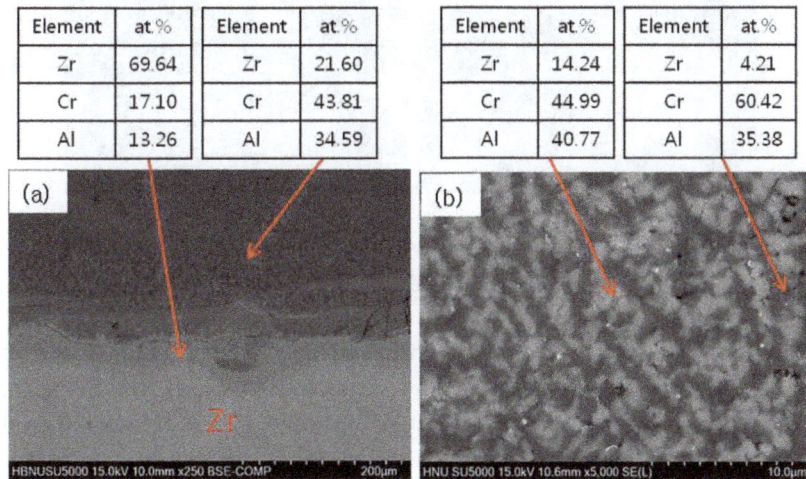

Figure 2. SEM-EDS (scanning electron microscopy–energy dispersive X-ray spectrometry) analysis results of CrAl laser-coated Zircaloy-4 substrate alloy: (**a**) near the Zr substrate; (**b**) Cr-rich coating layer.

If the mixture of CrAl coating powders and the top surface of substrate were completely melted and homogeneous, primary Cr phase would be formed first from the liquid, since the major component of the surface liquid pool was Cr. Normal solidification reactions to form phases for Cr-30 wt. % Al-20 wt. % Zr alloy (approximate chemical composition for the majority of the Cr-rich coating layer) predicted by Pandat thermo-calculation program are as follows [9]:

1. Liquid $\rightarrow$ Cr at 1492–1283 °C
2. Liquid $\rightarrow$ Cr + Al$_2$Zr at 1283–1255 °C
3. Liquid + Al$_2$Zr $\rightarrow$ Cr + Al$_3$Zr at 1255 °C
4. Liquid $\rightarrow$ Cr + Al$_3$Zr at 1255–1238 °C
5. Liquid $\rightarrow$ Cr + Al$_3$Zr + Al$_8$Cr$_5$ at 1238 °C

Cr dendrites and the Cr + AlZr eutectics should have formed upon solidification, however. The micrograph of the Cr-rich coating layer in Figure 2b reveals a somewhat different microstructure. Namely, dendritic primary Cr and AlCr phases were not observed. Additionally, in the case of the AlZr phase, it shows a dendritic morphology. This discrepancy seems to be because the growth occurred under a rapid solidification condition [10].

3.2. CrAl Laser-Coated Zr Alloy that Exposed to a High Temperature

Since the CrAl-coated Zr alloy is aimed to be resistant at high temperatures, the coated specimen was isothermally heated in inert atmosphere at 1100 °C for different times. Figure 3 indicates that inter-diffusion among phases in the coating layers apparently occurred after 2 h. The diffusion of aluminum appears to be significant so that the aluminum content can be detected in the Zr substrate. Generally, three distinct parts are shown: Zr-substrate, Zr-rich area, Cr-rich area (the majority of the coating layer). An intermediate area between the Zr-rich and Cr-rich area may be counted, but it was excluded as it can be regarded as a part of the Zr-rich area.

Element	at.%	Element	at.%	Element	at.%	Element	at.%
Zr	95.08	Zr	38.21	Zr	32.98	Zr	21.25
Cr	0.42	Cr	34.29	Cr	30.64	Cr	41.00
Al	4.50	Al	27.50	Al	36.38	Al	37.75

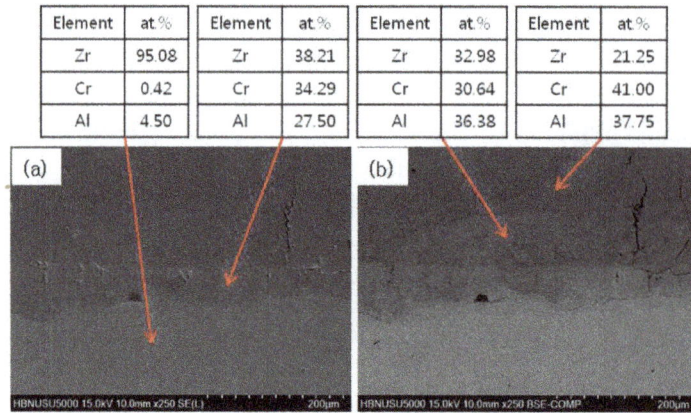

Figure 3. SEM-EDS analyses of laser-coated Zircaloy-4 substrate alloy after the isothermal heating at 1100 °C for 2 h in inert atmosphere: (**a**) bright image; (**b**) backscattered image.

As indicated in Figure 4, the isothermal heating clarified that the microstructure of the Zr-rich region near the substrate is composed of two regions: Zr-rich and Cr-rich. The Zr-rich area is Zr phase, and the Cr-rich area is postulated to be CrZr plus AlZr phases. Figure 5 also shows that the Cr-rich area is composed of three distinct phases. The main phase is Cr, containing about 40 at. % Al that is near the maximum solubility limit for Cr at 1100 °C. Others are Al-rich phases that include relatively large or very limited Zr content. According to the Pandat prediction and literature [9,11,12], the Al-rich phases with large Zr contents seem to be Al_3Zr. Meanwhile, the Al-rich phase with a very low Zr content should be Al_8Cr_5. The microstructure of the coated specimens isothermally heated for 10 h was also investigated. However, the coated specimens maintained for 10 h showed similar microstructural characteristics to those of specimens held for 2 h. It is believed that the microstructure of the coating layer could be converted into near the equilibrium structure even with a holding time of just 2 h, since 1100 °C is a very high temperature.

Element	at.%
Zr	87.26
Cr	2.02
Al	10.73
Element	at.%
Zr	40.48
Cr	48.90
Al	10.62

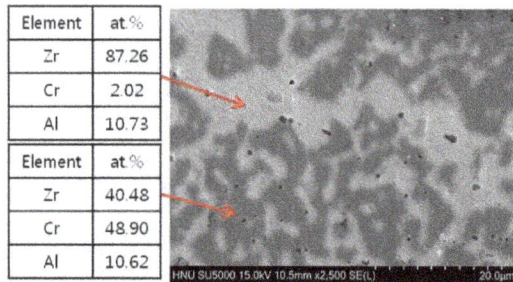

Figure 4. SEM-EDS analyses of Zr-rich area of CrAl-coated Zircaloy-4 alloy after the isothermal heating at 1100 °C for 2 h in inert atmosphere.

Element	at.%		Element	at.%
Zr	4.72		Zr	1.34
Cr	38.96		Cr	60.45
Al	56.32		Al	38.21
			Element	at.%
			Zr	23.17
			Cr	19.83
			Al	57.00

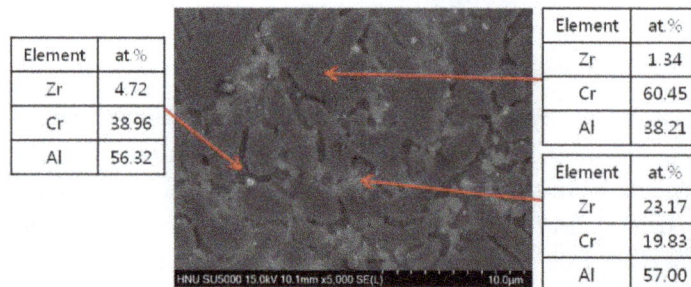

Figure 5. SEM-EDS analyses of CrAl coating layer on Zircaloy-4 alloy after the isothermal heating at 1100 °C for 2 h in inert atmosphere.

The chemical composition distribution as a function of depth for the CrAl-coated specimen exposed to a high temperature air is shown in Figure 6. Like Figure 3, three distinct parts are generally observed in the specimen, and it is clear that oxidation proceeded a little, only at the surface. If cracks are formed in the coating layer, they are undoubtedly undesirable, because the protective coating may be detached from the substrate. Vertically-formed cracks are supposed to be more detrimental to oxidation resistance, since oxygen ions can move easily through the cracks into the substrate. The formation of Zr oxides was observed on the Zr substrate near a vertical crack, as shown in Figure 7. However, a significant oxygen content was measured only at the top surface of CrAl coating layer in the sound region, and this suggests that the CrAl coating layer was effective in delaying the high temperature oxidation. It also appears that Al and Zr oxidize more preferentially than Cr in the coating layer. Unlike Al oxides, Zr oxides are not protective against oxidation toward the matrix at high temperatures [1,2]. Therefore, a mixing between CrAl coating layer and Zr substrate should be carefully controlled to minimize the Zr content at the top of the surface coating layer.

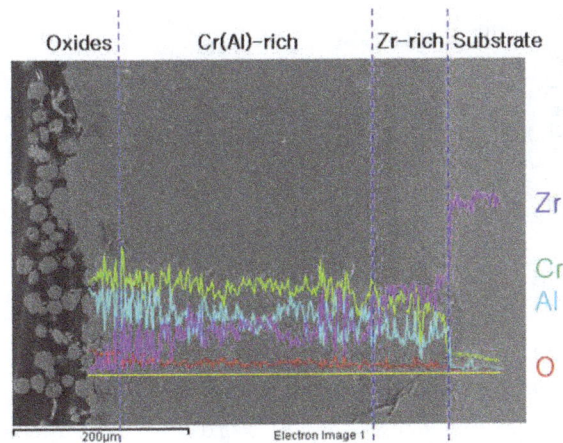

Figure 6. SEM micrographs with EDS profiles of the oxidized CrAl coating layer on Zircaloy-4 after isothermal heating at 1100 °C for 10 min in air.

Element	at.%	Element	at.%	Element	at.%	Element	at.%
O	67.12	O	67.62	O	11.15	O	66.84
Zr	30.89	Zr	11.48	Zr	0.75	Zr	3.15
Cr	-	Cr	16.18	Cr	68.45	Cr	2.31
Al	1.99	Al	4.72	Al	19.64	Al	27.70

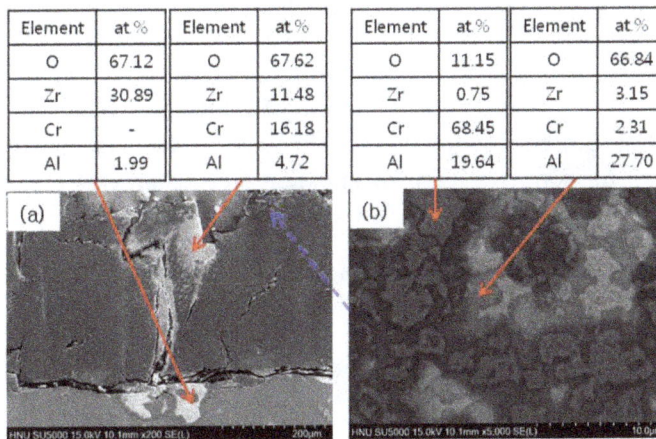

Figure 7. SEM-EDS analyses of laser coated Zircaloy-4 substrate alloy after isothermal heating at 1100 °C for 30 min in air: (**a**) near the substrate; (**b**) top surface of the coating layer.

3.3. Oxidation Behavior of CrAl Laser-Coated Zr Alloy at High Temperature

To clearly compare the oxidation resistance of CrAl coating layer with that of the Zr-incorporated CrAl layer, Cr-30 wt. % Al and Cr-30 wt. % Al-20 wt. % Zr alloy cast specimens were fabricated by vacuum arc remelting, and an oxidation test in steam at 1200 °C for 1 h was carried out. In the case of accident, the temperature for nuclear cladding can be extremely increased, and is expected to be still under steam atmosphere. The corrosion resistance under atmosphere containing moisture can be quite different from that under dry air. Although Si and SiO_2 are highly corrosion-resistant materials, they were quickly dissolved in a pressurized water condition at 360 °C in 18.9 MPa [2]. As shown in Figure 8, the coating layer without Zr possesses remarkably higher oxidation resistance than the Zr-mixed layer. Namely, much higher weight gain was observed for the Zr-containing alloy as compared to the CrAl alloy without Zr. The oxidation behavior for ZrCr30Al specimen (Zr alloy containing 30 wt. % Cr and 20 wt. % Al) was also compared for reference. Even though Cr and Al are contained in large amounts, it can be confirmed that the Zr alloy is seriously oxidized in a high temperature steam atmosphere.

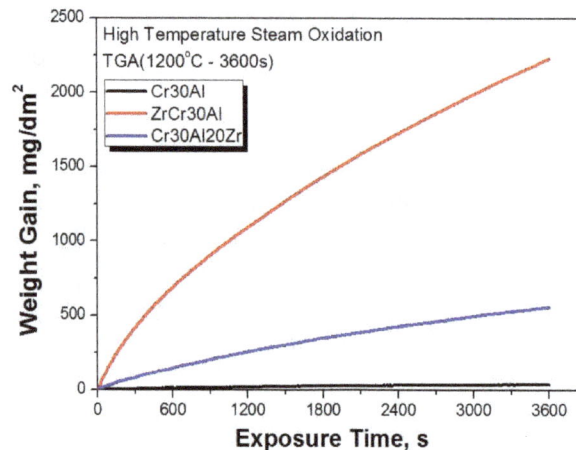

Figure 8. Oxidation behavior of Cr-30%Al cast alloys with and without Zr after the steam oxidation test at 1200 °C for 1 h (Zr alloy containing Cr and Al is also compared for reference).

Figure 9 indicates that a stable Al_2O_3 phase is observed a lot in the surface of the Cr-30%Al alloy. It is worth mentioning that a Cr_2O_3 phase was not found in that specimen. Although both Cr_2O_3 and Al_2O_3 phases are generally stable, the Al_2O_3 phase is believed to be more stable, resulting in a continuous external Al_2O_3 layer. This phenomenon has been known as transient oxidation [13]. If the content of aluminum in the coating layer is insufficient, it is considered that the Cr_2O_3 is observed on the coating surface. Meanwhile, a significant amount of ZrO_2 and Al_2O_3 phases were found in the case of the Zr-added alloy. Since the Zr oxide is not protective, the existence of a surface ZrO_2 layer should be responsible for the comparatively lower oxidation resistance.

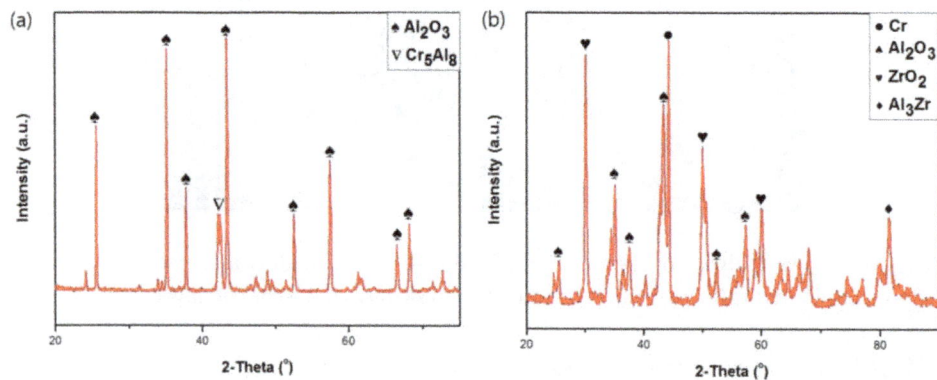

Figure 9. XRD analyses of cast specimens after the steam oxidation test at 1200 °C for 1 h: (**a**) Cr-30%Al; (**b**) Cr-30%Al-20%Zr alloy.

4. Conclusions

It was found that a significant mixing between the CrAl layer and the Zr substrate and the formation of rapidly solidified microstructure occurred during the laser surface coating process. Inter-diffusions among the solidified phases took place when the coated specimens were isothermally heated at 1100 °C, and resulted in the formation of equilibrium phases after just 2 h. Since Zr is easily oxidized and Zr oxides are not protective against oxidation, the Zr content at the top of coating layer should be minimized to avoid deteriorated oxidation resistance.

Acknowledgments: This work was supported by the National Research Foundation of Korea (NRF) grant funded by Korea government (MSIP) (NRF-2012M2A8A5025822).

Author Contributions: J.-M.K. designed the research and wrote the manuscript with help from the other authors; T.-H.H. and I.-H.K. performed experiments; J.-M.K. and H.-G.K. analyzed the data.

Conflicts of Interest: The authors declare no conflict of interest.

References

1. Kuprin, A.S.; Belous, V.A.; Voyevodin, V.N.; Bryk, V.V.; Vasilenko, R.L.; Ovcharenko, V.D.; Reshetnyak, E.N.; Tolmachova, G.N.; Vyugov, P.N. Vacuum-arc chromium-based coatings for protection of zirconium alloys from the high temperature oxidation in air. *J. Nucl. Mater.* **2015**, *465*, 400–406. [CrossRef]

2. Kim, H.G.; Kim, I.H.; Jung, Y.I.; Park, D.J.; Park, J.Y.; Koo, Y.H. Adhesion property and high-temperature oxidation behavior of Cr-coated Zircaloy-4 cladding tube prepared by 3D laser coating. *J. Nucl. Mater.* **2015**, *465*, 531–539. [CrossRef]

3. Terrani, K.A.; Parish, C.M.; Shin, D.; Pint, B.A. Protection of zirconium by alumina- and chromia-forming iron alloys under high-temperature steam exposure. *J. Nucl. Mater.* **2013**, *438*, 64–71. [CrossRef]

4. Jin, D.; Yang, F.; Zou, Z.; Gu, L.; Zhao, X.; Guo, F.; Xiao, P. A study of the zirconium alloy protection by Cr_2C_2-NiCr coating for nuclear reactor application. *Surf. Coat. Technol.* **2016**, *287*, 55–60. [CrossRef]

5. Kim, J.M.; Ha, T.H.; Park, J.S.; Kim, H.G. Effect of laser surface treatment on the corrosion behavior of FeCrAl-coated TZM alloy. *Metals* **2016**, *6*, 29. [CrossRef]

6. Chen, C.; Zhang, J.; Duan, C.; Feng, X.; Shen, Y. Investigation of Cr-Al composite coatings fabricated on pure Ti substrate via mechanical alloying method: Effects of Cr-Al ratio and milling time on coating, and oxidation behavior of coating. *J. Alloy. Compd.* **2016**, *660*, 208–219. [CrossRef]

7. Gonzalez, R.O.; Gribaudo, L.M. Analysis of controversial zones of the Zr-Cr equilibrium diagram. *J. Nucl. Mater.* **2005**, *342*, 14–19. [CrossRef]

8. Kim, H.G.; Kim, I.H.; Jung, Y.I.; Park, D.J.; Park, J.Y.; Koo, Y.H. High temperature oxidation behavior of Cr-coated zirconium. In Proceedings of the LWR fuel performance meeting, Charlotte, NC, USA, 15–19 September 2013; p. 840.

9. PANDAT. CompuTherm, LLC, Madison, WI, USA. Available online: http://www.computherm.com (accessed on 13 February 2017).

10. Yue, T.M.; Xie, H.; Lin, X.; Yang, H.O. Phase evolution and dendritic growth in laser cladding of aluminium on zirconium. *J. Alloy. Compd.* **2011**, *509*, 3705–3710. [CrossRef]

11. Okamoto, H. Phase diagrams for binary alloys. In *ASM Desk Handbook*; ASM International: Materials Park, OH, USA, 2000.

12. Zhang, M.; Xu, B.; Ling, G. Preparation and characterization of α-Al_2O_3 film by low temperature thermal oxidation of Al_8Cr_5 coating. *Appl. Surf. Sci.* **2015**, *331*, 1–7. [CrossRef]

13. Birks, N.; Meier, G.H.; Pettit, F.S. *Introduction to the High-Temperature Oxidation of Metals*; Cambridge University Press: Cambridge, UK, 2006; pp. 101–162.

13

One-Dimensional Constitutive Model for Porous Titanium Alloy at Various Strain Rates and Temperatures

Zhiqiang Liu *, Feifei Ji, Mingqiang Wang and Tianyu Zhu

School of Mechanical Engineering, Jiangsu University of Science and Technology, Zhenjiang 212003, China; jifeifei1990@163.com (F.J.); mqwang640526@sina.com (M.W.); zhuty_just@126.com (T.Z.)
* Correspondence: zhiqiangliu@just.edu.cn

Academic Editor: Hugo F. Lopez

Abstract: In this paper, the accurate description of the relationship between flow stress and strain of porous titanium alloys at various strain rates and temperatures were investigated with dynamic and quasistatic uniaxial compression tests for a further study on the processing mechanism of porous titanium material. Changes in their plastic flows were described through the one-dimensional Drucker-Prager (DP) constitutive model. Porous titanium alloys were micromilled in a DP simulation. After all parameters had been obtained in the DP model, the experimental and simulated true stress-strain curves and flow stress levels of two porous titanium alloys were compared to estimate the precision of the model. The findings were as follows. First, porous titanium alloys show deformation patterns characterized by pore collapse-induced deformation and have strong stress-hardening effects, but the patterns did not include noticeable plastic-flow plateaus. Second, porosity strongly affects the mechanical strength, strain-rate sensitivity, and temperature sensitivity of both alloys. Third, the DP model sufficiently describes the mechanical properties of both alloys at 25–300 °C and at strain rates of 1000–3000 s^{-1}, with a deviation of 10% or lower.

Keywords: constitutive model; porous titanium alloy; stress and strain

1. Introduction

Titanium alloys are increasingly used as structural material for aerospace, shipbuilding, petrochemical, power generation equipment, automotive industries and other fields because of their attractive properties such as high specific strength and high structural stiffness with excellent heat and corrosion resistance [1–4]. Compared with the dense titanium alloys, porous titanium alloys mainly obtained by spark plasma sintering, vacuum sintering and granulation loose sintering inherits general characteristics of titanium alloys [5–8]. In addition, low density, good adsorption as well as good biocompatibility are additional characteristics [9,10]. However, porous titanium alloys are recognized as difficult-to-cut materials due to their low thermal conductivity, high chemical activity and small elastic modulus [11–13]. The cutting process of porous titanium is that the shear slip of the workpiece occurs under the action of the cutting tool and the cutting chip flows out from the rack face of the cutter [14]. At the same time, the machining surface of the workpiece experienced the process of extrusion and friction, which occur in the flank face of the cutter [15,16]. During the process of machining, the material is subjected to strong elastic plastic deformation, which produces cutting resistance and a lot of cutting heat [17–19].

In order to further study the machining mechanism of porous titanium material, it is very important to obtain the accurate description of the relationship between the flow stress and strain of the material under high temperature and high strain rate. Constitutive model, which plays a vital role in the numerical simulation of material processing, is used to describe the flow stress varying with strain, temperature and strain rate [20,21]. Macroscopically, it reflects the relationship between material force and deformation. Nowadays, A number of constitutive equations were developed to predict constitutive behavior in a wide range of metals and alloys [22,23]. Particularly, two methods of phenomenological and physically-based constitutive models have been proposed widely [24]. Among them, Johnson-Cook (JC) constitutive model is the most popular due to its rich experience, simple form and availability of parameters. Drucker-Prager (DP) constitutive model, Zerilli-Armstrong (ZA) constitutive model, Artificial neural network (ANN) constitutive model and so on are also used [25,26].

Due to the influence of pore structure of the material, the mechanical properties of porous titanium alloys materials are more complicated than that of dense metal materials. It is mainly manifested in the density dispersion caused by the uneven microstructure. Furthermore, it also manifested in the uneven distribution of stress and discrete experimental results caused by uneven microstructure [27]. As a result, the relationship between the flow stress and strain of the material at various strain rates and temperatures can not be accurately described by the traditional constitutive model of titanium alloys.

Considering that the JC constitutive model is more suitable for the material with significant deformation, high strain rate and high temperature machining, but the description of the dynamic mechanical properties of porous titanium alloy material is limited [28]. It cannot describe the effect of temperature-strengthening. More seriously, the strain rate effect prediction enhancement is not accurate enough, and the deviation reaches about 10%–20%. Therefore, the DP constitutive model is proposed to describe the comprehensive mechanical properties of porous titanium alloys.

One-dimensional DP constitutive model proposed was based on the combination of quasistatic compression test, the SHPB tests, orthogonal experiment and micro-milling test in a DP simulation. A quasistatic compression test on the dynamic and static responses of porous titanium alloys with two porosity levels revealed their yield limits and elastic moduli and the effects of porosity on their static mechanical properties. The dynamic mechanical properties of the specimens were tested using a Split-Hopkinson pressure bar (SHPB). An orthogonal experiment was conducted to yield flow stress-strain curves at different strain rates and temperatures to analyze the strain-rate and temperature sensitivities of the specimens. Additionally, it can also determine the temperature and strain rate range for the DP constitutive model. Lastly, the parameters of the constitutive equation were determined by analyzing the experimental data and then micro-cutting simulation of DP constitutive model was carried out to compare with the experimental results in order to verify the rationality of the model.

2. Experimental Procedures

2.1. Testing Material

In this experiment, two kinds of porous titanium materials with different porosity were prepared by powder sintering. The particle sizes of two titanium powder raw materials are 500# and 200#. Besides that the diameter are less than 27 μm and 74 μm, respectively. The additive was 2 wt % polyvinyl alcohol aqueous solution during the sintering process and the titanium powder mixture was made into titanium body after compression molding whose forming pressure is 108 MPa. Taking into account the fact that the chemical properties of titanium is so active that can react with carbon, nitrogen, oxygen, hydrogen and other elements in the air, the vacuum sintering process was adopted and the vacuum degree was set at 10^{-4} Pa. Moreover, the sintering temperature of 200# titanium powder is 1100 °C, while the one of 500# titanium powder is 1200 °C. In addition, after sintering, do not have a natural cooling until the heat preservation for about 2 h have done. The specific parameters of the final porous titanium samples are shown in Table 1. In Table 1, it is easy to see that the porous titanium alloy samples not only have Ti elements, but also contain Fe, Cu, C, O, N and other elements. In addition, the porosities of preparations are mainly 26% and 36%, which is expressed as the ratio of the simple mass to volume [29].

Table 1. Chemical composition and structural properties of porous titanium alloy samples (wt %).

Type	Porosity (%)	Particle Size (μm)	Aperture (μm)	Ti (%)	Fe (%)	Cu (%)	C (%)	O (%)	N (%)
200#	26	≤27	15	≥99.7	≤0.15	≤0.005	≤0.05	≤0.2	≤0.03
500#	36	≤74	250	≥99.7	≤0.25	≤0.003	≤0.06	≤0.2	≤0.03

Moreover, the samples with different porosity should be machined as cylinders with the size of ϕ 8 mm × 3 mm, ϕ 8 mm × 4 mm, ϕ 8 mm × 6 mm and ϕ 8 mm × 8 mm by wire-electrode cutting. Where ϕ 8 represents a cylinder radius of 8 mm, and 3 mm, 4 mm, 6 mm and 8 mm is the height of cylinders. When processing, it is necessary to ensure that the level of parallelism and perpendicularity error is less than 0.01 mm and to polish the end face to adjust the contact surface roughness with the experimental equipment. Lastly, screening the samples to meet the requirement that the density difference of the test samples is no more than 0.01 g/cm^3 is also important. The final samples are listed in Figure 1.

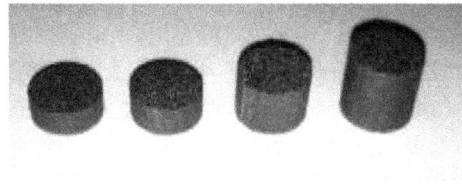

Figure 1. The final samples with the size of ϕ 8 mm × 3 mm, ϕ 8 mm × 4 mm, ϕ 8 mm × 6 mm and ϕ 8 mm × 8 mm.

2.2. Experimental Database

A uniaxial quasistatic compression test was performed with the parameters shown in Table 2. An electronic digital controller was switched to the PC-control mode for automated control, and the quasistatic compression test was conducted after the digital signal stabilized.

Table 2. Parameters of the quasistatic compression test.

Temperature	Strain Rate	Crosshead Speed	Number of Repetitions
25 °C	0.001	0.48 mm/min	2

A set of SHPB experiments was designed on the basis of micromachining parameters, the mechanical properties of the porous titanium alloys, and the performance of the bar (Table 3). The experiments were conducted at a maximum temperature of 300 °C and a maximum strain rate of 4000 s^{-1}.

Table 3. Parameters of the employed Split-Hopkinson pressure bar experiments.

Temperature	25 °C		100 °C		200 °C		300 °C		Specimen Size/mm
Porosity/%	26	36	26	36	26	36	26	36	
Strain rate/s^{-1}	1000	1000	1000	1000	1000	1000	1000	1000	ϕ 8 × 6
	2000	2000	2000	2000	2000	2000	2000	2000	ϕ 8 × 4
	3000	3000	3000	3000	3000	3000	3000	3000	ϕ 8 × 3
	—	—	4000	4000	4000	—	4000	4000	ϕ 8 × 3

The strut of SHPB with a diameter of 15 mm was used during the experiments. The impact velocity of the striker bar of the SHPB was adjusted to achieve different strain rates; each experiment was repeated two to three times to ensure sufficient reproducibility. The means of valid data from the experiments were calculated to reduce the experimental error.

3. Results and Discussion

3.1. *Static Mechanical Response of Porous Titanium Alloys*

Considerable deformation occurred on specimens during compression and induced crushing. By the law of constant volume, true stress-strain curves did not accurately reflect the mechanical properties of the specimens. Subsequently, engineering stress-strain curves were used to represent these mechanical properties. Figure 2 presents the engineering stress-strain curves of porous titanium alloys with porosities of 26% and 36% under quasistatic compression. These curves are classified into elastic, plastic, and densification stages. Generally, the interface between the elastic and plastic stages is defined as the yield limit, whereas the interface between the plastic and densification stages is defined as densification (crushing). The stress and strain values that correspond with the interface between the plastic and densification stages are termed as the densification stress and densification strain, respectively. Figure 3 shows that the stress-strain curves of the specimens exhibit an overall linear relationship in the elastic stage and that, in the plastic stage, stress values begin to level off as strain values increase, resulting in a plastic yield plateau. After crushing occurs at the end of the plastic stage, the curves begin to rise. These curves are highly reproducible before crushing, but separate after crushing because of internal defects in the specimens.

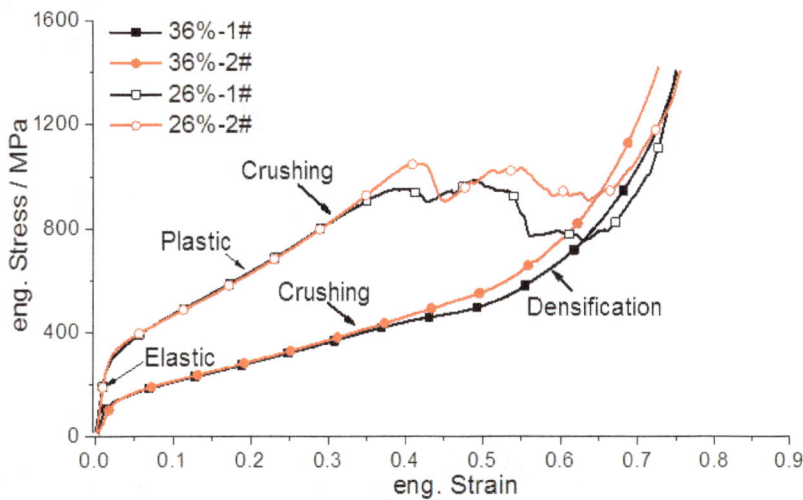

Figure 2. Engineering stress-strain curves of porous titanium alloys with porosities of 26% and 30% under quasistatic compression.

Table 4 tabulates the mechanical parameters of the porous titanium alloy specimens under quasistatic compression. These parameters were obtained from a quasistatic experiment. The yield limits of the specimens were estimated at a nonproportional compressive strain of 0.2%, and the elastic moduli were measured using the graphical method. As the table shows, the mechanical performance of the specimen with a higher porosity was lower than that of the specimen with a lower porosity.

Table 4. Parameters of the porous titanium alloy specimens under quasistatic compression.

Porosity	Elastic Modulus	Yield Limit	Densification Stress	Densification Strain
26%	20.13 GPa	250 MPa	850 MPa	0.31
36%	7.67 GPa	115 MPa	375 MPa	0.32

Equation (1), which had been derived through a finite element analysis of a spherical cavity model to estimate elastic modulus and porosity, was used to measure the simulated elastic modulus values of the 26% and 36% porous titanium alloys; those values were 43.6 and 30.1 GPa, respectively. These elastic modulus values were larger than their corresponding experimental values because the porous structures of the specimens were not uniformly spherical; the majority of their pores were irregular; those irregular pores facilitated stress concentration, fissure induction, and mechanical performance reduction in these alloys.

$$E = E_0 \left[\frac{1 - \varepsilon_p}{1 + 1.1\varepsilon_p} \right] \tag{1}$$

where E is the elastic modulus of the porous titanium alloy, ε_p is the porosity, E_0 is the elastic modulus of the substrate (E_0 of a pure titanium alloy is approximately 102.5), and the Poisson's ratio of a pure titanium alloy is approximately 0.3.

The specimen with a porosity of 26% experienced a short densification stage (Figure 2). As loading continued to increase, the specimen absorbed enough energy to induce severe damage to its structure; it was crushed after the short densification stage (Figure 3a), indicating the brittleness of the specimen. By contrast, the specimen with a porosity of 36% experienced a longer densification stage, showing less stress variations and no signs of crushing (Figure 3b). This specimen exhibited a relatively high toughness, implying a high capability to absorb plastic energy. Clear gullies, cracks can be seen in the microstructure (Figure 3c) with small porosity while the large porosity has the fewer (Figure 3d).

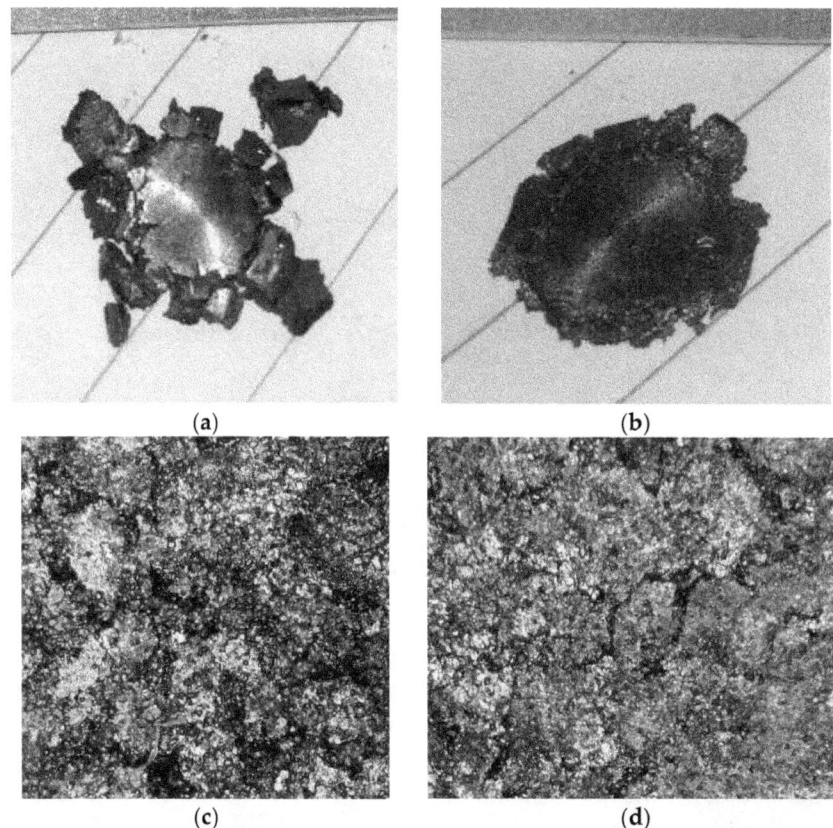

Figure 3. The porous titanium alloy specimens experiencing various degrees of crushing: (a) $\varepsilon_p = 26\%$; (b) $\varepsilon_p = 36\%$; (c) $\varepsilon_p = 26\%$, $\times 100$; and (d) $\varepsilon_p = 36\%$, $\times 100$.

3.2. Dynamic Mechanical Response of Porous Titanium Alloys

SHPB experiments were conducted to yield the true stress–strain curves of the porous titanium alloy specimens (porosity: 26% and 36%) at different strain rates and temperatures. These curves were grouped for the same temperatures and for different strain rates to examine the strain rate effects on alloys under dynamic loading. The true stress and strain values are translated from the engineering value by Equations (2) and (3).

$$\sigma_T = \sigma_E(1 + \varepsilon_E) \tag{2}$$

$$\varepsilon_T = \ln(1 + \varepsilon_E) \tag{3}$$

where σ_T is the true stress, ε_T is the true strain; σ_E is the engineering stress, and ε_E is the engineering strain.

Figure 4 shows the true stress-strain curves of the specimens at $T = 25\,°C$, $T = 100\,°C$, $T = 200\,°C$ and $T = 300\,°C$ for various strain rates. Compared with the stress-strain curves obtained under quasistatic compression (Figure 2), the stress-strain curves under dynamic loading suggest that the yield limits and flow stress values of the alloys increased at high strain rates and their mechanical strength values were higher at high strain rates than at low strain rates. This indicates that the alloy specimens exhibit certain degrees of sensitivity to strain rates. Moreover, the true stress-strain data suggested that the higher the strain rate was, the more discrete the data were. As Figure 4 shows, when the strain rate reached 4000 s^{-1} or higher, the distributions of plastic flow stress were irregular at different temperatures. Thus, the data for both specimens were nonreproducible.

Figure 4 also depicts the deformation of both alloys under dynamic loading during elastic and plastic stages. Unlike compacted materials, these alloy specimens had no plastic flow plateaus in the plastic stage and their stress levels increased when strain levels rose, indicating that strain-hardening rates ($\partial\sigma/\partial\varepsilon$) did not converge to zero as strain levels increased. Moreover, under the same conditions, the strain-hardening rates increased at higher porosity levels and were slightly affected by strain rates, whereas the yield stress and flow stress levels decreased at higher porosity levels. This contrast indicated that high porosity levels led to poor mechanical performance, conforming to the mechanical properties obtained under quasistatic compression.

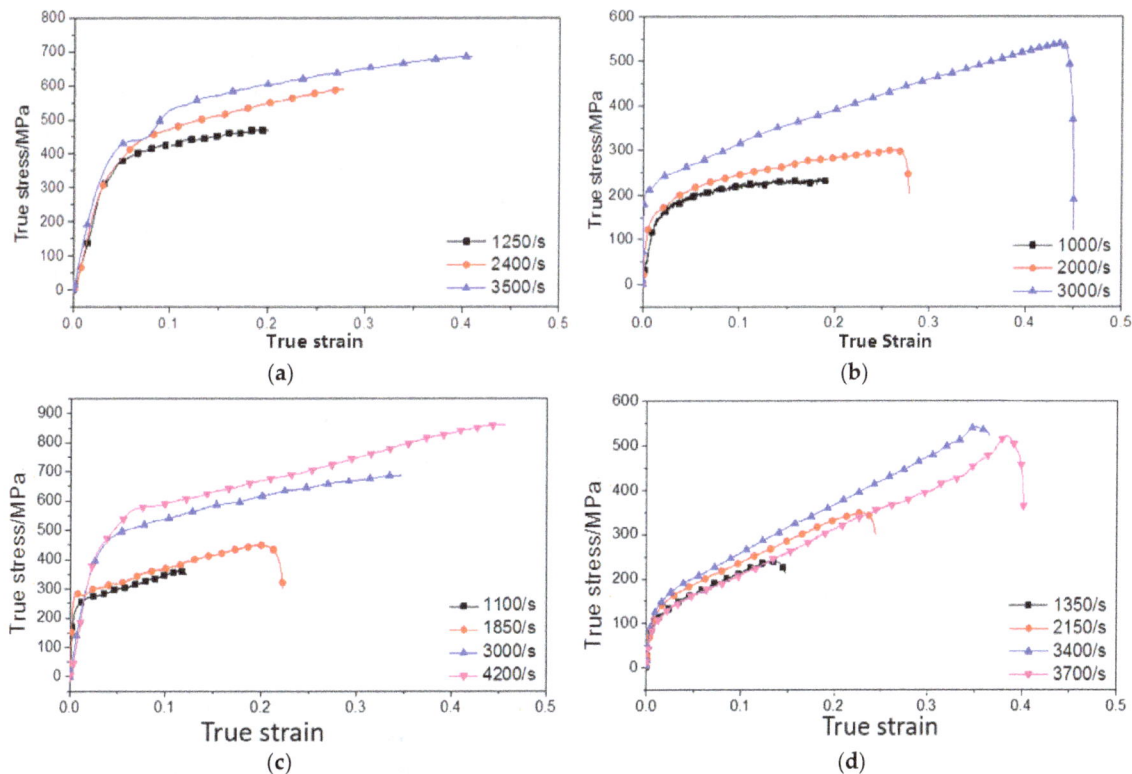

Figure 4. *Cont.*

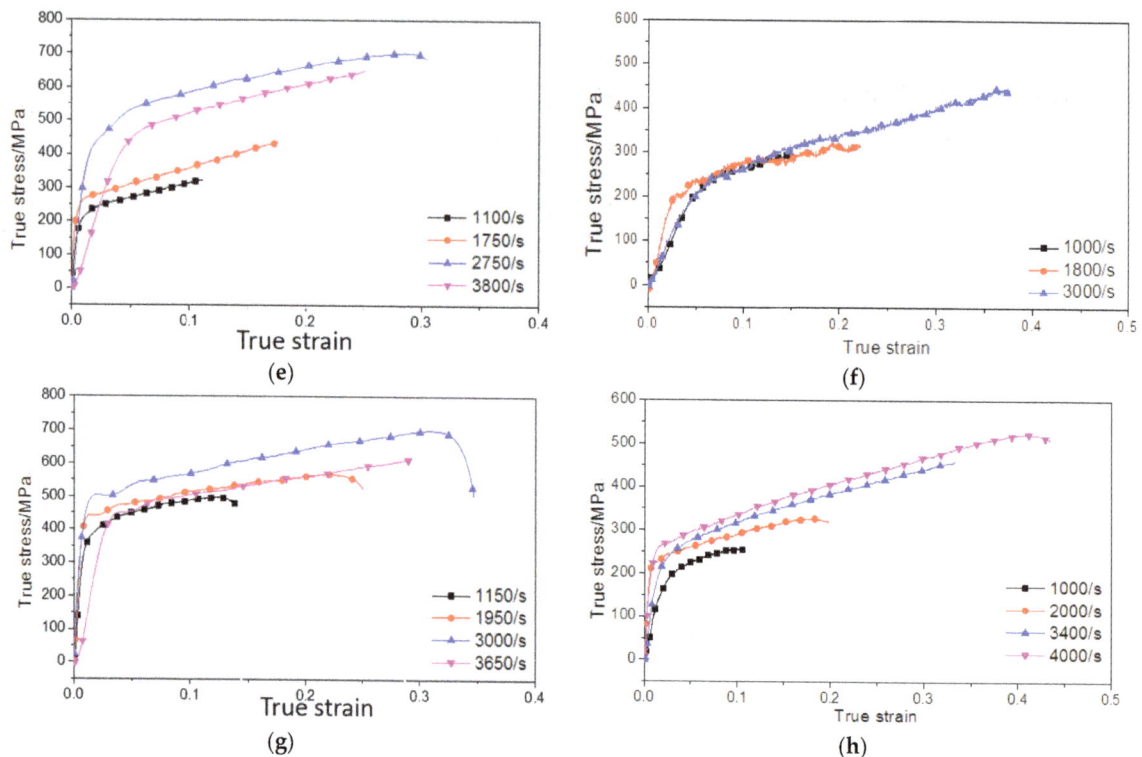

Figure 4. Stress–strain curves of the porous titanium alloy specimens (porosity: 26% and 36%) at the same temperatures but different strain rates: (a) $\varepsilon_p = 26\%$, $T = 25\ °C$; (b) $\varepsilon_p = 36\%$, $T = 25\ °C$; (c) $\varepsilon_p = 26\%$, $T = 100\ °C$; (d) $\varepsilon_p = 36\%$, $T = 100\ °C$; (e) $\varepsilon_p = 26\%$, $T = 200\ °C$; (f) $\varepsilon_p = 36\%$, $T = 200\ °C$; (g) $\varepsilon_p = 26\%$, $T = 300\ °C$; and (h) $\varepsilon_p = 36\%$, $T = 300\ °C$.

The strain-rate sensitivity of a porous titanium alloy is typically determined according to the strain-rate effect in the alloy's microscopic structure and that in the alloy's substrate. Plastic bending in a microscopic structure under loading leads to a strain-rate effect. Because specimens with open-textured structures were used in this study, plastic bending was likely to occur to pore walls and edges under loading. An analysis of the strain-effects in the specimens' substrates using the dynamic mechanical properties of pure titanium (TA2) showed that these substrates were highly sensitive to strain rates. Therefore, because the specimens exhibited strain-rate effects, the effects of air in the pores were not considered.

Higher strain rates were achieved at higher impact velocities of the striker bar. However, a high impact velocity resulted in crushing in the pore structures, inducing irregular stress. As such, measurement results were subject to the axial-inertia effect of the specimens, and the axial-inertia effect and strain-rate effect were coupled to the extent that these effects were indistinguishable from one another. These results did not accurately reflect the dynamic mechanical properties of the specimens under strain-rate effects but indicated irregular stress-strain curve distributions at a strain rate of 4000 s^{-1}.

That the strain-hardening rate ($\partial\sigma/\partial\varepsilon$) increases at high porosity levels can be elucidated by investigating the deformation of porous materials at high strain rates. Extrusions generally occur in a compacted material under impact loading, exposing high-density defects. Thus, such materials exhibit two impact-strengthening effects, i.e., strain-strengthening and strain rate-strengthening effects. A porous material under impact loading is prone to pore collapse-induced deformation, leading to compaction toward the support end. During compaction, this material increases rapidly in density, exhibiting pronounced strain-strengthening effects. Subsequently, substantial strain-strengthening effects can be observed from porous titanium alloys because of their low densities.

To examine the specimens' temperature sensitivity, stress-strain curves obtained from the experiments were classified at the same strain rates but different temperatures. Figure 5 shows the true stress–strain curves obtained from the specimens (porosity: 26% and 36%) at different temperatures and at $\dot{\varepsilon} = 1000$ s^{-1}, $\dot{\varepsilon} = 2000$ s^{-1}, and $\dot{\varepsilon} = 3000$ s^{-1}. At temperatures of less than 200 °C, the yield limit and flow stress of the 26%-porosity specimen declined as the temperature increased, indicating that the presence of temperature-softening effects in the specimen. However, at 300 °C, the yield limit and flow stress increased to levels higher than their room-temperature levels, indicating changes in the specimen's mechanical properties or microscopic structure at this temperature; these changes were unlikely to be caused by strain-rate effects or strain-hardening effects. In addition, the flow stress of the 36%-porosity specimen only fluctuated slightly as the temperature increased. Thus, its flow-stress variations at various temperatures could not be easily generalized. However, two features of stress-strain curves for the 36%-porosity specimen were observed: the specimen exhibited limited temperature sensitivity at temperatures under 300 °C, and it showed stronger strain-hardening effects at 100 °C than at other temperatures. By contrast, the strain-hardening effects of the 26%-porosity specimen at the same strain rates did not change much despite temperature variations.

Figure 6 depicts the changes in flow stress in relation to temperature at a stress of 0.1 ($\varepsilon = 0.1$) and at consistent strain rates. This figure illustrates the temperature-softening effects of the 26%-porosity specimen under 200 °C and the substantial increases in these effects at 300 °C. However, the flow stress of the 36%-porosity specimen showed a downward trend at temperatures lower than 100 °C but an upward trend at temperatures higher than 100 °C. The stress-rate sensitivity of the 26%-porosity specimen increased with rising temperatures and decreased once the temperature reached 300 °C, whereas that of the 36%-porosity counterpart decreased with rising temperatures and increased once the temperature reached 300 °C.

Flow stress exhibited similar trends at strain rates of 1000 s^{-1} and 2000 s^{-1}, but the discreteness of the strain-rate data collected from the specimens increased at a strain rate of 3000 s^{-1}, at which these trends were less noticeable.

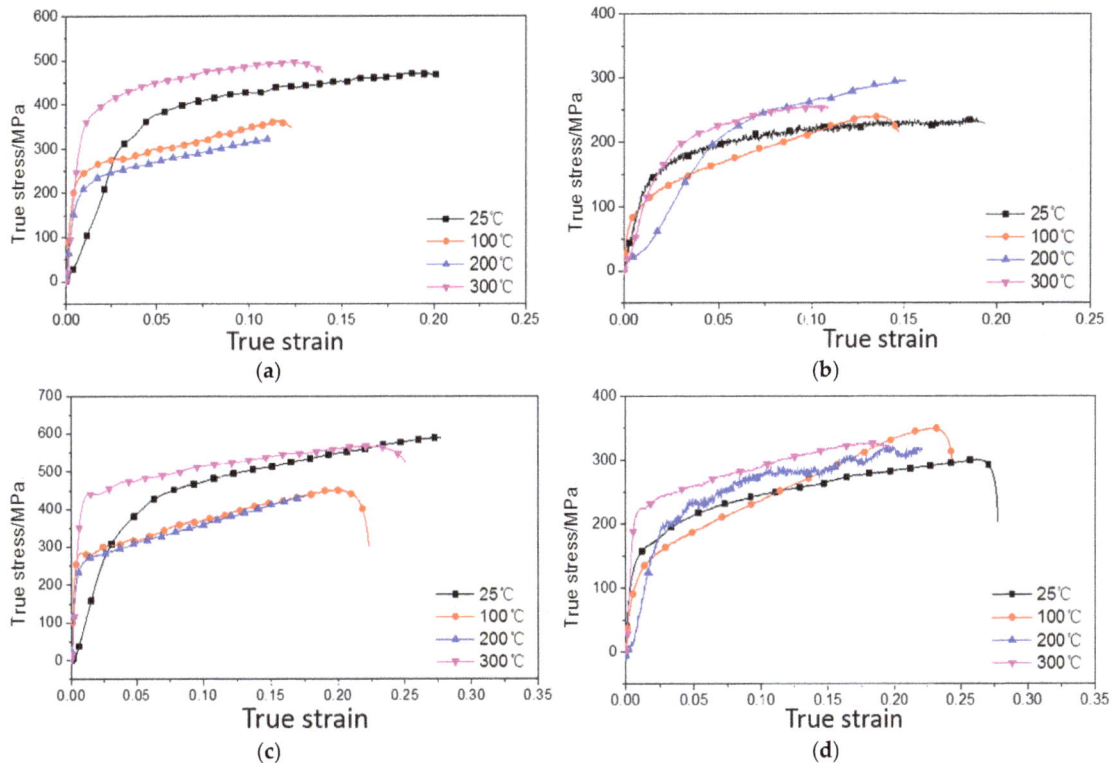

Figure 5. *Cont.*

(e)

(f)

(g)

(h)

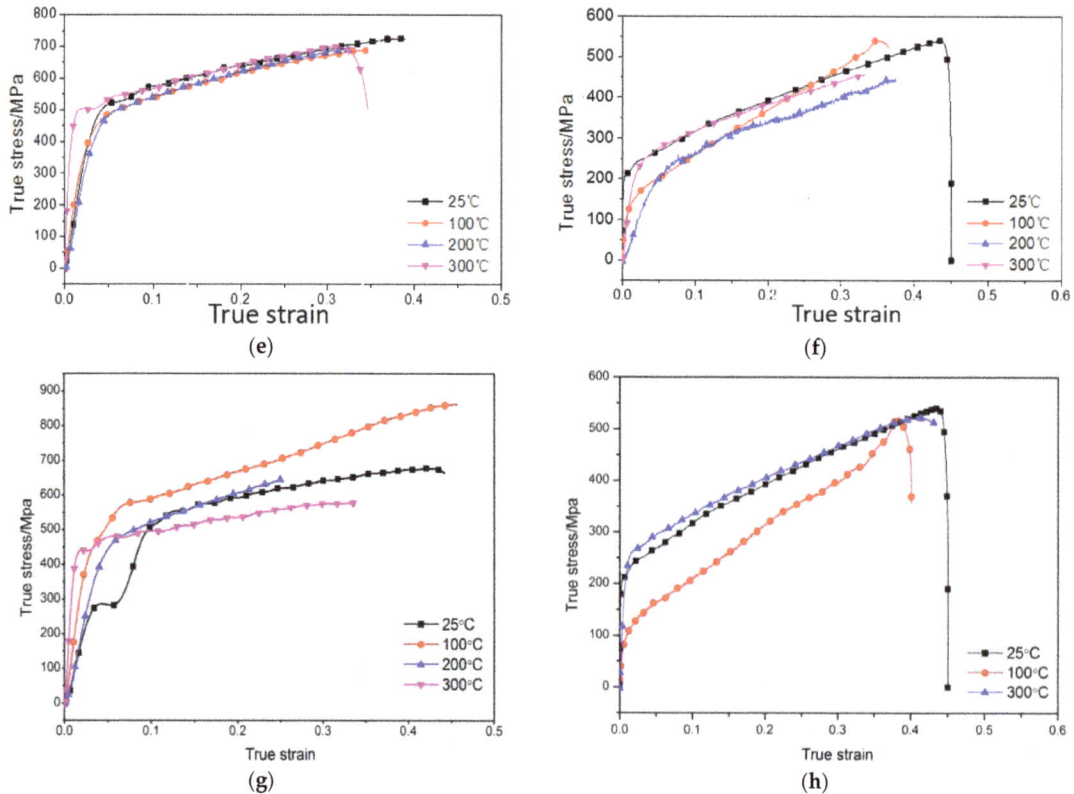

Figure 5. Stress–strain curves obtained from the 26%- and 36%-porosity specimens at different temperatures but the same strain rates: (a) $\varepsilon_p = 26\%$, $\dot{\varepsilon} = 1000$ s^{-1}; (b) $\varepsilon_p = 36\%$, $\dot{\varepsilon} = 1000$ s^{-1}; (c) $\varepsilon_p = 26\%$, $\dot{\varepsilon} = 2000$ s^{-1}; (d) $\varepsilon_p = 36\%$, $\dot{\varepsilon} = 2000$ s^{-1}; (e) $\varepsilon_p = 26\%$, $\dot{\varepsilon} = 3000$ s^{-1}; (f) $\varepsilon_p = 36\%$, $\dot{\varepsilon} = 3000$ s^{-1}; (g) $\varepsilon_p = 26\%$, $\dot{\varepsilon} = 4000$ s^{-1}; and (h) $\varepsilon_p = 36\%$, $\dot{\varepsilon} = 4000$ s^{-1}.

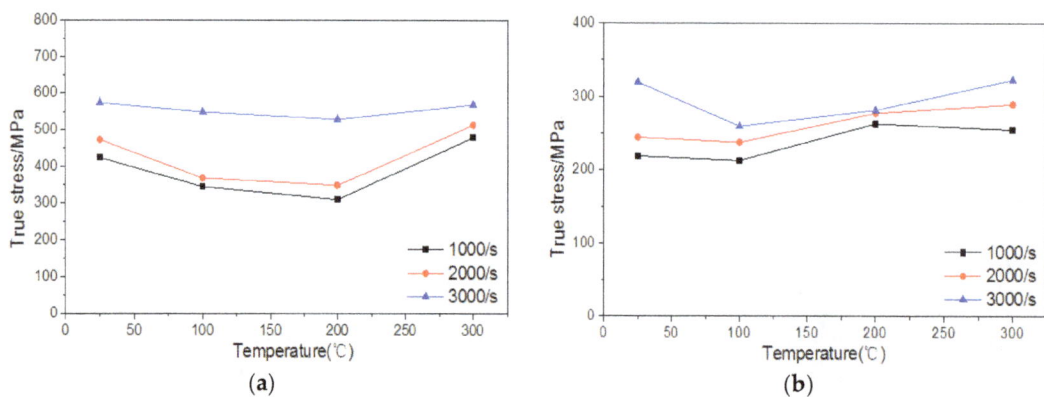

(a)

(b)

Figure 6. Flow stress of the 26%- and 36%-porosity specimens in response to temperature variations at different strain rates: (a) $\varepsilon_p = 26\%$; and (b) $\varepsilon_p = 36\%$.

Different temperature effects were absorbed between the porous titanium alloy specimens used in this study and compact titanium alloys because the specimens' mechanical strength levels increased, rather than monotonically decreased, with rising temperatures.

3.3. The Drucker-Prager (DP) Model

The DP model is a constitutive model applied to the finite element simulation of micromachining processes; it describes stress-strain relationships in detail. Each term in the model contains piecewise functions, thus corresponding closely with the specimens' plastic mechanical properties and ensuring selectivity for fitting. The DP model was used to examine the patterns of plastic flow variations.

The DP model is expressed by

$$\sigma(\varepsilon, \dot{\varepsilon}, T) = G(\varepsilon) \times \Gamma(\dot{\varepsilon}) \times \Theta(T) \tag{4}$$

$$G(\varepsilon) = \begin{cases} \sigma_0 \left(1 + \frac{\varepsilon}{\varepsilon_0}\right)^{\frac{1}{n}}, & \text{if } \varepsilon \leq \varepsilon_{\text{cut}}; \\ \sigma_0 \left(1 + \frac{\varepsilon_{\text{cut}}}{\varepsilon_0}\right)^{\frac{1}{n}}, & \text{if } \varepsilon > \varepsilon_{\text{cut}}; \end{cases} \tag{5}$$

$$\Gamma(\dot{\varepsilon}) = \begin{cases} \left(1 + \frac{\dot{\varepsilon}}{\dot{\varepsilon}_0}\right)^{\frac{1}{m_1}}, & \text{if } \dot{\varepsilon} \leq \dot{\varepsilon}_t; \\ \left(1 + \frac{\dot{\varepsilon}}{\dot{\varepsilon}_0}\right)^{\frac{1}{m_2}} \left(1 + \frac{\dot{\varepsilon}_t}{\dot{\varepsilon}_0}\right)^{\left(\frac{1}{m_1} - \frac{1}{m_2}\right)}, & \text{if } \dot{\varepsilon} > \dot{\varepsilon}_t; \end{cases} \tag{6}$$

$$\Theta(T) = \begin{cases} c_0 + c_1 T + c_2 T^2 + c_3 T^3 + c_4 T^4 + c_5 T^5, & \text{if } \dot{\varepsilon} \leq \dot{\varepsilon}_t; \\ \Theta(T_{\text{cut}}) - \frac{T - T_{\text{cut}}}{T_{\text{melt}} - T_{\text{cut}}}, & \text{if } \dot{\varepsilon} > \dot{\varepsilon}_t; \end{cases} \tag{7}$$

where $G(\varepsilon)$ is the strain-hardening term, $\Gamma(\dot{\varepsilon})$ is the strain-rate strengthening term, $\Theta(T)$ is the temperature-softening term, σ_0 is the yield limit, ε is the plastic strain, ε_0 is the reference plastic strain, ε_{cut} is the breaking strain, n is the hardening exponent, m_1 and m_2 are strain-rate effect exponents, $\dot{\varepsilon}$ is the strain rate, $\dot{\varepsilon}_0$ is the reference strain rate, $\dot{\varepsilon}_t$ is the critical strain rate, c_0–c_5 are polynomial coefficients, T is the temperature, T_{melt} is the melting point of the material, and T_{cut} is the critical temperature.

Fitting was conducted through the method of separating variables for each term in the DP model. Experimental data were fitted through linear regression and polynomial fitting to yield constitutive equation parameters.

Fitting was first performed on the strain-hardening term. A DP model with reference strain rates and under room-temperature conditions was used to describe the stress-strain relationship under quasistatic conditions. No noticeable plastic-flow plateaus were generated in the specimens because deformation and densification occurred simultaneously. Thus, the fitting was performed using equations with strain-hardening terms smaller than the breaking strain (ε_{cut}).

The strain-hardening term was converted into Equation (7). In a log–log graph, this equation is a straight line with an intercept of $\ln \sigma_0$ and a slope of $1/n$, in which σ_0 is defined as the yield limit under quasistatic conditions and the reference plastic strain (ε_0) is defined as the nonproportional compressive strain (which was set to be 0.2% in all experiments of the present study). The hardening exponent (n) was obtained through the fitting of quasistatic true stress-strain data.

$$\ln \sigma = \ln \sigma_0 + \frac{1}{n} \ln \left(1 + \frac{\varepsilon}{\varepsilon_0}\right) \tag{8}$$

This fitting yielded the strain-hardening parameters of the 26%- and 36%-porosity titanium alloy specimens (Table 5).

Table 5. Strain-hardening parameters of the DP model.

Porosity	σ_0/MPa	ε_0	n
26%	250	0.002	8.331
36%	115	0.002	8.105

Second, fitting was performed on strain-rate strengthening terms. The DP model was employed under room-temperature conditions to describe the stress-strain relationship at different strain rates at room temperature. Comparing flow stress values at the same room temperature but various strain rates with flow stress values under quasistatic conditions ($\varepsilon = 0.1$) yielded strain-rate strengthening coefficients for different strain rates. In Figure 7, the straight line indicates changes in strain-rate strengthening coefficients under dynamic conditions; the intersection between the line and coordinate axis is the lower limit of the critical-strain-rate range (whereas its upper limit is $1000\ \mathrm{s}^{-1}$); and the critical-strain-rate ranges of the 26%- and 36%-porosity specimens are $800–1000\ \mathrm{s}^{-1}$ and $700–1000\ \mathrm{s}^{-1}$, respectively. From these ranges, adequate critical strain rates ($\dot{\varepsilon}_t$) were selected.

To conduct fitting using equations with strain-hardening terms larger than the selected threshold critical strain rates ($\dot{\varepsilon}_t$), the strain-rate strengthening term was converted into Equation (9). In a log–log graph, this equation is a straight line with a slope of $(1/m_1 - 1/m_2)\ln(1 + \dot{\varepsilon}_t/\dot{\varepsilon}_0)$ and an intercept of $1/m_2$, in which the reference strain rate ($\dot{\varepsilon}_0$) was set to be 0.001. The strain-rate strengthening coefficients of both alloys at the strain rates of $1000\ \mathrm{s}^{-1}$, $2000\ \mathrm{s}^{-1}$ and $3000\ \mathrm{s}^{-1}$ at room temperature were fitted to yield strain-rate sensitivity exponents m_1 and m_2.

$$\ln[\Gamma(\dot{\varepsilon})] = \begin{cases} \frac{1}{m_1}\ln\left(1 + \frac{\dot{\varepsilon}}{\dot{\varepsilon}_0}\right), & \text{if } \dot{\varepsilon} \le \dot{\varepsilon}_t; \\ \frac{1}{m_2}\ln\left(1 + \frac{\dot{\varepsilon}}{\dot{\varepsilon}_0}\right) + \left(\frac{1}{m_1} - \frac{1}{m_2}\right)\ln\left(1 + \frac{\dot{\varepsilon}_t}{\dot{\varepsilon}_0}\right) & \text{if } \dot{\varepsilon} > \dot{\varepsilon}_t; \end{cases} \tag{9}$$

This fitting yielded strain-rate strengthening term parameters from the DP model for both specimens (Table 6).

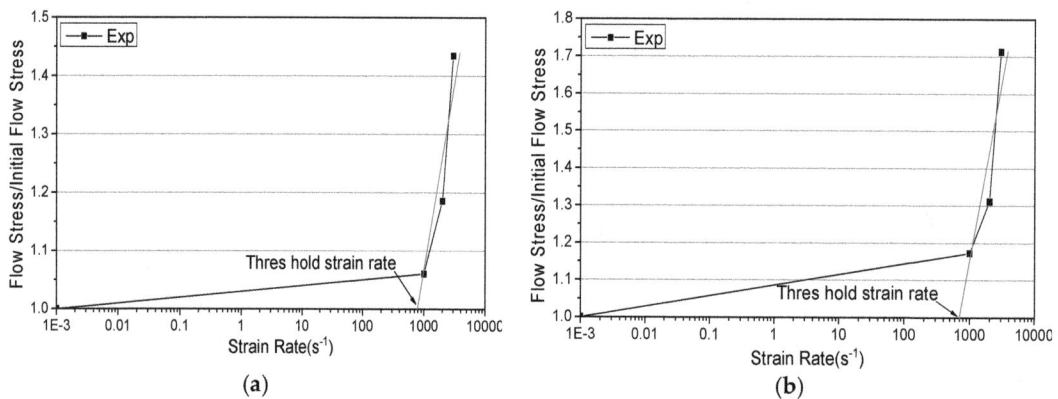

(a)   (b)

Figure 7. Selection ranges of critical strain rates in the DP model: (a) $\varepsilon_p = 26\%$, $T = 25\ ^\circ\mathrm{C}$; and (b) $\varepsilon_p = 36\%$, $T = 25\ ^\circ\mathrm{C}$.

Table 6. Strain-rate strengthening term parameters of the DP model.

Porosity	ε_0	ε_t	m_1	m_2
26%	0.001	900	7402	3.611
36%	0.001	800	427.5	2.821

Fitting was performed on temperature-softening terms. The flow stress values of the alloy specimens changed nonlinearly with temperature; thus, linear decrease equations with temperature-softening terms greater than the critical temperature (T_{cut}) were not adequate for the fitting. Instead, polynomial equations obtained from the temperature-softening terms in the model were used to describe the temperature effects of the specimens.

During the fitting, flow stress ($\varepsilon = 0.1$) values for identical strain rates at $25\ ^\circ\mathrm{C}$, $100\ ^\circ\mathrm{C}$, $200\ ^\circ\mathrm{C}$ and $300\ ^\circ\mathrm{C}$ was divided by the initial flow stress and the corresponding strain-rate strengthening coefficients to derive temperature-effect coefficients for the alloys at these different temperatures (Figure 8). Polynomial fitting was subsequently performed on these temperature-effect coefficients to yield polynomial coefficients c_0–c_5 (Table 7).

Table 7. Temperature-softening parameters of the DP model.

Porosity	c_0	c_1	c_2	c_3
26%	1.057	-2.241×10^{-3}	-9.996×10^{-7}	2.867×10^{-8}
36%	1.105	-5.048×10^{-3}	3.933×10^{-5}	-7.447×10^{-8}

Note: Polynomial coefficients excluded from this table were set as zero.

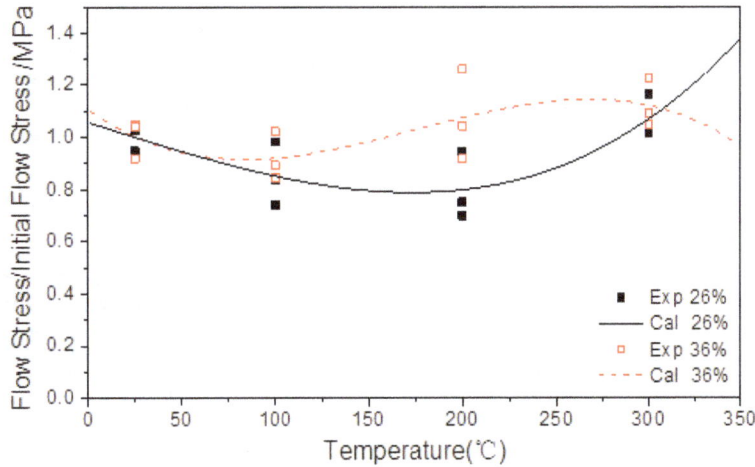

Figure 8. Polynomial fitting conducted using temperature-effect terms in the DP model.

3.4. True Stress-Strain Curves and Flow Stress: Comparing Experimental and Simulated Results

The 26%- and 36%-porosity specimens were micromilled in the DP simulation. Figure 9 shows true stress-strain curves under dynamic compression and DP simulation. The simulated true stress-strain curves at room temperatures exhibited favorable consistency with the experimental ones, indicating the ability of the DP model to sufficiently describe flow-stress changes from the specimens at room temperature. Moreover, for the 36%-porosity specimen, the simulated and experimental results of strain-rate strengthening effects were practically identical, indicating the ability of the DP model to sufficiently describe the strain-rate strengthening effects of the specimen. The DP model was also applied to describing the strain-rate strengthening effects of the 26%-porosity specimen, although the collected simulated and experimental data showed certain degrees of deviation at considerably high deformation levels.

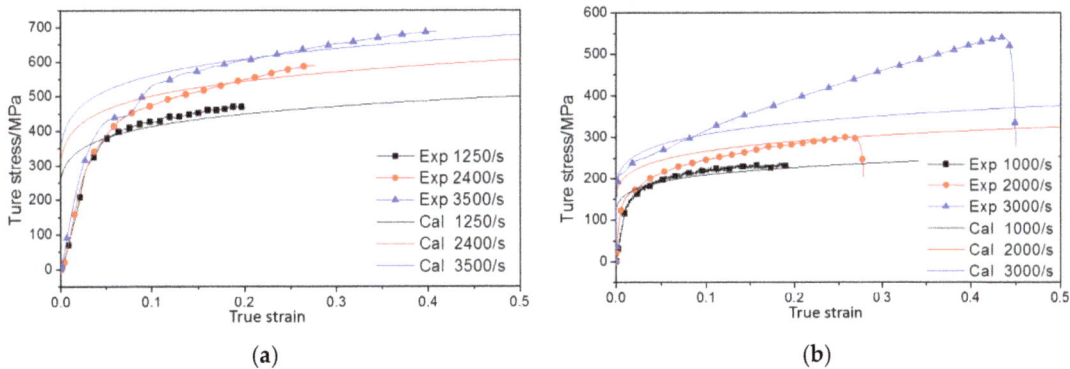

(a)

(b)

Figure 9. True stress–strain curves (dynamic compression vs. DP simulation): (a) $\varepsilon_p = 26\%$, $T = 25\,°C$; and (b) $\varepsilon_p = 36\%$, $T = 25\,°C$.

Figure 10 compares the experimental and simulated results of the flow stress phenomena of both specimens at different strain rates and at a flow stress value of 0.1. The simulated and experimental results of flow stress were practically identical with a deviation of 10% or lower, indicating the ability of the DP model to sufficiently describe the dynamic mechanical properties of the specimens. The polynomial equations obtained using temperature-related terms in the model were able to characterize the temperature-softening and temperature-strengthening effects of both specimens.

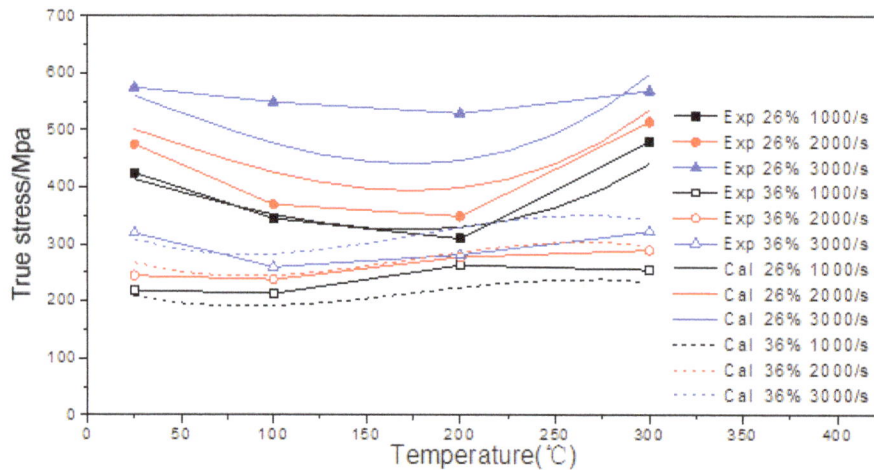

Figure 10. Comparison of the simulated and experimental flow stress values of the DP model ($\varepsilon = 0.1$).

Figure 11 provides a comparison between the simulated stress and experimental stress values. Eliminating the data at beginning, relatively stable data were selected for analysis. As can be seen, a good agreement has been obtained which shows the applicability of the DP constitutive model in describing the relationship between flow stress and strain of porous titanium alloys at various strain rates and temperatures. The error of the model was examined using average absolute relative error parameter (AARE) [30,31]. The AARE was found to be 5.36% and 4.609%, respectively.

$$AARE = \frac{1}{N} \sum_{i=1}^{n} \left| \frac{E_i - S_i}{E_i} \right| \times 100 \qquad (10)$$

where E_i is the experimental value, S_i is the simulated value, and N is the number of data.

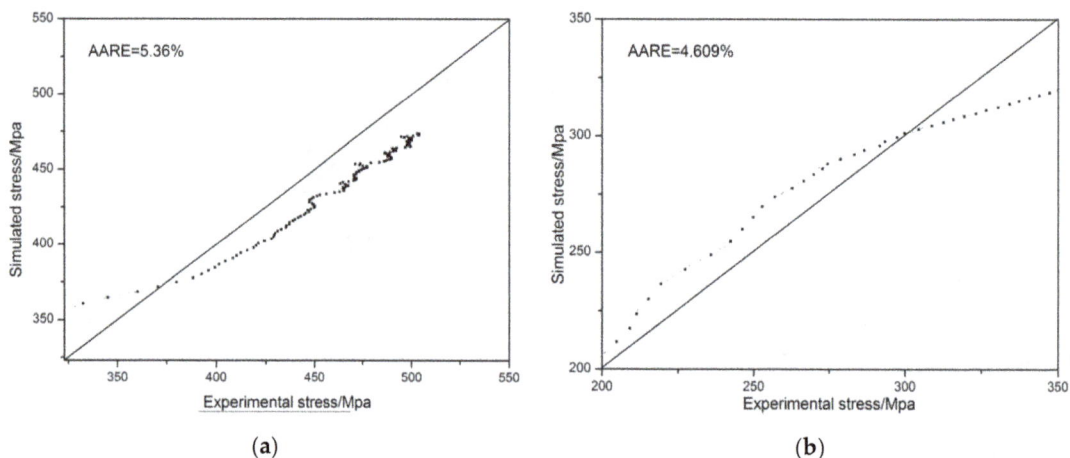

(a)   (b)

Figure 11. Error analysis between simulation stress and experimental stress: (a) $\varepsilon_p = 26\%$, $T = 25\,°C$, $\dot{\varepsilon} = 1200\ s^{-1}$; and (b) $\varepsilon_p = 36\%$, $T = 25\,°C$, $\dot{\varepsilon} = 3000\ s^{-1}$.

4. Conclusions

Several conclusions can be drawn from the dynamic and quasistatic mechanical response testing of the 26%- and 36%-porosity titanium alloy specimens and the establishment and validation of the DP constitutive model:

(1) As with a conventional porous material, the deformation patterns of porous titanium alloys under quasistatic loading can be divided into elastic, plastic, and densification stages.

(2) The employed dynamic and quasistatic mechanical response tests showed that the presence of pores substantially reduced the mechanical strength levels of the alloy specimens. The compressive elastic moduli of compacted pure titanium and the 26%- and 36%-porosity titanium alloy specimens were 102.5 GPa, 20.13 GPa, and 7.67 GPa, respectively. The yield limits of compacted pure titanium and the 26%- and 36%-porosity specimens were 275 MPa, 250 MPa and 115 MPa, respectively.

(3) Dissimilar to the deformation patterns of compacted titanium, the deformation patterns of porous titanium alloys are characterized by pore collapse-induced deformation. Thus, the alloy specimens show strong stress-hardening effects, although no noticeable plastic flow plateaus were observed.

(4) Porous titanium alloys exhibit stress-rate effects at strain rates of 1000–3000 s^{-1}. However, no statistical regularity was observed in the strain-rate data at a strain rate of 4000 s^{-1}, indicating that the discreteness of these data increased with rising strain rates.

(5) A 26%-porosity titanium has notable temperature sensitivity, whereas a 36%-porosity titanium has weak temperature sensitivity. Moreover, the strain-rate sensitivity levels of both alloys vary at different temperatures.

(6) The DP model accurately describes the mechanical properties of both alloy specimens at 25–300 °C and at strain rates of 1000–3000 s^{-1}, with a deviation of 10% or lower.

Acknowledgments: The work is supported by the National Natural Science Foundation of China (No. 51305174).

Author Contributions: Zhiqiang Liu and Tianyu Zhu conceived and designed the experiments; Tianyu Zhu performed the experiments; Feifei Ji and Tianyu Zhu analyzed the data; Zhiqiang Liu and Mingqiang Wang contributed analysis tools; Tianyu Zhu and FeiFei Ji wrote the paper.

Conflicts of Interest: The authors declare no conflict of interest.

References

1. Ezugwu, E.O.; Wang, Z.M. Titanium alloys and their machinability—A review. *J. Mater. Process. Technol.* **1997**, *68*, 262–274. [CrossRef]

2. Liu, Y.; Jiang, G.F.; He, G. Enhancement of entangled porous titanium by BisGMA for load-bearing biomedical applications. *Mater. Sci. Eng. C* **2016**, *61*, 37–41. [CrossRef] [PubMed]

3. Yetim, A.F. Investigation of wear behavior of titanium oxide films, produced by anodic oxidation, on commercially pure titanium in vacuum conditions. *Surf. Coat. Technol.* **2010**, *205*, 1757–1763. [CrossRef]

4. Ossama, M.B.; Frédéric, B.B.R.; Lee, M.G.; Peter, H.; Matthias, W. Constitutive modelling of high strength titanium alloy Ti-6Al-4 V for sheet forming applications at room temperature. *Int. J. Solids Struct.* **2016**, *80*, 334–347.

5. Zhang, L.; He, Z.Y.; Zhang, Y.Q.; Jiang, Y.H.; Zhou, R. Enhanced in vitro bioactivity of porous NiTi–HA composites with interconnected pore characteristics prepared by spark plasma sintering. *Mater. Des.* **2016**, *101*, 170–180. [CrossRef]

6. Wu, C.C.; Chang, T.P.; Peng, K.Y.; Chang, W.C. Study on the properties of WC-10Co alloys adding Cr_3C_2 powder *via* various vacuum sintering temperatures. *J. Alloy. Compd.* **2016**, *686*, 810–815. [CrossRef]

7. Yao, Y.C.; Qu, P.W.; Gan, X.K.; Huang, X.P.; Zhao, Q.F.; Liang, F. Preparation of porous-structured $LiFePO_4/C$ composite by vacuum sintering for lithium-ion battery. *Ceram. Int.* **2016**, *42*, 12726–12734. [CrossRef]

8. Chen, H.; Wang, C.; Zhu, K.; Fan, Y.J.; Zhang, X.D. Fabrication of porous titanium scaffolds by stack sintering of microporous titanium spheres produced with centrifugal granulation technology. *Mater. Sci. Eng. C* **2014**, *43*, 182–188. [CrossRef] [PubMed]

9. Taniguchi, N.; Fujibayashi, S.; Takemoto, M.; Sasaki, K.; Otsuki, B.; Nakamura, T.; Matsushita, T.; Kokubo, T.; Matsuda, S. Effect of pore size on bone ingrowth into porous titanium implants fabricated by additive manufacturing: An in vivo experiment. *Mater. Sci. Eng. C* **2016**, *59*, 690–701. [CrossRef] [PubMed]

10. Quan, H.X.; Gao, S.S.; Zhu, M.H.; Yu, H.Y. Comparison of the torsional fretting behavior of three porous titanium coatings for biomedical applications. *Tribol. Int.* **2015**, *92*, 29–37. [CrossRef]

11. Yang, Y.; Zhang, C.; Wang, Y.; Dai, Y.J.; Luo, J.B. Friction and wear performance of titanium alloy against tungsten carbide lubricated with phosphate ester. *Tribol. Int.* **2015**, *95*, 27–34. [CrossRef]

12. Rahman, M.; Wang, Z.G.; Wong, Y.S. A review on high-speed machining of tita-nium alloys. *JSME Int. J. Ser. C Mech. Syst. Mach. Elem. Manuf.* **2006**, *49*, 11–20. [CrossRef]

13. Budinski, K.G. Tribological properties of titanium alloys. *Wear* **1991**, *151*, 203–217. [CrossRef]

14. Lauro, C.H.; Filhob, S.L.M.R.; Brandãob, L.C.; Davin, J.P. Analysis of behaviour biocompatible titanium alloy (Ti-6Al-7Nb) in the micro-cutting. *Measurement* **2016**, *96*, 529–540. [CrossRef]

15. Schoop, J.; Jawahir, I.S.; Balk, T.J. Size effects in finish machining of porous powdered metal for engineered surface quality. *Precis. Eng.* **2015**, *44*, 180–191. [CrossRef]

16. Wang, X.H.; Li, J.S.; Hu, R.; Kou, H.C. Mechanical properties and pore structure deformation behavior of biomedical porous titanium. *Trans. Nonferr. Met. Soc. Chin.* **2015**, *25*, 1543–1550. [CrossRef]

17. Dai, J.; Chen, C.; Weng, F. High temperature oxidation behavior and research status of modifications on improving high temperature oxidation resistance of titanium alloys and titanium aluminides: A review. *J. Alloy. Compd.* **2016**, *685*, 784–798. [CrossRef]

18. Revankar, G.D.; Shetty, R.; Rao, S.S.; Gaitonde, V.N. Wear resistance enhancement of titanium alloy (Ti–6Al–4V) by ball burnishing process. *J. Mater. Res. Technol.* **2016**, *5*, 190–197. [CrossRef]

19. Yang, D.; Liu, Z. Surface topography analysis and cutting parameters optimization for peripheral milling titanium alloy Ti–6Al–4V. *Int. J. Refract. Met. Hard Mater.* **2015**, *51*, 192–200. [CrossRef]

20. Lin, Y.C.; Chen, X.M. A critical review of experimental results and constitutive descriptions for metals and alloys in hot working. *Mater. Des.* **2011**, *32*, 1733–1759. [CrossRef]

21. Zhang, C.; Zhang, L.; Shen, W.F.; Liu, C.R.; Xia, Y.N.; Li, R.Q. Study on constitutive modeling and processing maps for hot deformation of medium carbon Cr–Ni–Mo alloyed steel. *Mater. Sci. Eng. A* **2014**, *590*, 255–261. [CrossRef]

22. Chen, L.; Zhao, G.; Yu, J. Hot deformation behavior and constitutive modeling of homogenized 6026 aluminum alloy. *Mater. Des.* **2015**, *74*, 25–35. [CrossRef]

23. Roters, F.; Raabe, D.; Gottstein, G. Work hardening in heterogeneous alloys—A microstructural approach based on three internal state variables. *Acta Mater.* **2000**, *48*, 4181–4189. [CrossRef]

24. Samantaray, D.; Mandal, S.; Bhaduri, A.K.; Venugopal, S.; Sivaprasad, P.V. Analysis and mathematical modelling of elevated temperature flow behaviour of austenitic stainless steels. *Mater. Sci. Eng. A* **2011**, *528*, 1937–1943. [CrossRef]

25. Li, P.; Li, F.G.; Cao, J.; Ma, X.K.; Li, J.H. Constitutive equations of 1060 pure aluminum based on modified double multiple nonlinear regression model. *Trans. Nonferr. Met. Soc. Chin.* **2016**, *26*, 1079–1095. [CrossRef]

26. Haghdadi, N.; Zarei-Hanzaki, A.; Khalesian, A.R.; Abedi, H.R. Artificial neural network modeling to predict the hot deformation behavior of an A356 aluminum alloy. *Mater. Des.* **2013**, *49*, 386–391. [CrossRef]

27. Rubshtein, A.P.; Trakhtenberg, S.; Makarova, E.B.; Bliznets, D.G.; Yakovenkova, L.I.; Vladimirov, V.B. Porous material based on spongy titanium granules: Structure, mechanical properties, and osseointegration. *Mater. Sci. Eng. C* **2014**, *35*, 363–369. [CrossRef] [PubMed]

28. Wang, F.; Zhao, J.; Zhu, N.; Li, Z. A comparative study on Johnson-Cook constitutive modeling for Ti–6Al–4V alloy using automated ball indentation (ABI) technique. *J. Alloy. Compd.* **2015**, *633*, 220–228. [CrossRef]

29. Fakhri, M.A.; Bordatchev, E.V.; Tutunea-Fatan, O.R. An image-based methodology to establish correlations between porosity and cutting force in micromilling of porous titanium foams. *Int. J. Adv. Manuf. Technol.* **2012**, *60*, 841–851. [CrossRef]

30. Haghdadi, N.; Zarei-Hanzaki, A.; Abedi, H.R. The flow behavior modeling of cast A356 aluminum alloy at elevated temperatures considering the effect of strain. *Mater. Sci. Eng. A* **2012**, *535*, 252–257. [CrossRef]

31. Haghdadi, N.; Martin, D.; Hodgson, P. Physically-based constitutive modelling of hot deformation behavior in a LDX 2101 duplex stainless steel. *Mater. Des.* **2016**, *106*, 420–427. [CrossRef]

14

The Improvement of Dehydriding the Kinetics of NaMgH$_3$ Hydride via Doping with Carbon Nanomaterials

Zhong-Min Wang *, Song Tao, Jia-Jun Li, Jian-Qiu Deng, Huaiying Zhou and Qingrong Yao

School of Material Science and Engineering, Guilin University of Electronic Technology, Guilin 541004, China; taosongaa@163.com (S.T.); ljjdzn@hotmail.com (J.-J.L.); jqdeng@guet.edu.cn (J.-Q.D.); zhy@guet.edu.cn (H.Z.); qingry96@guet.edu.cn (Q.Y.)
* Correspondence: zmwang@guet.edu.cn

Academic Editor: Jacques Huot

Abstract: NaMgH$_3$ perovskite hydride and NaMgH$_3$–carbon nanomaterials (NH-CM) composites were prepared via the reactive ball-milling method. To investigate the catalytic effect of CM on the dehydriding kinetic properties of NaMgH$_3$ hydride, multiwall carbon nanotubes (MWCNTs) and graphene oxide (GO) were used as catalytic additives. It was found that dehydriding temperatures and activation energies (ΔE_1 and ΔE_2) for two dehydrogenation steps of NaMgH$_3$ hydride can be greatly reduced with a 5 wt. % CM addition. The NH–2.5M–2.5G composite presents better dehydriding kinetics, a lower dehydriding temperature, and a higher hydrogen-desorbed amount (3.64 wt. %, 638 K). ΔE_1 and ΔE_2 can be reduced by about 67 kJ/mol and 30 kJ/mol, respectively. The results suggest that the combination of MWCNTs and GO is a better catalyst as compared to MWCNTs or GO alone.

Keywords: NaMgH$_3$ hydride; doping; carbon nanomaterial (CM) composite; catalytic effect; dehydriding kinetics

---

1. Introduction

Perovskite-type hydride, NaMgH$_3$, as a potential candidate for on-board applications, has received considerable attention for its high gravimetric and volumetric hydrogen densities (6 wt. % and 88 kg/m^3, respectively), as well as its reversible hydriding and dehydriding reactions [1–7]. NaMgH$_3$ is characterized by an orthorhombic perovskite structure comprising [MgH$_6$] octahedra and [NaH$_{12}$] cubo-octahedra, which is analogous to GdFeO$_3$-type perovskite (space group Pnma). This space group is typical for low-tolerance-factor oxide perovskites, where the singly charged Na cation occupies eight-fold coordinated voids [8]. Sheppard et al. [9] reported the desorption enthalpy and entropy are 86.6 $\pm$ 1.0 kJ/(mol·H$_2$) and 132.2 $\pm$ 1.3 kJ/(mol·H$_2$) for the decomposition of NaMgH$_3$ into NaH and Mg, respectively, indicating that NaMgH$_3$ is thermodynamically more stable than magnesium hydride (MgH$_2$). One of the most promising approaches to improve its poor dehydrogenation property is incorporating with other metals or complex hydrides (e.g., Na$_{1-x}$Li$_x$MgH$_3$ [10–12], NaMgH$_3$-g-C$_3$N$_4$ [13], and NaMgH$_3$-MgH$_2$ [14,15]).

The addition of small amounts of catalytic material (transition metals, carbon materials, etc.) to MgH$_2$ has successfully reduced the time taken to absorb or desorb hydrogen [16–22]. Recently, carbon nanomaterials have been investigated as catalytic materials for enhancing the uptake and release of hydrogen from MgH$_2$ due to its lightweight nature. Alsabawi et al. [23] investigated the catalytic effect of up to 10 wt. % carbon buckyballs (C$_{60}$) on the kinetics of hydrogen desorption and the subsequent absorption of MgH$_2$. They found that a 1–2 wt. % C$_{60}$ additive with 2 h of milling time are the optimum conditions for the best desorption kinetics. Imamura et al. reported the enhanced kinetics of a ball-milled Mg–graphite composite with organic additives (tetrahydrofuran, cyclohexane, or benzene), and Raman characterization indicated that the organic additives allowed the graphite to shear along planes rather than grind into small particles, as it did without organic additives [24,25]. Thiangviriya et al. showed that the improvement of dehydrogenation kinetics of the $2LiBH_4$–MgH_2 composite by doping with activated carbon nanofibers is due to the increase in the hydrogen diffusion pathway and thermal conductivity [26]. Kadri et al. demonstrated that the presence of both a V-based catalyst and carbon nanotubes reduces the enthalpy and entropy of MgH_2, and partially destroyed CNTs are better at enhancing the hydrogen sorption performance [27]. Other forms of carbon, such as amorphous carbon [28,29], carbon black [28,30], activated carbons [31,32], and carbon nanotubes [28,33–39], have also been studied for their catalytic effect on the magnesium hydrogen system. $NaMgH_3$ can be synthesized via reactive mechanochemical means. Indeed, mechanochemical approaches provide not only less energy intensive routes to the hydrides, but also ensure that particles sizes are minimized, improving the dehydrogenation kinetics of the ternary hydrides compared to those prepared at high temperature [10,40].

In this work, multiwall carbon nanotubes (MWCNTs) and graphite (G) were used as catalytic additives. The reactive ball-milling method was employed to prepare $NaMgH_3$ perovskite hydride and $NaMgH_3$-carbon nanomaterials (NH-CM) composites. The effects of different additives on the structure, thermal stability, and dehydriding kinetic properties of $NaMgH_3$ hydride were investigated.

2. Materials and Methods

All sample handlings were undertaken in an argon atmosphere glovebox (Mbraun, Labstar, 1 ppm H_2O, 1 ppm O_2). $NaMgH_3$ hydride was prepared by reactive milling of stoichiometric mixtures of MgH_2 (>98% purity, Alfa-Aesar, Ward Hill, MA, USA) and NaH (99.9% purity, Sigma-Aldrich, Billerica, MA, USA). The mixed powders were milled with stainless steel balls, with a ball-to-powder weight ratio of 80:1, under a H_2 atmosphere (0.8 MPa) for 45 h at 320 rpm using a QM-3SP2 planetary mill at ambient temperature. Multiwall carbon nanotubes (MWCNTs, tube length: 0.5–2 μm, tube diameter: 30–50 nm, >95% purity, Chengdu Organic Chemistry Co., Ltd., Chengdu, China) and graphene oxide (GO, <10 sheets, >95% purity, Chengdu Organic Chemistry Co., Ltd., Chengdu, China) were introduced to $NaMgH_3$ hydride as catalysts. NH-CM composites were prepared by reactive milling the mixture of MgH_2, NaH, and catalytic additives under the same milling conditions. The composites of $NaMgH_3$ with different loading of carbon nanomaterials were labeled as follows: NH–5M ($NaMgH_3$ + 5 wt. % MWCNT composite), NH–5G ($NaMgH_3$ + 5 wt. % GO composite), and NH–2.5M–2.5G ($NaMgH_3$ + 2.5 wt. % MWCNTs + 2.5 wt. % G composite).

X-ray diffraction (XRD) samples were prepared in a glove box. To avoid exposure to air during the measurement, the sample was spread uniformly on the sample holder and covered with Scotch tape. XRD analysis was performed on Empyrean PIXcel 3D (PANalytical B.V., Almelo, The Netherlands) (Cu Kα radiation) with a scanning speed of 5°/min. The mean crystallite size was determined by the Scherer formula ($D = k\lambda/(\beta\_(hkl)\cos\theta)$), where D is the crystallite size, k is the shape factor (0.89), λ is the wavelength of the Cu Kα radiation (0.154056 nm), β_hkl is the FWHM (full width of peak at half maximum), and θ is the diffraction angle. The evaluations of the hydrogen-desorbed amount and the dehydriding kinetic properties of the samples were carried out in an automatic Sievert-type apparatus (PCTpro2000, Setaram Co., Caluire, France). Thermal properties of the $NaMgH_3$ and NH-CM composites were investigated by differential scanning calorimetry (DSC, NETZSCH STA 449F3, Selb, Germany) at different ramping rates (5, 10, and 15 K/min) under a continuous argon flow (20 mL/min) from 298 K to 725 K. The sample was loaded into an alumina crucible in the glove box. The crucible was then placed in a sealed glass bottle in order to prevent oxidation during transportation from the glove box to the TGA/DSC apparatus. An empty alumina crucible was used as a reference.

3. Results and Discussion

3.1. Structure Characterization of As-Prepared NaMgH$_3$ Hydride and NH-CM Composites

Figure 1 shows the XRD patterns of as-prepared NaMgH$_3$ hydride and NH-CM composites (NH–5M, NH–5G, and NH–2.5M–2.5G). The red long string is the standard line of NaMgH$_3$ phase. As can been seen, typical peaks of NaMgH$_3$ phase can be observed in all samples, which indicate an orthorhombic perovskite structure, similar to those of the GdFeO$_3$ type perovskite (space group Pnma) [41,42]. The sharp peaks are more prominent for NaMgH$_3$ hydride without a catalyst addition. A slight peak shift to a lower angle also is observed in these NH-CM composites. The calculated crystallite size are 15.8 nm for the as-synthesized NaMgH$_3$ sample, 14.9 nm for the NH–5M sample, 14.0 nm for the NH–5G sample, and 13.4 nm for the NH–2.5M–2.5G sample, respectively. The results indicate that the addition of 5 wt. % catalyst will reduce the crystallite size, which may improve dehydriding kinetics. The peaks of the catalysts are absent, probably because only a small amount of CM were added to the composites, and these additives were appeared in the form of an amorphous phase after ball-milling [11]. The existence of MgO can be attributed to the slight oxidization of samples in the handling process [43].

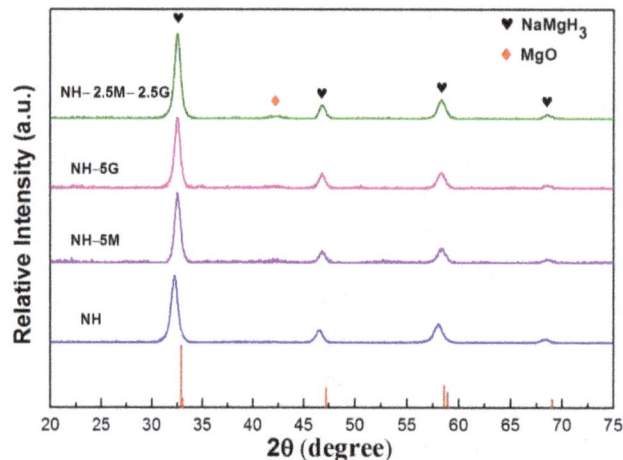

Figure 1. The XRD patterns of as-prepared NaMgH$_3$ hydride and NH-CM composites.

3.2. Thermal Stabilities of NaMgH$_3$ Hydride and NH-CM Composites

Figure 2 presents the DSC curves of NaMgH$_3$ hydride and three NH-CM composites at different heating rates (5, 10, and 20 K/min) under a continuous argon flow from 298 K to 750 K, where T_x and T_p represent the start and peak temperature of dehydrogenation, respectively. From the DSC curves, there are two endothermic peaks for all four samples, which are related to two decomposition steps of NaMgH$_3$ hydride. The decomposition steps can be expressed as follows [2]:

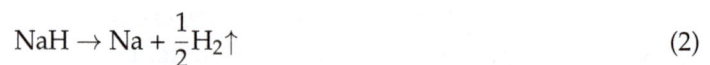

$$NaMgH_3 \rightarrow NaH + Mg + H_2\uparrow \tag{1}$$

$$NaH \rightarrow Na + \frac{1}{2}H_2\uparrow \tag{2}$$

In comparison with NaMgH$_3$ hydride without a catalyst addition, an obvious decrease of dehydriding temperature is observed in all composites. At a heating rate of 5 K/min, NH–2.5M–2.5G has the largest dehydriding temperatures reduction with temperature difference values of ΔT_{x1}, ΔT_{p1}, ΔT_{x2}, and ΔT_{p2} are 52.2 K, 51.1 K, 44.1 K, and 20.4 K, respectively. The next sample with the biggest reduction temperature differences was NH–5G, followed by NH–5M.

In order to estimate the dehydriding activation energy (ΔE) of NaMgH$_3$ hydride, the Kissinger plot is used for non-isothermal DSC analysis and can be expressed in the form as follows [44]:

$$\ln (T^2/\upsilon) = \Delta E/RT + \ln (\Delta E/R u_0) \tag{3}$$

where T is the peak temperature, R is the gas constant (8.3145 J/(K·mol)), and υ and u_0 are the heating rate and frequency factor, respectively.

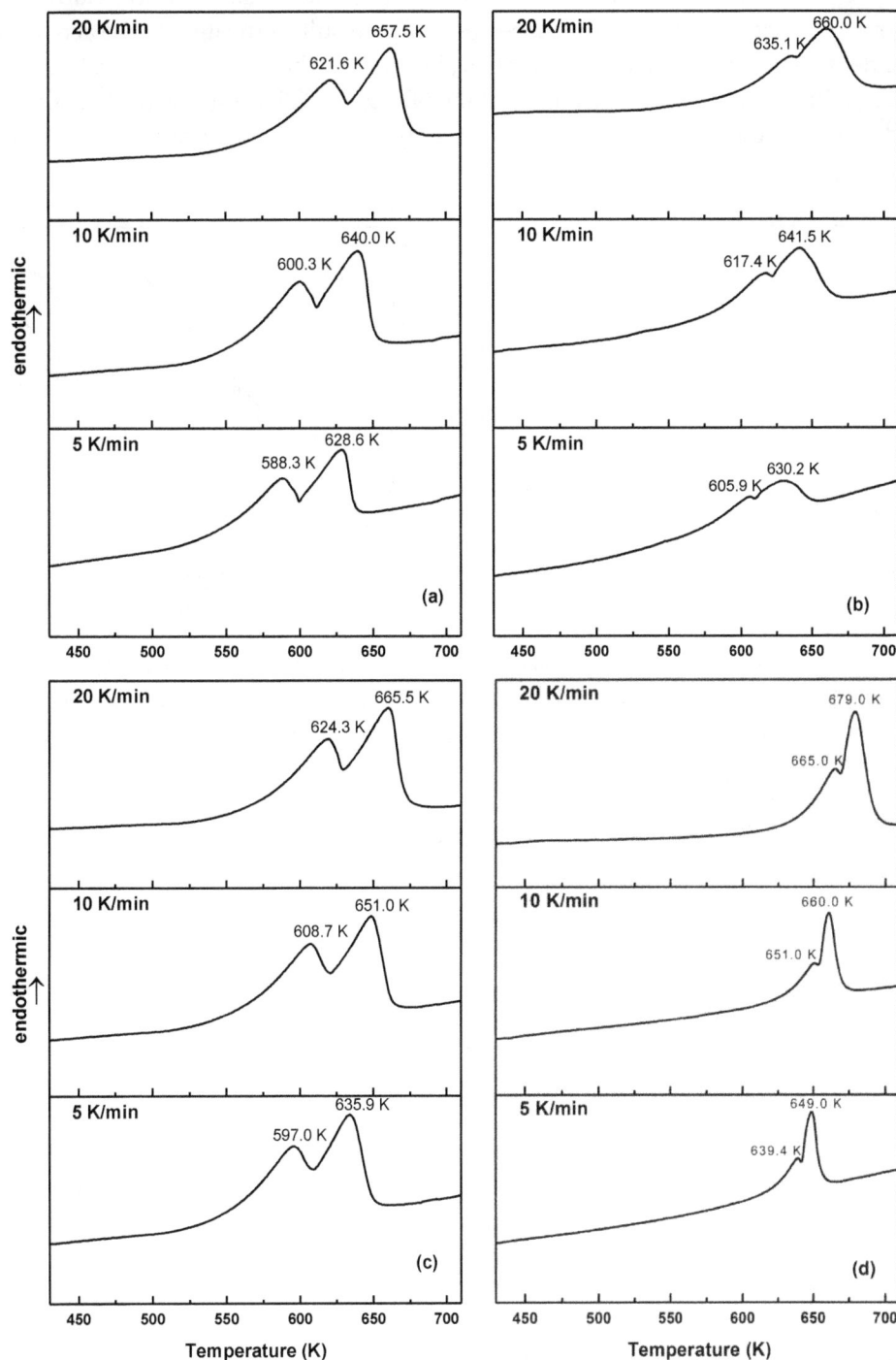

Figure 2. DSC curves of NaMgH$_3$ hydride and NH-CM composites at different heating rates. (**a**) NH–2.5M–2.5G; (**b**) NH–5G; (**c**) NH–5M; (**d**) NaMgH$_3$ hydride.

In Figure 3, both datasets at T_x and T_p show a good linear relation between $\ln(T^2/\upsilon)$ and $(1000/T)$ with a slope of $\Delta E/R$. The calculated ΔE values are listed in Table 1. An obvious decrease of ΔE for both dehydrogenation steps of NaMgH$_3$ hydride is observed for those samples milled with the CM additive. The calculated ΔE_1 (the first decomposition step) and ΔE_2 (the second decomposition step) are 113.8 kJ/mol and 126.6 kJ/mol for the NH–2.5M–2.5G sample, 139.8 kJ/mol and 147.6 kJ/mol for the NH–5G sample, and 146.4 kJ/mol and 153.9 kJ/mol for the NH–5M sample, respectively. In comparison with NaMgH$_3$ hydride without a CM addition (ΔE_1 = 180.3 kJ/mol, ΔE_2 = 156.2 kJ/mol), the NH–2.5M–2.5G sample has the highest reduction of activation energy, ΔE, where the deviation values of ΔE_1 and ΔE_2 are about 67 kJ/mol and 30 kJ/mol, respectively. The results indicate that the activation energy (ΔE) for the dehydrogenation steps of the NaMgH$_3$ hydride can be greatly decreased by milling with a 5 wt. % CM addition, especially in the case of the NH–2.5M–2.5G composite. These observations correspond well with the DSC results, such that the dehydriding temperatures are lowered by the activation energy.

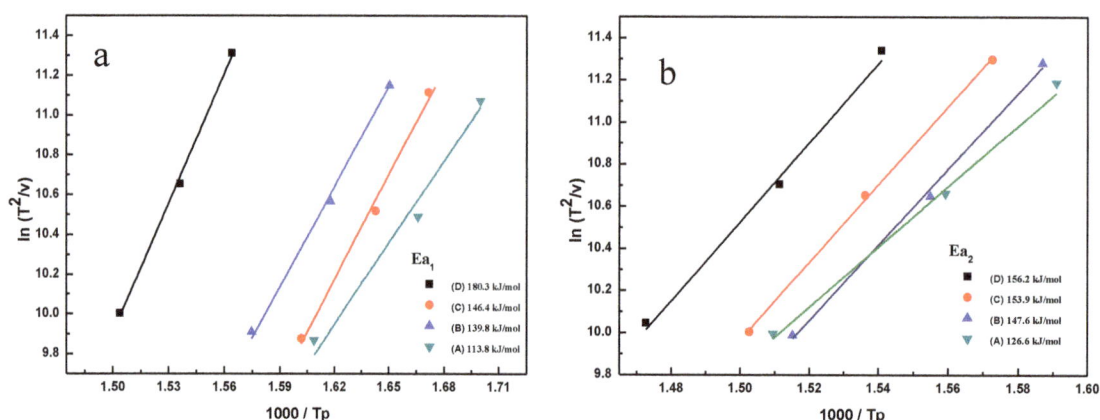

Figure 3. The relationship curves between $\ln(T_p^2/\upsilon)$ and $(1000/T_p)$ for NaMgH$_3$ hydride and composites fitted by Kissinger method. (A) NH–2.5M–2.5G; (B) NH–5G; (C) NH–5M; (D) NaMgH$_3$ hydride (**a**) for the first decomposition step of NaMgH$_3$ hydride and (**b**) for the second decomposition step of NaMgH$_3$ hydride.

Table 1. Calculated value of ΔE of the decomposition of the NaMgH$_3$ hydride.

Sample	1st Step ΔE_1 (kJ/mol)	2nd Step ΔE_2 (kJ/mol)
NH–2.5M–2.5G	113.8	126.6
NH–5G	139.8	147.6
NH–5M	146.4	153.9
NaMgH$_3$ hydride	180.3	156.2

3.3. Dehydriding Kinetic Properties of NaMgH$_3$ Hydride and NH-CM Composites

The isothermal dehydriding properties of the four samples at different temperatures (593 K, 613 K, and 638 K) are shown in Figures 4–6, respectively. In comparison with NaMgH$_3$ hydride, all NH-CM composites present better dehydriding kinetic properties. The hydrogen-desorbed amount increases with the increase in temperature. Among these three NH-CM composites, the NH–2.5M–2.5G composite has the best catalytic effect in improving the dehydriding kinetic properties of the NaMgH$_3$ hydride, where 90% of the maximum theoretical capacity (3.64 wt. % hydrogen) is released within 20 min at 638 K. Table 2 shows the maximum amount of hydrogen desorbed from the NaMgH$_3$ hydride and the NH-CM composites at different temperatures. These results agree well with that observed in Figures 2 and 3, indicating that dehydriding kinetics and dehydriding temperatures can be effectively reduced by a combined catalytic addition of MWCNTs and GO.

Table 2. The maximum amount of hydrogen desorbed from NaMgH$_3$ hydride and NH-CM composites at different temperatures.

Sample/Temperature	593 K (wt. %)	613 K (wt. %)	638 K (wt. %)
NaMgH$_3$	1.13	2.03	3.42
NH–5M	1.36	2.46	3.40
NH–5G	1.21	2.49	3.64
NH–2.5G–2.5M	1.46	2.56	3.64

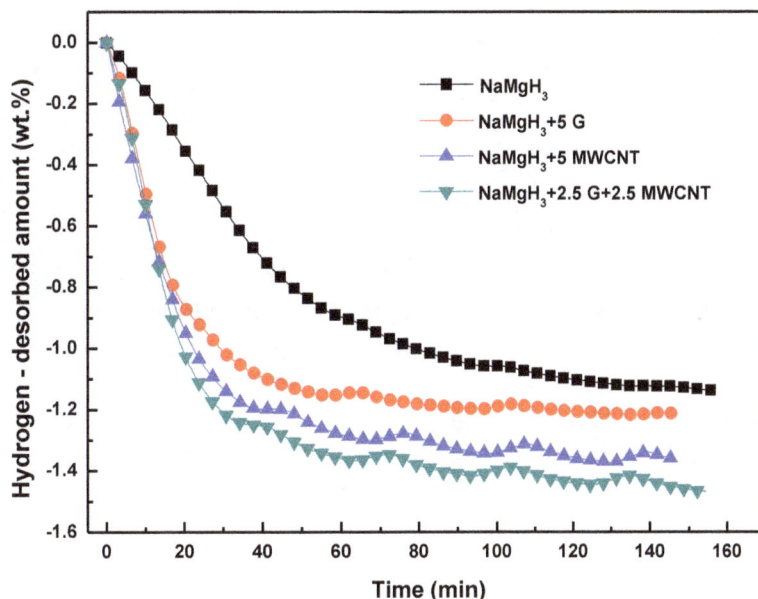

Figure 4. Dehydriding kinetic curves of the NaMgH$_3$ hydride and the NH-CM composites at 593 K.

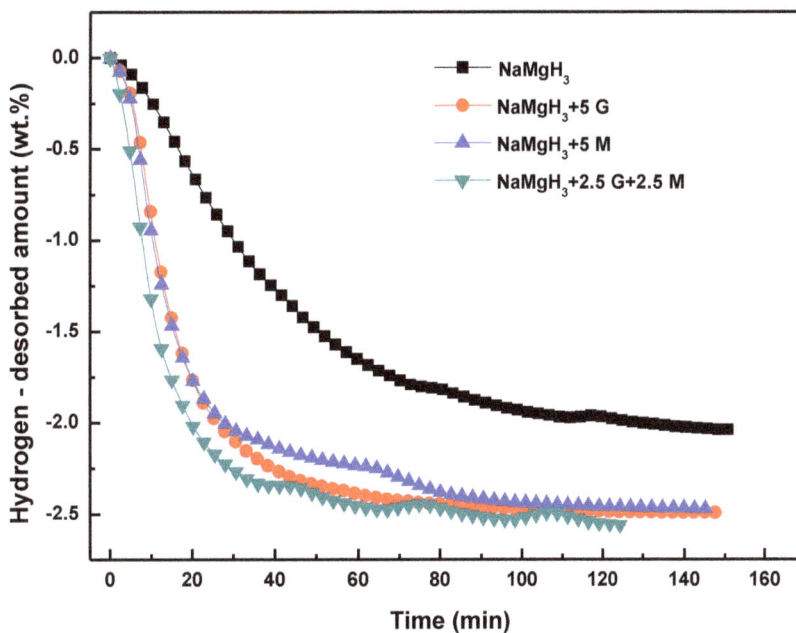

Figure 5. Dehydriding kinetic curves of the NaMgH$_3$ hydride and the NH-CM composites at 613 K.

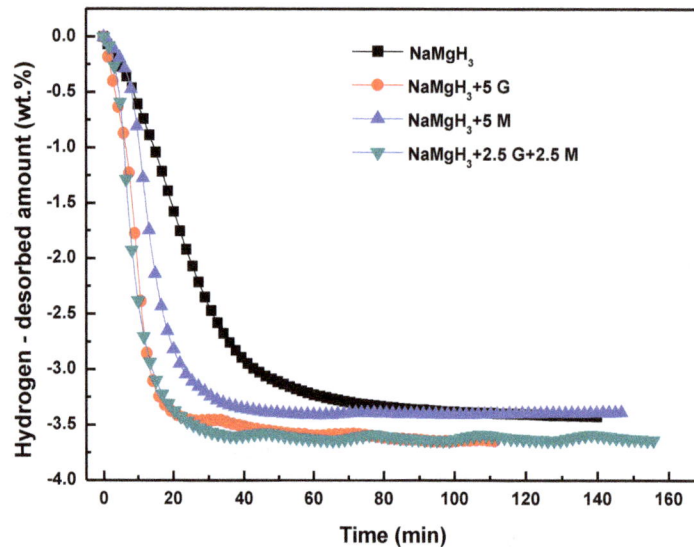

Figure 6. Dehydriding kinetic curves of the $NaMgH_3$ hydride and the NH-CM composites at 638 K.

To illustrate the decomposition mechanism of the NH-CM hydride composites, the XRD patterns of the $NaMgH_3$ + 2.5 G + 2.5 M sample after dehydriding at different temperatures are shown in Figure 7. With the increase in temperature from 593 K to 638 K, peaks of the $NaMgH_3$ phase become weakened, and peaks of the NaH phase and Mg phase become strengthened, such results agree well with our previous work for pristine $NaMgH_3$ hydride reported in [12]. In another word, the decomposition of $NaMgH_3$ is a two-step reaction: $NaMgH_3 \rightarrow NaH + Mg + H_2 \rightarrow Na + Mg + 3/2H_2$. The dopping with NM cannot change the decomposition process of $NaMgH_3$, but contribute to its dehydriding kinetics.

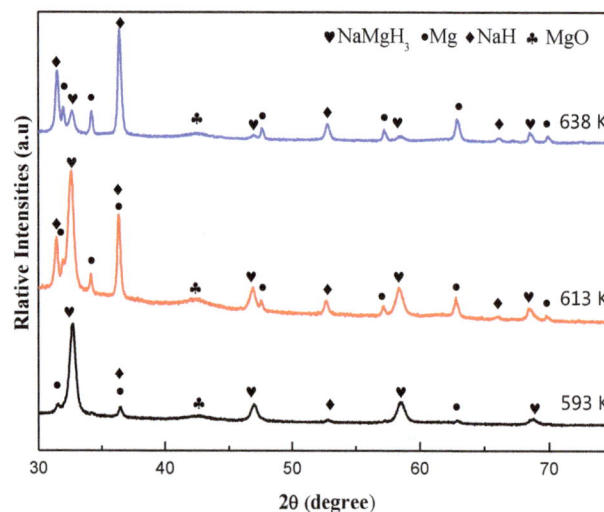

Figure 7. XRD patterns of $NaMgH_3$ + 2.5 G + 2.5 M hydride composite after dehydriding at different temperatures (593 K, 613 K, and 638 K).

One of the possible reasons for the enhancement of dehydriding kinetics of $NaMgH_3$ hydride is because of the ball milling process with CM additives, which creates more defects, a refined grain size, and a distorted crystal structure. Such structural features and observed improvements have also been reported in the case of the reactive ball milling of magnesium hydride with carbon additives in hydrogen gas [10,11,41–43]. A synergetic effect may exist in the NH–2.5M–2.5G composite, where the presence of MWCNTs may hinder the restacking of GO; hence, improving the dehydriding kinetics. Bhatnagar et al. reported this synergetic effect in MgH_2–$NaAlH_4$ composite with the addition of 1.5 wt. % of graphene nanosheets and 0.5 wt. % of single wall carbon nanotube [45]. Further work is still needed to illustrate its possible mechanism.

4. Conclusions

$NaMgH_3$ perovskite hydride and NH-CM composites were prepared via the reactive ball-milling method under a H_2 atmosphere. MWCNTs and GO were used as a catalyst to improve the dehydriding kinetic properties of $NaMgH_3$ hydride. Dehydriding temperature and activation energy (ΔE) for two dehydrogenation steps of $NaMgH_3$ hydride can be greatly reduced with a 5 wt. % CM addition, especially the composite with a combined addition of 2.5 wt. % MWCNTs + 2.5 wt. % GO (NH–2.5M–2.5G). In comparison with $NaMgH_3$ hydride, the ΔE_1 and ΔE_2 of the NH–2.5M–2.5G composite are reduced by about 67 kJ/mol and 30 kJ/mol, respectively. The maximum amount of hydrogen desorbed is 3.64 wt. % at 638 K, and about 90% of the maximum amount was released within 20 min. This can be attributed to the synergetic effect between MWCNTs and GO, indicating that the combination of MWCNTs and GO is a better catalyst as compared to MWCNTs or GO alone.

Acknowledgments: This work was financially supported by the National Natural Science Foundation of China (51261003, 51471055) and the Guangxi Natural Science Foundation (2016GXNSFGA380001).

Author Contributions: Wang and Tao conceived and designed the study. Tao and Li performed the experiments. Deng provided the DSC analyses. Yao contributed significantly to XRD refinement and analysis; Wang wrote the manuscript. Zhou helped perform the analysis with constructive discussions. All authors read and approved the manuscript.

Conflicts of Interest: The authors declare no conflict of interest.

References

1. Bououdina, M.; Grant, D.; Walker, G. Review on hydrogen absorbing materials—Structure, microstructure, and thermodynamic properties. *Int. J. Hydrog. Energy* **2006**, *31*, 177–182. [CrossRef]

2. Pottmaier, D.; Pinatel, E.R.; Vitillo, J.G.; Garroni, S.; Orlova, M.; Baro, M.D.; Vaughan, G.B.M.; Fichtner, M.; Lohstroh, W.; Baricco, M. Structure and thermodynamic properties of the $NaMgH_3$ perovskite: A comprehensive study. *Chem. Mater.* **2011**, *23*, 2317–2326. [CrossRef]

3. Bouhadda, Y.; Fenineche, N.; Boudouma, Y. Hydrogen storage: Lattice dynamics of orthorhombic $NaMgH_3$. *Phys. B* **2011**, *406*, 1000–1003. [CrossRef]

4. Vajeeston, P.; Ravindran, P.; Kjekshus, A.; Fjellvag, H. First-principles investigations of the $MMgH_3$ (M = Li, Na, K, Rb, Cs) series. *J. Alloy. Compd.* **2008**, *450*, 327–337. [CrossRef]

5. Bouhadda, Y.; Boudouma, Y.; Fenineche, N.; Bentabet, A. Ab initio calculations study of the electronic, optical and thermodynamic properties of $NaMgH_3$, for hydrogen storage. *J. Phys. Chem. Solids* **2010**, *71*, 1264–1268. [CrossRef]

6. Wu, H.; Zhou, W.; Udovic, T.J.; Rush, J.J.; Yildirim, T. Crystal chemistry of perovskite-type hydride $NaMgH_3$: Implications for hydrogen storage. *Chem. Mater.* **2008**, *20*, 2335–2342. [CrossRef]

7. Bouhadda, Y.; Bououdina, M.; Fenineche, N.; Boudouma, Y. Elastic properties of perovskite-type hydride $NaMgH_3$ for hydrogen storage. *Int. J. Hydrog. Energy* **2013**, *38*, 1484–1489. [CrossRef]

8. Bouamrane, A.; Laval, J.P.; Soulie, J.P.; Bastide, J.P. Structural characterization of $NaMgH_2F$ and $NaMgH_3$. *Mater. Res. Bull.* **2000**, *35*, 545–549. [CrossRef]

9. Sheppard, D.A.; Paskevicius, M.; Buckley, C.E. Thermodynamics of hydrogen desorption from $NaMgH_3$ and its application as a solar heat storage medium. *Chem. Mater.* **2011**, *23*, 4298–4300. [CrossRef]

10. Ikeda, K.; Nakamori, Y.; Orimo, S. Formation ability of the perovskite-type structure in $Li_xNa_{1-x}MgH_3$ ($x = 0$, 0.5 and 1.0). *Acta Mater.* **2005**, *53*, 34–53. [CrossRef]

11. Martínez-Coronado, R.; Sánchez-Benítez, J.; Retuerto, M.; Fernández-Díaz, M.T.; Alonso, J.A. High-pressure synthesis of $Na_{1-x}Li_xMgH_3$ perovskite hydrides. *J. Alloy. Compd.* **2012**, *522*, 101–105. [CrossRef]

12. Wang, Z.; Li, J.; Tao, S.; Deng, J.; Zhou, H.; Yao, Q. Structure, thermal analysis and dehydriding kinetic properties of $Na_{1-x}Li_xMgH_3$ hydrides. *J. Alloy. Compd.* **2016**, *660*, 402–406. [CrossRef]

13. Tao, S.; Wang, Z.; Li, J.; Deng, J.; Zhou, H.; Yao, Q. Improved dehydriding properties of NaMgH$_3$ pervoskite hydride by addition of graphitic carbon nitride. *Mater. Sci. Forum* **2016**, *852*, 502–508. [CrossRef]

14. Li, Y.; Zhang, L.; Zhang, Q.; Fang, F.; Sun, D.; Li, K.; Wang, H.; Ouyang, L.; Zhu, M. In situ embedding of Mg$_2$NiH$_4$ and YH$_3$ nanoparticles into bimetallic hydride NaMgH$_3$ to inhibit phase segregation for enhanced hydrogen storage. *J. Phys. Chem. C* **2014**, *118*, 23635–23644. [CrossRef]

15. Chaudhary, A.; Paskevicius, M.; Sheppard, D.A.; Buckley, C.E. Thermodynamic destabilization of MgH$_2$ and NaMgH$_3$ using Group IV elements Si, Ge or Sn. *J. Alloy. Compd.* **2015**, *623*, 109–116. [CrossRef]

16. Webb, C.J. A review of catalyst-enhanced magnesium hydride as a hydrogen storage material. *J. Phys. Chem. Solids* **2014**, *84*, 96–106. [CrossRef]

17. Bogdanović, B.; Hartwig, T.H.; Spliethoff, B. The development, testing and optimization of energy storage materials based on the MgH$_2$-Mg system. *Int. J. Hydrog. Energy* **1993**, *18*, 575–589. [CrossRef]

18. Sakintuna, B.; Lamari-Darkrim, F.; Hirscher, M. Metal hydride materials for solid hydrogen storage: A review. *Int. J. Hydrog. Energy* **2007**, *32*, 1121–1140. [CrossRef]

19. Bogdanović, B.; Spliethoff, B. Active MgH$_2$-Mg-systems for hydrogen storage. *Int. J. Hydrog. Energy* **1987**, *12*, 863–873. [CrossRef]

20. Bobet, J.-L.; Akiba, E.; Nakamura, Y.; Darriet, B. Study of Mg-M (M = Co, Ni and Fe) mixture elaborated by reactive mechanical alloying—Hydrogen sorption properties. *Int. J. Hydrog. Energy* **2000**, *25*, 987–996. [CrossRef]

21. Reiser, A.; Bogdanović, B.; Schlichte, K. The application of Mg-based metal-hydrides as heat energy storage systems. *Int. J. Hydrog. Energy* **2000**, *25*, 425–430. [CrossRef]

22. Fakioğlu, E.; Yürüm, Y.; Veziroğlu, T.N. A review of hydrogen storage systems based on boron and its compounds. *Int. J. Hydrog. Energy* **2004**, *29*, 1371–1376. [CrossRef]

23. Alsabawi, K.; Webb, T.A.; Gray, E.M.; Webb, C.J. The effect of C$_{60}$ additive on magnesium hydride for hydrogen storage. *Int. J. Hydrog. Energy* **2015**, *40*, 10508–10515. [CrossRef]

24. Imamura, H.; Takesue, Y.; Akimoto, T.; Tabata, S. Hydrogen absorbing magnesium composites prepared by mechanicalgrinding with graphite: Effects of additives on composite structures and hydriding properties. *J. Alloy. Compd.* **1999**, *293–295*, 564–568. [CrossRef]

25. Imamura, H.; Tabata, S.; Shigetomi, N.; Takesue, Y.; Sakata, Y. Composites for hydrogen storage by mechanical grinding ofgraphite carbon and magnesium. *J. Alloy. Compd.* **2002**, *330–332*, 579–583. [CrossRef]

26. Thiangviriya, S.; Utke, R. Improvement of dehydrogenation kinetics of 2LiBH$_4$-MgH$_2$ composite by doping with activated carbon nanofibers. *Int. J. Hydrog. Energy* **2016**, *41*, 2797–2806. [CrossRef]

27. Kadri, A.; Jia, Y.; Chen, Z.; Yao, X. Catalytically enhanced hydrogen sorption in Mg-MgH$_2$ by coupling vanadium-based catalyst and carbon nanotubes. *Materials* **2015**, *8*, 3491–3507. [CrossRef]

28. Spassov, T.; Zlatanova, Z.; Spassova, M.; Todorova, S. Hydrogen sorption properties of ball-milled Mg-C nanocomposites. *Int. J. Hydrog. Energy* **2010**, *35*, 10396–10403. [CrossRef]

29. Rud, A.D.; Lakhnik, A.M. Effect of carbon allotropes on the structure and hydrogen sorption during reactive ball-milling of Mg-C powder mixtures. *Int. J. Hydrog. Energy* **2012**, *37*, 4179–4187. [CrossRef]

30. Huang, Z.G.; Guo, Z.P.; Calka, A.; Wexler, D.; Liu, H.K. Effects of carbon black, graphite and carbon nanotube additives on hydrogen storage properties of magnesium. *J. Alloy. Compd.* **2007**, *427*, 94–100. [CrossRef]

31. Mandzhukova, T.; Grigorova, E.; Khristov, M.; Tsyntsarski, B. Investigation of the effect of activated carbon and 3d-metalcontaining additives on the hydrogen sorption properties of magnesium. *Mater. Res. Bull.* **2011**, *46*, 1772–1776. [CrossRef]

32. Grigorova, E.; Mandzhukova, T.; Tsyntsarski, B.; Budinova, T.; Khristov, M.; Tzvetkov, P.; Petrova, B.; Petrov, N. Effect of activated carbons derived from different precursors on the hydrogen sorption properties of magnesium. *Fuel Process. Technol.* **2011**, *92*, 1963–1969. [CrossRef]

33. Wu, C.Z.; Wang, P.; Yao, X.; Liu, C.; Chen, D.M.; Lu, G.Q.; Cheng, H.M. Effect of carbon/noncarbon addition on hydrogen storage behaviors of magnesium hydride. *J. Alloy. Compd.* **2006**, *414*, 259–264. [CrossRef]

34. Wu, C.Z.; Wang, P.; Yao, X.; Liu, C.; Chen, D.M.; Lu, G.Q.; Cheng, H.M. Hydrogen storage properties of MgH$_2$/SWNT composite prepared by ball milling. *J. Alloy. Compd.* **2006**, *420*, 278–282. [CrossRef]

35. Skripnyuk, V.M.; Rabkin, E.; Bendersky, L.A.; Magrez, A.; Carreño-Morelli, E.; Estrin, Y. Hydrogen storage properties of as-synthesized and severely deformed magnesium multiwall carbon nanotubes composite. *Int. J. Hydrog. Energy* **2010**, *35*, 5471–5478. [CrossRef]

36. Schaller, R.; Mari, D.; dos Santos, S.M.; Tkalcec, I.; Carreño-Morelli, E. Investigation of hydrogen storage in carbonnanotube-magnesium matrix composites. *Mater. Sci. Eng. A* **2009**, *521–522*, 147–150. [CrossRef]

37. Singh, R.K.; Raghubanshi, H.; Pandey, S.K.; Srivastava, O.N. Effect of admixing different carbon structural variants on the decomposition and hydrogen sorption kinetics of magnesium hydride. *Int. J. Hydrog. Energy* **2010**, *35*, 4131–4137. [CrossRef]

38. Ranjbar, A.; Ismail, M.; Guo, Z.P.; Yu, X.B.; Liu, H.K. Effects of CNTs on the hydrogen storage properties of MgH_2 and MgH_2-BCC composite. *Int. J. Hydrog. Energy* **2010**, *35*, 7821–7826. [CrossRef]

39. Chen, B.-H.; Kuo, C.-H.; Ku, J.-R.; Yan, P.-S.; Huang, C.-J.; Jeng, M.-S.; Tsau, F.-H. Highly improved with hydrogen storage capacity and fast kinetics in Mg-based nanocomposites by CNTs. *J. Alloy. Compd.* **2013**, *568*, 78–83. [CrossRef]

40. Ikeda, K.; Kato, S.; Shinzato, Y.; Okuda, N.; Nakamori, Y.; Kitano, A.; Yukawa, H.; Morinaga, M.; Orimo, S. Thermodynamical stability and electronic structure of aperovskite-type hydride, $NaMgH_3$. *J. Alloy. Compd.* **2007**, *446–447*, 162–165. [CrossRef]

41. RÖnnebro, E.; Noréus, D.; Kadir, K.; Reiser, A.; Bogdanovic, B. Investigation of the perovskite related structures of $NaMgH_3$, $NaMgF_3$ and Na_3AlH_6. *J. Alloy. Compd.* **2000**, *299*, 101–106. [CrossRef]

42. Tao, S.; Wang, Z.; Deng, J.; Zhou, H.; Yao, Q. Improvement of dehydrogenation kinetics of $NaMgH_3$ hydride by introducing K_2TiF_6 as Dopant. *Int. J. Hydrog. Energy* **2016**, in press.

43. Reardon, H.; Mazur, N.; Gregory, D.H. Facile synthesis of nanosized sodium magnesium hydride, $NaMgH_3$. *Prog. Nat. Sci. Mater. Int.* **2013**, *23*, 343–350. [CrossRef]

44. Kissinger, H.E. Reaction kinetics in differential thermal analysis. *Anal. Chem.* **1957**, *29*, 1702–1706. [CrossRef]

45. Bhatnagar, A.; Pandey, S.K.; Dixit, V.; Shukla, V.; Shahi, R.R.; Shaza, M.A.; Srivastava, O. Catalytic effect of carbon nanostructures on the hydrogen storage properties of MgH_2–$NaAlH_4$ composite. *Int. J. Hydrog. Energy* **2014**, *39*, 14240–14246. [CrossRef]

15

The Effects of Cr and Al Addition on Transformation and Properties in Low-Carbon Bainitic Steels

Junyu Tian, Guang Xu *, Mingxing Zhou, Haijiang Hu and Xiangliang Wan

The State Key Laboratory of Refractories and Metallurgy, Hubei Collaborative Innovation Center for Advanced Steels, Wuhan University of Science and Technology, 947 Heping Avenue, Qingshan District, Wuhan 430081, China; 13164178028@163.com (J.T.); kdmingxing@163.com (M.Z.); hhjsunny@sina.com (H.H.); wanxiangliang@wust.edu.cn (X.W.)
* Correspondence: xuguang@wust.edu.cn

Academic Editor: Robert Tuttle

Abstract: Three low-carbon bainitic steels were designed to investigate the effects of Cr and Al addition on bainitic transformation, microstructures, and properties by metallographic method and dilatometry. The results show that compared with the base steel without Cr and Al addition, only Cr addition is effective for improving the strength of low-carbon bainitic steel by increasing the amount of bainite. However, compared with the base steel, combined addition of Cr and Al has no significant effect on bainitic transformation and properties. In Cr-bearing steel, Al addition accelerates initial bainitic transformation, but meanwhile reduces the final amount of bainitic transformation due to the formation of a high-temperature transformation product such as ferrite. Consequently, the composite strengthening effect of Cr and Al addition is not effective compared with individual addition of Cr in low-carbon bainitic steels. Therefore, in contrast to high-carbon steels, bainitic transformation in Cr-bearing low-carbon bainitic steels can be finished in a short time, and Al should not be added because Al addition would result in lower mechanical properties.

Keywords: bainitic transformation; microstructure; property; Cr; Al

1. Introduction

Low-carbon bainitic steels are commonly designed with alloying elements added to achieve favorable properties [1–11]. It is well known that the mechanical properties of bainitic steel are significantly influenced by the volume fraction of bainite, the amount of retained austenite (RA), the precipitation of cementite, among other factors [2]. The main purpose of the addition of alloying elements is to promote bainitic transformation and control the microstructures. For example, the addition of manganese, copper, and nickel, among others, are strong austenite stabilizers and can result in high stability of austenite and higher amounts of RA [3–5]. Silicon is used to restrain the formation of cementite during bainitic transformation [6]. In addition, a higher strength can be achieved due to precipitation hardening and grain refinement effects by adding vanadium, titanium, molybdenum, or niobium [7–11].

Chromium (Cr) and aluminium (Al) are also very important alloying elements, which are commonly added to low-carbon bainitic steels. Some researchers have investigated the effect of Al addition on bainitic transformation [5,12–17]. For example, Jimenez-Melero et al. [5] and Zhao et al. [12] reported that the addition of Al significantly increases the chemical driving force for the formation of bainitic ferrite plates and shortens the austenite-to-bainite transformation time. Similar results were also obtained by Garcia-Mateo et al. [13] and Hu et al. [14,15]. They claimed that the addition of Co and Al accelerates the bainitic transformation by increasing the free energy change accompanying the austenite-to-bainite ferrite transformation. Moreover, Monsalve et al. [16] and Meyer et al. [17] proved that Al promotes the formation of ferrite.

As to Cr, You et al. [18] and Chance et al. [19] studied the effect of Cr on continuous cooling transformation (CCT) diagrams, indicating that a single C-curve is separated by a bay of austenite stability due to the presence of Cr. They claimed that the addition of Cr may delay the bainitic transformation. Kong et al. [20] and Zhang et al. [21] investigated the influence of Cr on the transformation kinetics, demonstrating that the addition of Cr enhances the hardenability of metastable austenite. Moreover, Zhou et al. [22] investigated the effect of Cr on transformation and microstructure, and showed that Cr appreciably restrains ferrite transformation.

In summary, some investigations have been conducted on the effects of individual Cr or Al addition on microstructure and properties of bainitic steels. Cr and Al, as important alloying elements, are often compositely added to many bainitic steels [12–15]. However, few studies have reported on the composite effects of the combined addition of Cr and Al on the transformation, microstructure, and properties in low-carbon bainitic steels. Therefore, three kinds of low-carbon bainitic steels were designed in the present study to investigate the effects of Cr and Al addition on bainitic transformation, microstructure, and properties. Heat treatment experiments were performed on ThermecMaster-Z hot thermal–mechanical simulator followed by microstructure and property analyses, as well as quantitative characterization of bainitic transformation with dilation data. The purpose of the present study is to clarify the effects of the combined addition of Cr and Al on the transformation, microstructure, and properties in low-carbon bainitic steels. The results are useful for optimizing the composition design of Cr–Al alloying low-carbon steels.

2. Experimental Procedure

Three low-carbon bainitic steels with different chemical compositions were designed and their compositions are given in Table 1. Silicon (Si), manganese (Mn), and molybdenum (Mo), as important alloying elements, are often added in bainitic steels. The addition of Si prevents the formation of carbide, as carbide is detrimental to mechanical properties. The addition of Mn and Mo can enhance the hardenability of metastable austenite, which increases the bainite amount. Therefore, some Si, Mn, and Mo were added to the three steels. In addition, only Cr addition to steel B was used to study the effects of Cr on transformation and properties in the low-carbon bainitic steel, while the combined addition of Cr and Al in steel C was used to investigate the effects of Al and the composite addition of Cr and Al. The three steels were refined using a laboratory-scale 50 kg vacuum furnace. Cast ingots were heated at 1280 °C for 2 h and then hot-rolled to 12 mm thick plates in seven passes. The start and finish rolling temperatures were 1180 °C and 915 °C, respectively. After rolling the plates, they were cooled to 550 °C at a cooling rate of 20 °C/s followed by final air-cooling to room temperature.

Table 1. The chemical compositions of steels (wt %).

Steels	C	Si	Mn	Cr	Mo	Al	N	P	S
A (base)	0.218	1.831	2.021	/	0.227	/	<0.003	<0.006	<0.003
B (Cr)	0.221	1.792	1.983	1.002	0.229	/	<0.003	<0.006	<0.003
C (Cr + Al)	0.219	1.824	2.041	1.021	0.230	0.502	<0.003	<0.006	<0.003

Samples for dilatometric tests were machined to a cylinder of 8 mm diameter and 12 mm height. The top and bottom surfaces of samples were polished conventionally to keep the measurement surface level. The experiments were conducted according to the procedure shown in Figure 1 on a ThermecMaster-Z hot thermal–mechanical simulator equipped with a light-emitting diode (LED) dilatometer to quantitatively analyze the bainitic transformation of the three steels. The specimens were heated to 1000 °C at a rate of 10 °C/s and held for 900 s to achieve a homogeneous austenitic microstructure, followed by cooling to 350 °C at a rate of 10 °C/s. The austenitizing temperature of 1000 °C is larger than Ac_3 (the temperature at which transformation of ferrite to austenite is completed during heating). The small grain size can be obtained at a lower heating temperature. In addition, there are few inclusions and precipitates in steels, so the holding time of 900 s was chosen. The cooling rate of 10 °C/s refers to the cooling ability in the central area of thick plates in industrial production. After isothermal holding for 3600 s at 350 °C for bainite precipitation, the samples were air-cooled to room temperature.

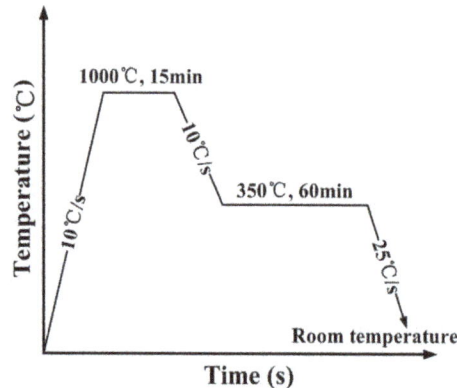

Figure 1. Experimental procedure.

According to the empirical Equations (1) in Ref. [23] and (2) in Ref. [24], the bainite starting temperature (B_S) and martensite starting temperature (M_S) of base steel are 471 °C and 336 °C, respectively. Therefore, the isothermal transformation temperature is designed to be 350 °C. It can be seen in Equation (2) that the Ms decreases by the addition of Cr, so the M_S temperature for steels B and C are smaller than 350 °C. It is known that finer bainite laths—and more of them—can be obtained at the lower phase transition temperature, which is beneficial for the mechanical properties of bainitic steel [11]. Moreover, martensite transformation will occur at the lower isothermal temperature (e.g., 300 °C). Bainite laths become coarse at the higher isothermal temperature (e.g., 400 °C), which is harmful to the mechanical properties of bainitic steel.

$$B_S \ (^{\circ}C) = 630 - 45Mn - 40V - 35Si - 30Cr - 25Mo - 20Ni - 15W \tag{1}$$

$$M_S \ (^{\circ}C) = 498.9 - 333.3C - 33.3Mn - 27.8Cr - 16.7Ni - 11.1(Si + Mo + W) \tag{2}$$

The time–temperature-transformation (TTT) curves of the three steels are plotted by MUCG83 software developed by Bhadeshia at Cambridge University (Figure 2). Chromium can enhance the stability and hardenability of austenite, so that the TTT curve moves to bottom-right with Cr addition, indicating that it is easy to obtain bainite at a certain cooling rate. In addition, higher temperature transformation may occur by Al addition, which is harmful to bainitic transformation.

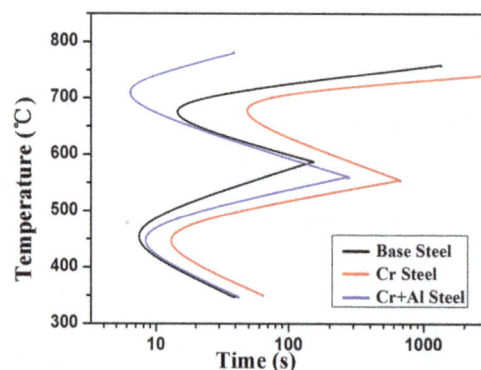

Figure 2. The time–temperature-transformation (TTT) curves of three steels.

Additionally, in order to investigate the properties of the tested steels, $140 \times 20 \times 10$ mm blocks were cut from hot-rolled sheets and heat-treated using the same procedure shown in Figure 1. The specimens were mechanically polished and etched with a 4% nital solution for microstructure examination. Both bainite morphology and fracture surfaces were examined using a Nova 400 Nano field-emission scanning electron microscope (FE-SEM) operated at an accelerating voltage of 20 kV. The volume fraction of bainite in the three specimens was calculated by Image-Pro Plus software, and tensile tests were carried out on UTM-4503 electronic universal tensile tester at room temperature. Tensile specimens were prepared according to the ASTM standard and the beam displacement rate was 1 mm/min. Four tensile tests were conducted for each kind of tested steel and the corresponding average values were calculated in this work. In order to determine the volume fraction of RA, X-ray diffraction (XRD) experiments were conducted on BRUKER D8 ADVANCE diffractometer, using unfiltered Cu Kα radiation and operating at 40 kV and 40 mA.

3. Results

3.1. Microstructure

Figure 3 presents the microstructures of the three steels before heating treatment. It can be observed that the microstructures of the three steels mainly consist of lath-like bainite. The grain sizes of prior-austenite before heating treatment are measured by Image-Pro Plus software to be 34.6 ± 8.5 μm, 31.5 ± 8.5 μm, and 39.2 ± 8.5 μm for steels A, B, and C, respectively. In addition, few inclusions and precipitates are observed in the three steels due to very small amounts of N, P, and S (Table 1 and Figure 3). At the same time, the same processing routes are utilized for all three steels. Therefore, the influences of inclusions and precipitates in the three steels are small and similar.

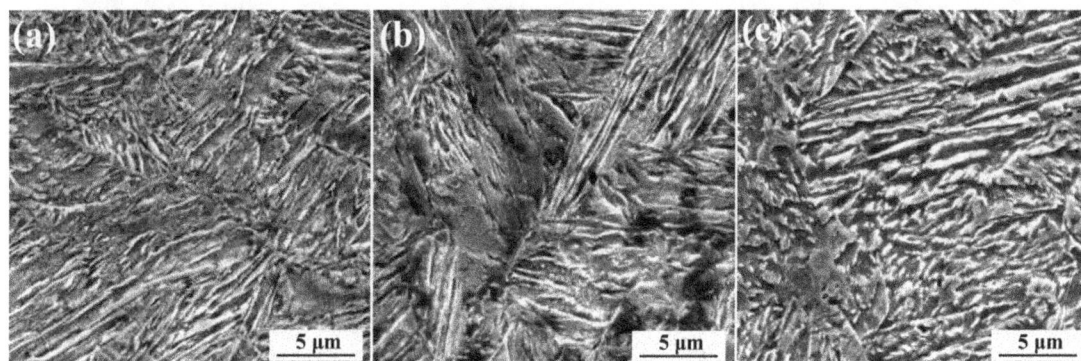

Figure 3. SEM micrographs of three low-carbon bainitic steels before heating treatment: (**a**) base, (**b**) Cr addition, and (**c**) Cr and Al addition.

Figure 4 shows the typical SEM microstructures of the three steels after isothermal holding for 3600 s at 350 °C following austenization at 1000 °C for 900 s. The classification method proposed in Ref. [11,25] is used in the present work to identify the microstructure in low-carbon bainitic steels: the microstructure is classified as ferrite (F), bainite ferrite (BF), and martensite (M). It can be observed that the microstructures of the three specimens mainly consist of lath-like BF and martensite/austenite (M/A) islands as shown in Figure 4. Prior-austenite grain boundaries (AGB) are well defined, as shown by arrows in Figure 4b. The original austenite grain size of the three tested steels, which influences the bainite morphology [11], was calculated by Image-Pro Plus software (Table 2). The prior-austenite grain sizes of the three steels have no significant difference. In addition, some ferrite is observed in steel A (base steel) and steel C (Cr + Al steel), as marked by the arrow in Figure 4c, but there is no ferrite in steel B (Figure 4b). According to the micrographs, the volume fraction of bainite was calculated by Image-Pro Plus software using the method in Ref. [11,26]. Further, the volume fractions of RA were calculated based on XRD results using the method in Ref. [4,27]. The results are shown in Table 3.

It reveals that the sample with only Cr addition has the largest amount of bainite, while the base steel without Cr and Al addition has the smallest percentage of bainite. It is clear that, compared with base steel, Cr addition obviously increases the amount of bainite. However, the combined addition of Cr and Al decreases the bainite amount to the level of base steel, indicating that Al addition in Cr-bearing low-carbon bainitic steel obviously retards the bainitic transformation. It should be pointed out that the M_S of base steel is calculated to be 374 °C and 386 °C by empirical Equation (3) in Ref. [28] and (4) in Ref. [29], respectively. However, the microstructures of the three steels after isothermal transformation at 350 °C mainly consist of bainite rather than martensite (Figure 4). It indicates that the empirical Equations (3) and (4) for M_S are not suitable for the three tested steels in the present study. The calculated results of B_S and M_S with different equations are not the same. The M_S temperatures of the base steel calculated by some equations are higher than 350 °C and others are lower than 350 °C, indicating that the different equations were obtained with different steel grades and experimental conditions. Moreover, the B_S and M_S in these equations may be corresponding to equilibrium conditions. The B_S and M_S are also affected by cooling rate. Therefore, B_S and M_S calculated by theoretical equations can only be used as theoretical reference. The real B_S and M_S at a certain cooling rate for a steel grade should be measured by experiments.

$$M_S\ (^\circ C) = 537.8 - 361.1C - 38.9(Mn + Cr) - 19.4Ni - 27.8Mo \tag{3}$$

$$M_S\ (^\circ C) = 561.1 - 473.9C - 33Mn - 16.7(Ni + Cr) - 21.1Mo \tag{4}$$

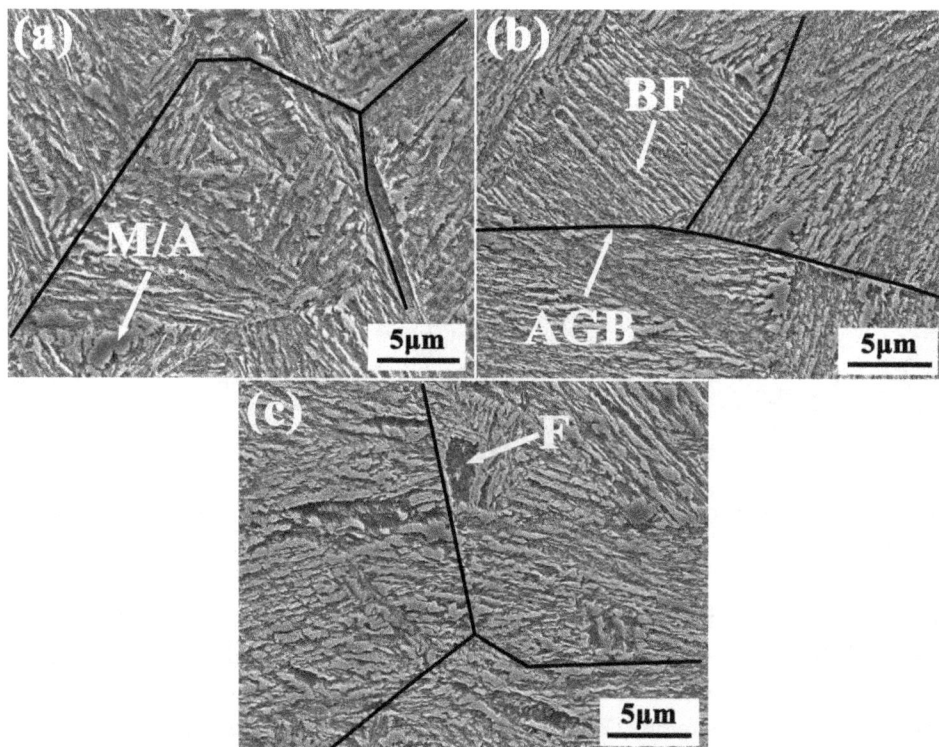

Figure 4. SEM micrographs of three low-carbon bainitic steels after isothermal holding at 350 °C for 3600 s: (a) base, (b) Cr addition, and (c) Cr and Al addition.

Table 2. Prior-austenite grain sizes of three steels (μm).

	Base	Cr	Cr + Al
Prior-austenite grain size	30.8 ± 9.4	29.4 ± 9.1	32.6 ± 8.5

Table 3. The volume fractions of bainite ferrite (BF) and retained austenite (RA) in three steels.

Steels	$V_{(BF)}$ (%)	$V_{(RA)}$ (%)
A (base)	45.6	3.5
B (Cr)	68.4	11.5
C (Cr + Al)	48.6	10.7

3.2. Mechanical Properties

The engineering strain–stress curves of the three tested steels are presented in Figure 5, and the corresponding mechanical properties are given in Table 4. Both the strength and the elongation are improved by only Cr addition, while the combined addition of Cr and Al has no significant effect on strength and elongation. Compared with steel A (base steel), the ultimate tensile strength (UTS) of Cr addition steel (steel B) increases by 135 MPa, while the UTS and yield strength (YS) increments of Cr and Al additional steel (steel C) are only 21 MPa and 12 MPa, respectively. Additionally, the strength and total elongation (TE) of steel C (Cr + Al) is unexpectedly smaller than that of steel B (only Cr), suggesting that no further improvement of mechanical properties occurs by Al addition in Cr-bearing bainitic steel.

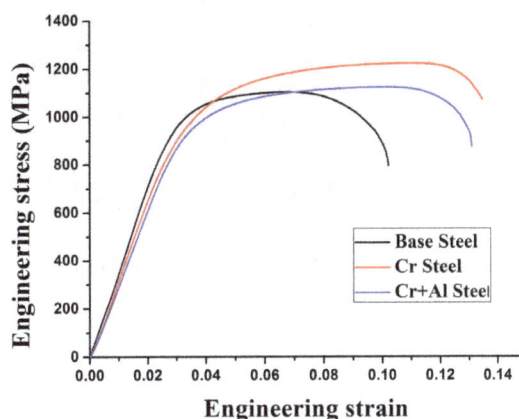

Figure 5. Engineering strain–stress curves of three tested steels with different compositions.

Table 4. Mechanical Properties of steels with different compositions.

Steels	UTS (MPa)	YS (MPa)	TE (%)	UTS × TE (GPa%)
A (base)	1103 ± 18	867 ± 22	10.2 ± 0.4	11.25 ± 0.007
B (Cr)	1238 ± 21	889 ± 18	13.1 ± 0.8	16.22 ± 0.017
C (Cr + Al)	1124 ± 15	873 ± 16	12.8 ± 0.5	14.39 ± 0.008

UTS: ultimate tensile strength; YS: yield strength; TE: total elongation.

Figure 6 displays the tensile fracture morphologies of the three steels. The mix of quasi-cleavage and dimples is exhibited in steel A (base steel), as shown by arrows in Figure 6a. A small number of quasi-cleavage fractures with a river pattern are observed in steel A, without Cr and Al addition (Figure 6a), indicating a portion of brittle fracture [19]. Nevertheless, this kind of brittle fracture microstructure rarely appears in steels B (only Cr) and C (Cr + Al). Additionally, it is observed that the diameters of dimples in the fracture of the steels with only Cr addition and combined addition of Cr and Al are larger than that of steel A without Cr and Al addition. This means that the toughness of steel is improved by Cr addition. The result is consistent with the mechanical properties of steels listed in Table 4.

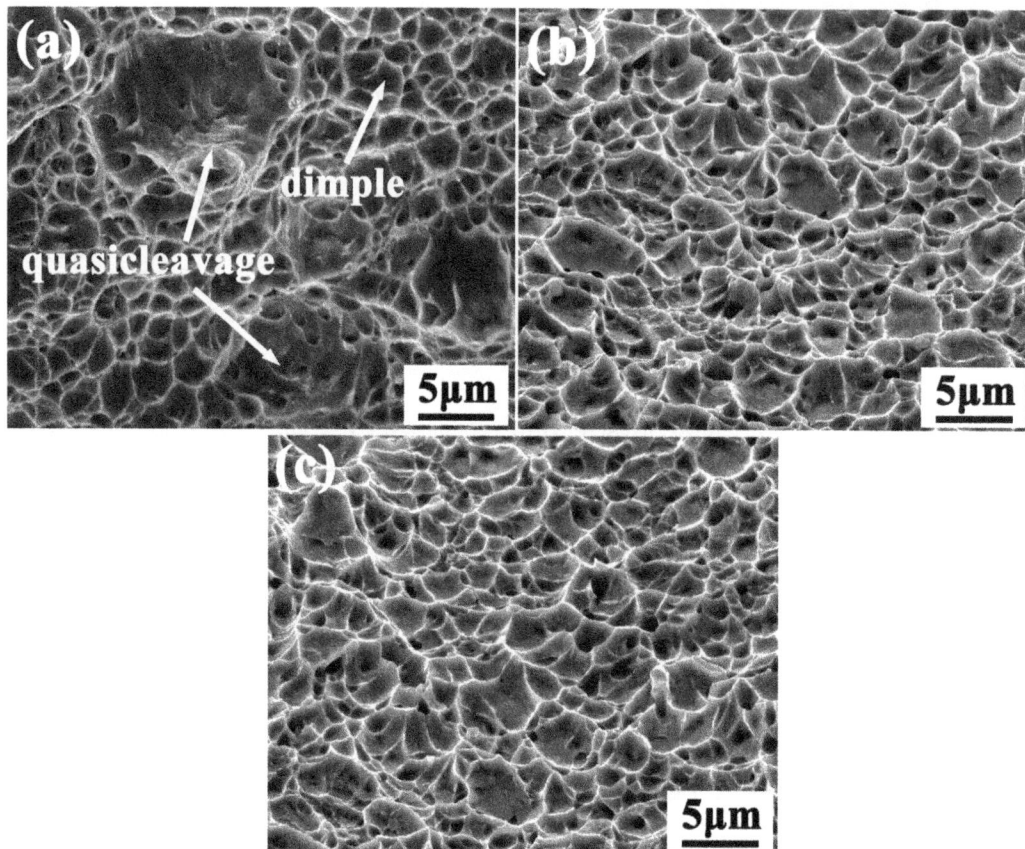

Figure 6. Fractographs of three tested steels with different compositions: (**a**) base, (**b**) Cr addition, and (**c**) Cr and Al addition.

3.3. Thermal Dilatometry

In order to quantitatively and accurately investigate the influence of combined Cr and Al addition on bainite transformation, dilatometric experiments were conducted on the thermo–mechanical simulator. According to the recorded dilatometric data, dilatation curves of the three steels are plotted. Figure 7 shows the recorded dilatation curves and transformation rates of the three steels during isothermal holding at 350 °C. Figure 7a shows dilatations as a function of holding time during isothermal holding at 350 °C, where the beginning of isothermal holding was selected as the zero point of abscissa and ordinate axes. The transformation temperature was constant and no extra force was applied on the sample during isothermal holding, thus the dilatation in Figure 7a represents the real bainite transformation amount. It can be observed that compared with the base steel, the final amount of bainite obviously increases with the addition of Cr, but it merely slightly rises with the combined addition of Cr and Al. Moreover, the amount of bainite transformation of steel C (Cr + Al steel) is obviously smaller than that of steel B (Cr steel), indicating that the addition of Al in Cr-bearing bainitic steel has a negative effect on the amount of bainite.

In addition, dilatation rates versus holding time during isothermal holding at 350 °C are given in Figure 7b. It shows that bainite transformation in steel C with combined Cr + Al addition and steel A without Cr and Al is completed prior to steel B with individual Cr addition. Although the maximum amount of bainite appears in steel B (Cr steel), the fastest transformation rate shows up in steel C (Cr + Al steel). It indicates that the addition of Al accelerates initial bainitic transformation, while the addition of Cr delays bainitic transformation. Besides, it is noticed that there is no distinguishable difference in the transformation processes between steel A (base steel) and steel C (Cr + Al steel). The time consumed to complete bainitic transformation is basically equal for the two steels, indicating that the combined Cr and Al addition has an ignorable effect on bainitic transformation rate compared with the base steel. Moreover, bainitic transformation completes quickly in the three low-carbon steels in the present study. This is contrast to high-carbon steels in which several hours or days are needed to finish bainitic transformation [13,15].

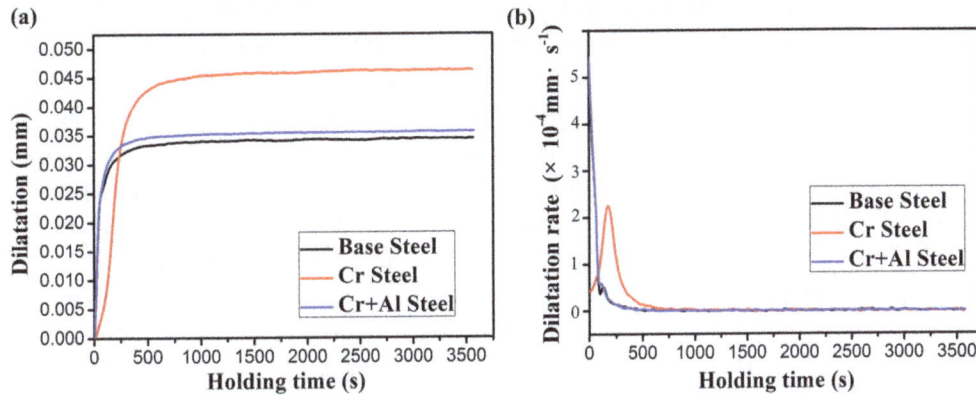

Figure 7. (a) Dilation curves and (b) transformation rates of three steels recorded by dilatometer on thermal simulator during bainitic transformation at 350 °C.

4. Discussions

4.1. Influence of Cr Addition

SEM micrographs (Figure 4) show that steel A mainly consists of bainite sheaves, a small amount of ferrite, and M/A, while no ferrite exists in steel B. As reported by some researches [30–32], bainitic transformation is characterized by incomplete reaction, thus some untransformed austenite after isothermal holding could transform into martensite. The results of dilatometric test (Figure 7) and microstructural determination (Table 3) both indicate that the bainite amount obviously increases with about 1% Cr addition. For bainitic steels, more bainite amount and RA fractions can improve mechanical properties [3,33,34]. In the present work, compared with steel A, the volume fractions of bainite and RA in steel B increase by 22.8% and 8%, respectively, resulting in an increment of 135 MPa in UTS and about 3% in TE. It can also be explained from the viewpoint of wetting of grain boundaries. It can be observed that the M/A particles with lath morphology clearly wet the austenite grain boundary (AGB) as shown in the bottom-right in Figure 4b, while they wet few AGBs in Figure 4a,c. It was reported by Straumal et al. [35] that the transition from incomplete to complete surface wetting is a phase transformation. It indicates that more M/A particles distribute in steel B (Cr steel) than steel A (base steel) and steel C (Cr + Al steel). The increased surface area contact of martensite particles with ferrite facilitates stress transfer from ductile to hard phase, which contributes to its high strength [36]. It demonstrates that a small amount of Cr addition in low-carbon bainitic steel improves strength of steel with a better total elongation. The existence of ferrite in steel A indicates that high-temperature phase transition happens during the cooling process before bainite transformation. However, with the Cr addition, the ferrite transformation is avoided, resulting in an increased bainite fraction. It is reported that Cr causes a separation of the bainite C-curve and extends the bainite formation field [18,19]. Similar results can be obtained from Equations (1) and (2). The addition of Cr decreases the B_S and M_S, which contributes to the greater amount of bainite. Moreover, Cr addition enhances the hardenability of metastable austenite [22] and increases the stability of austenite to ferrite [37]. Therefore, more undercooled austenite can transform into lath-like BF in the Cr addition steel, which leads to the improvement of the mechanical properties of steel B.

4.2. Influence of Al Addition

Figure 7 shows that bainite transformation is accelerated by Al addition, which is consistent with the results by Hu et al. [14,15], Caballero et al. [37] and others. SEM micrographs (Figure 4) indicate that steel B (Cr steel) mainly consists of bainite sheaves without ferrite, while a small amount of ferrite presents in steel C (Cr +Al steel), showing that Al addition promotes the formation of ferrite. According to the dilation diagram (Figure 7), the total dilatation of steel C (Cr + Al steel) is only 0.0368 mm, reduced by 19.5% compared to steel B (Cr steel, 0.0457 mm), which is consistent with the result listed in Table 3. It indicates that 0.5% Al addition in Cr-bearing steel obviously reduces the final bainite amount. In addition, the product of tensile strength and total elongation for steel C (Cr + Al steel) slightly decreases from 16.22 GPa% for steel B (Cr steel) to 14.39 GPa%. Fonstein et al. [38] and Meyer et al. [17] reported that Al addition can promote the formation of a high-temperature transformation product such as ferrite. Therefore, the amount of bainite transformation decreases because of less supercooled austenite. On the other hand, the results by Caballero and Bhadeshia [37] indicate that Al addition has no significant effect on final bainite amount. This depends on whether the ferrite transformation occurs. There is no ferrite transformation in their study, while ferrite appears in the present study, which leads to the decrease of bainite amount. Therefore, the mechanical properties of steel C (Cr + Al steel) are slightly smaller compared to steel B (Cr steel).

4.3. Influence of Composited Addition of Cr and Al

It can be observed from Table 3 that the volume fractions of bainite in steels A and C are 45.6% and 48.6%, respectively. Also, from the dilation curves, the dilatation of steel C (Cr and Al) is 0.0368 mm, increased by 0.0022 mm compared to the 0.0346 mm expansion of base steel, demonstrating that the effect of combined addition of Cr and Al on bainite transformation is very small. In addition, the UTS of steel C (Cr and Al) increases by only 21 MPa compared with base steel, indicating that the strengthening effect of combined Cr and Al addition has no significant improvement over that of steel A (base steel). As mentioned previously, although the only Cr addition increases the final bainite transformation amount, Al addition in Cr-bearing low-carbon steel reduces the amount of bainite transformation because of the formation of high-temperature transformation product (F). This means that the addition of Al weakens the promotion function of Cr on bainite transformation. Therefore, the composite strengthening effect of Cr and Al addition has no significant improvement over that of base steel. Al addition is very effective at shortening the bainitic transformation time in high-carbon bainitic steel [13,37]. However, for the low-carbon bainitic steels, bainitic transformation can finish in a short time even without Al addition (Figure 7), and Al should not be added in Cr-bearing low-carbon bainitic steels because Al addition would result in lower mechanical properties.

5. Conclusions

Three low-carbon bainitic steels were designed in the present work. Metallographic method and dilatometry were used to investigate the effects of Cr and Al addition on bainitic transformation, microstructures, and properties. The results show that the individual addition of Cr is effective for improving the strength of low-carbon bainite steel by increasing the amount of bainite transformation. In addition, in Cr-bearing low-carbon steel, Al addition leads to lower mechanical properties due to decreased amount of bainite transformation, although the addition of Al accelerates the initial bainitic transformation and shortens the austenite-to-bainite transformation time. Moreover, the addition of Al can effectively shorten the bainitic transformation time in high-carbon bainitic steels. However, bainitic transformation in Cr-bearing low-carbon bainitic steels can finish in a short time. Therefore, it is not necessary to add Al for the acceleration of bainitic transformation because Al addition would result in lower mechanical properties.

Acknowledgments: The authors gratefully acknowledge the financial supports from National Natural Science Foundation of China (NSFC) (No. 51274154), the National High Technology Research and Development Program of China (No. 2012AA03A504). State Key Laboratory of Development and Application Technology of Automotive Steels (Baosteel Group).

Author Contributions: Guang Xu, supervisor, conceived and designed the experiments; Junyu Tian conducted experiments, analyzed the data and wrote the paper; Mingxing Zhou, conducted experiments; Haijiang Hu conducted experiments; Xiangliang Wan, conducted experiments. All authors participated in the discussion of experimental results.

Conflicts of Interest: The authors declare no conflict of interest. The founding sponsors had no role in the design of the study; in the collection, analyses, or interpretation of data; in the writing of the manuscript, and in the decision to publish the results.

References

1. Zhou, M.X.; Xu, G.; Wang, L.; Yuan, Q. The varying effects of uniaxial compressive stress on the bainitic transformation under different austenitization temperatures. *Metals* **2016**, *6*, 119. [CrossRef]

2. He, J.G.; Zhao, A.M.; Yao, H.; Zhi, C.; Zhao, F.Q. Effect of ausforming temperature on bainite transformation of high carbon low alloy steel. *Mater. Sci. Forum.* **2015**, *817*, 454–459. [CrossRef]

3. Hu, H.J.; Xu, G.; Wang, L.; Zhou, M.X.; Xue, Z.L. Effect of ausforming on the stability of retained austenite in a C-Mn-Si bainitic steel. *Met. Mater. Int.* **2015**, *21*, 929–935. [CrossRef]

4. Zhou, M.X.; Xu, G.; Wang, L.; He, B. Effects of austenitization temperature and compressive stress during bainitic transformation on the stability of retained austenite. *T. Indian. I. Metals.* **2016**, 1–7. [CrossRef]

5. Jimenez-Melero, E.; Dijk, N.H.V.; Zhao, L.; Sietsma, J.; Offerman, S.E.; Wright, J.P. The effect of aluminium and phosphorus on the stability of individual austenite grains in trip steels. *Acta Mater.* **2009**, *57*, 533–543. [CrossRef]

6. Girault, E.; Mertens, A.; Jacques, P.; Houbaert, Y.; Verlinden, B.; Humbeeck, J.V. Comparison of the effects of silicon and aluminium on the tensile behaviour of multiphase trip-assisted steels. *Scripta Mater.* **2001**, *44*, 885–892. [CrossRef]

7. Heller, T.; Nuss, A. Effect of alloying elements on microstructure and mechanical properties of hot rolled multiphase steels. *Ironmak. Steelmak.* **2005**, *32*, 303–308. [CrossRef]

8. Pereloma, E.V.; Timokhina, I.B.; Russell, K.F.; Miller, M.K. Characterization of clusters and ultrafine precipitates in Nb-containing C–Mn–Si steels. *Scripta Mater.* **2006**, *54*, 471–476. [CrossRef]

9. Shi, W.; Li, L.; Yang, C.X.; Fu, R.Y.; Wang, L.; Wollants, P. Strain-induced transformation of retained austenite in low-carbon low-silicon trip steel containing aluminum and vanadium. *Mater. Sci. Eng. A* **2006**, *429*, 247–251. [CrossRef]

10. Wang, X.D.; Huang, B.X.; Wang, L.; Rong, Y.H. Microstructure and mechanical properties of microalloyed high-strength transformation-induced plasticity steels. *Metall. Mater. Trans. A* **2008**, *39*, 1–7. [CrossRef]

11. Hu, H.J.; Xu, G.; Wang, L.; Xue, Z.L.; Zhang, Y.L.; Liu, G.H. The effects of Nb and Mo addition on transformation and properties in low carbon bainitic steels. *Mater. Des.* **2015**, *84*, 95–99. [CrossRef]

12. Zhao, J.; Wang, T.S.; Lv, B.; Zhang, F.C. Microstructures and mechanical properties of a modified high-C–Cr bearing steel with nano-scaled bainite. *Mater. Sci. Eng. A* **2015**, *628*, 327–331. [CrossRef]

13. Garcia-Mateo, C.; Caballero, F.G.; Bhadeshia, H.K.D.H. Acceleration of low-temperature bainite. *ISIJ Int.* **2003**, *43*, 285–288. [CrossRef]

14. Hu, F.; Wu, K.M.; Zheng, H. Influence of Co and Al on pearlitic transformation in super bainitic steels. *Ironmak. Steelmak.* **2012**, *39*, 535–539. [CrossRef]

15. Hu, F.; Wu, K.M.; Zheng, H. Influence of Co and Al on bainitic transformation in super bainitic steels. *Steel Res. Int.* **2013**, *84*, 1060–1065. [CrossRef]

16. Monsalve, A.; Guzmán, A.; Barbieri, F.D.; Artigas, A.; Carvajal, L.; Bustos, O. Mechanical and microstructural characterization of an aluminum bearing trip steel. *Metall. Mater. Trans. A* **2016**, *47*, 3088–3094. [CrossRef]

17. Meyer, M.D.; Mahieu, J.; Cooman, B.C.D. Empirical microstructure prediction method for combined intercritical annealing and bainitic transformation of trip steel. *Mater. Sci. Technol.* **2002**, *18*, 1121–1132. [CrossRef]

18. You, W.; Xu, W.H.; Liu, Y.X.; Bai, B.Z.; Fang, H.S. Effect of chromium on CCT diagrams of novel air-cooled bainite steels analyzed by neural network. *J. Iron. Steel Res. Int.* **2007**, *14*, 39–42.

19. Chance, J.; Ridley, N. Chromium partitioning during isothermal transformation of a eutectoid steel. *Metall. Mater. Trans. A* **1981**, *12*, 1205–1213. [CrossRef]

20. Kong, L.; Liu, Y.; Liu, J.; Song, Y.; Li, S.; Zhang, R. The influence of chromium on the pearlite-austenite transformation kinetics of the Fe–Cr–C ternary steels. *J. Alloy. Compd.* **2015**, *648*, 494–499. [CrossRef]

21. Zhang, G.H.; Chae, J.Y.; Kim, K.H.; Dong, W.S. Effects of Mn, Si and Cr addition on the dissolution and coarsening of pearlitic cementite during intercritical austenitization in Fe-1mass%C alloy. *Mater. Charact.* **2013**, *81*, 56–67. [CrossRef]

22. Zhou, L.Y.; Liu, Y.Z.; Yuan, F.; Huang, Q.W.; Song, R.B. Effect of Cr on transformation of ferrite and bainite dual phase steels. *J. Iron Steel Res. Int.* **2009**, *21*, 37–41.

23. Zhao, Z.; Cheng, L.; Liu, Y.; Northwood, D.O. A new empirical formula for the bainite upper temperature limit of steel. *J. Mater. Sci.* **2001**, *36*, 5045–5056. [CrossRef]

24. Rowland, E.S.; Lyle, S.R. The application of M_S points to case depth measurement. *ASM Trans.* **1946**, *37*, 26–47.

25. Xiao, F.; Liao, B.; Ren, D.; Shan, Y.; Yang, K. Acicular ferritic microstructure of a low-carbon Mn–Mo–Nb microalloyed pipeline steel. *Mater. Charact.* **2005**, *54*, 305–314. [CrossRef]

26. Zhou, M.X.; Xu, G.; Wang, L.; Hu, H.J. Combined effect of the prior deformation and applied stress on the bainite transformation. *Met. Mater. Int.* **2016**, *22*, 956–961. [CrossRef]

27. Lindström, A. Austempered High Silicon Steel: Investigation of Wear Resistance in A Carbide Free Microstructure. Master's Thesis, Luleå Tekniska Universitet, Sweden, 2006.

28. Grange, R.A.; Stewart, H.M. The temperature range of martensite formation. *Trans. AIME* **1946**, *167*, 467–472.

29. Steven, W.; Haynes, A.G. The temperature formation of martensite and bainite in low-alloy steels. *J. Iron Steel Inst.* **1956**, *183*, 349–359.

30. Wang, X.; Zurob, H.S.; Xu, G.; Ye, Q.; Bouaziz, O.; Embury, D. Influence of microstructural length scale on the strength and annealing behavior of pearlite, bainite, and martensite. *Metall. Mater. Trans. A* **2013**, *44*, 1454–1461. [CrossRef]

31. Wang, X.L.; Wu, K.M.; Hu, F.; Yu, L.; Wan, X.L. Multi-step isothermal bainitic transformation in medium-carbon steel. *Scripta Mater.* **2014**, *74*, 56–59. [CrossRef]

32. Cornide, J.; Garcia-Mateo, C.; Capdevila, C.; Caballero, F.G. An assessment of the contributing factors to the nanoscale structural refinement of advanced bainitic steels. *J. Alloy. Compd.* **2012**, *577*, S43–S47. [CrossRef]

33. Hu, H.J.; Xu, G.; Zhou, M.X.; Yuan, Q. Effect of Mo content on microstructure and property of low-carbon bainitic steels. *Metals* **2016**, *6*, 173. [CrossRef]

34. Shi, J.; Sun, X.; Wang, M.; Hui, W.; Dong, H.; Cao, W. Enhanced work-hardening behavior and mechanical properties in ultrafine-grained steels with large-fractioned metastable austenite. *Scripta Mater.* **2010**, *63*, 815–818. [CrossRef]

35. Straumal, B.B.; Baretzky, B.; Kogtenkova, O.A.; Straumal, A.B.; Sidorenko, A.S. Wetting of grain boundaries in Al by the solid Al_3Mg_2 phase. *J. Mater. Sci.* **2010**, *45*, 2057–2061. [CrossRef]

36. Ahmad, E.; Manzoor, T.; Ziai, M.M.A.; Hussain, N. Effect of martensite morphology on tensile deformation of dual-phase steel. *J. Mater. Eng. Perform.* **2012**, *21*, 1–6. [CrossRef]

37. Caballero, F.G.; Bhadeshia, H.K.D.H. Very strong bainite. *Curr. Opin. Solid State Mater. Sci.* **2004**, *8*, 251–257. [CrossRef]

38. Fonstein, N.; Yakubovsky, O.; Bhattacharya, D.; Siciliano, F. Effect of niobium on the phase transformation behavior of aluminum containing steels for trip products. *Mater. Sci. Forum.* **2005**, *500–501*, 453–460. [CrossRef]

16

Microstructural and Mechanical Characteristics of Novel 6% Cr Cold-Work Tool Steel

Singon Kang [1], Minwook Kim [2] and Seok-Jae Lee [2],*

[1] Advanced Steel Processing and Products Research Center, Department of Metallurgical and Materials Engineering, Colorado School of Mines, Golden, CO 80401, USA; sikang@mines.edu
[2] Division of Advanced Materials Engineering, Research Center for Advanced Materials Development, Chonbuk National University, Jeonju 561-756, Korea; kmw0717@jbnu.ac.kr
* Correspondence: seokjaelee@jbnu.ac.kr

Academic Editor: Klaus-Dieter Liss

Abstract: We investigated a new cold-work tool steel with a low Cr content of 6 wt. % which was designed based on thermodynamic calculation to minimize the formation of primary carbide. A smaller particle size and a smaller volume fraction of carbides were observed in this 6% Cr steel. Superior mechanical properties in terms of hardness, impact toughness, tensile strength, and total elongation were achieved in this steel, due to fine secondary carbides precipitated during tempering. These carbide particles were M_6C and $(Mo,V)C$ carbides with a diameter below 100 nm.

Keywords: cold-work tool steel; alloy design; carbide formation; tempering

1. Introduction

The recent trend in the automotive industry is to enhance the use of advanced high-strength steel (AHSS), which reduces the weight of vehicles and improves fuel efficiency. Simultaneously, the use of high-strength tool steels is required for various metal working processes, e.g., press forming or drawing. For an effective metal working process, basically, the strength of tool steels should be higher than that of AHSS which is fabricated using the tool steels. This means that the development of tool steels tends to take precedence over the development of AHSS as there is a huge demand for AHSS with a strength higher than 1 GPa.

The tool steels mainly consist of high C, high Cr, and other carbide-forming elements such as Mo, V, and W [1–3]. The mechanical properties of the tool steels can be influenced by altering the tempering condition, which is related to the type and amount of carbides formed during tempering. For instance, a tempered STD11 steel, widely used as a representative of cold-work tool steel, has a high hardness over 58 HRC and a good wear resistance due its high carbide content in the tempered martensite matrix. However, some transition carbides can act as crack initiation sites, which lower the toughness of the product [4,5]. It has been reported that the primary carbide formed and coarsened can be elongated during hot deformation, which causes vacancy formation and crack initiation due to the anisotropy of the elongated shape of the carbide [6].

Recently, the development trend of the cold-work tool steels is to reduce the amounts of C and Cr, to decrease the volume fraction and size of the primary carbides [7–11]. These steels with a lower Cr content have better mechanical properties, because the formation of primary carbide is suppressed, compared to STD11 steel. Kim et al. developed a cold-work tool steel containing 8 wt. % Cr, which was designed based on thermodynamic calculations to avoid the primary carbide [11]. However, they observed some amounts of primary carbide in the proposed 8% Cr steel specimen.

Therefore, in the present work, we developed a new cold-work tool steel with a Cr content of 6 wt. %. The reduced Cr content was designed to inhibit the formation of the primary carbide as much as possible. The expected strength decrease due to the reduction of Cr content was complemented by increasing the amounts of carbide-forming elements, such as Mo, V, and W. The mechanical properties and microstructural features of the developed 6% Cr cold-work steel were evaluated in comparison with those of STD11 steel after a conventional quenching and tempering treatment.

2. Materials and Methods

The chemical composition of the steels used in this study is shown in Table 1. The base contents of C and Cr in the newly designed steel alloy were 0.9 wt. % and 6.0 wt. %, respectively (hereinafter referred to as "Cr6C9" steel). The commercial STD11 steel was used to evaluate the Cr6C9 steel. Then 30 kg of the Cr6C9 steel ingot was fabricated by vacuum induction melting (VIM). The ingot was homogenized at 1150 °C for 1 h and hot-forged to remove the cast structure. The forged plate was annealed at 870 °C for 4 h, followed by furnace cooling for spheroidization. Charpy U-notched specimens and ASTM E8 sub-size tensile specimens with the gauge length of 25.4 mm and the thickness of 1.5 mm were prepared from the spheroidized plate. The machined specimens were austenitized at 1030 °C for 30 min in a vacuum tube furnace under the vacuum condition and then air-cooled to room temperature. The tempering treatment was carried out at 520 °C for 2 h twice, indicating 4 h tempering in total. The microstructure of the tempered specimens was observed under an optical microscope (OM, Leica Co., DM ILM, Wetzlar, Germany) and a field emission emitter scanning electron microscope (FE-SEM, Hitachi Co., SU-70, Tokyo, Japan). The samples were mechanically polished to 1 μm and etched in the Vilella etchant (0.6 g picric acid + 3 mL HCl + 97 mL ethanol). Carbon replica samples were prepared to observe small carbides below 500 nm in diameter by means of a transmission electron microscope (TEM, JEM-2010, JEOL, Ltd, Tokyo, Japan) with an energy dispersive spectrometer (EDS). The volume fraction of retained austenite was determined using X-ray diffraction (XRD, RIGAKU MAX-2500, Rigaku Co., Tokyo, Japan) with Cu Kα radiation. The equation to calculate the volume fraction of retained austenite is mentioned in detail in reference [10]. Room temperature tensile test was carried out with a strain rate of 1×10^{-3} s^{-1} using Instron 5982 testing machine (Instron, Norwood, MA, USA). The hardness of the tempered specimens was measured on the Rockwell hardness C scale. The Charpy U-notch impact test was carried out to evaluate the impact toughness.

Table 1. Chemical composition of two steels used in the present study in wt. %.

Steels	C	Si	Mn	Cr	Mo	W	V	Fe
Cr6C9	0.88	0.56	0.56	6.08	3.27	0.52	0.99	Bal.
STD11	1.48	0.23	0.26	11.17	0.74	-	0.23	Bal.

3. Results and Discussion

Figure 1 shows the temperature-dependent equilibrium phase volume fractions for the Cr6C9 and STD11 steels determined by means of the MatCalc software [12]. The database of "mc_fe_v2.057.tdb" is used for the thermodynamic calculations in MatCalc. The solidification temperature and the initial temperature required to form the primary carbide were compared. In the STD11 steel, the solidification temperature was 1248 °C, but the primary M$_7$C$_3$ carbide appeared at the higher temperature of 1252 °C. It indicates that the M$_7$C$_3$ carbide can be precipitated before the full solidification and the precipitates easily coarsened during the following thermo-mechanical processing, which had a damaging influence on the mechanical properties. The solidification in the Cr6C9 steel was completed at 1244 °C, similar to the STD11 steel. As the temperature decreased, M$_6$C carbide was precipitated at 1196 °C and M$_7$C$_3$ carbide was precipitated at 1110 °C, successively. The carbide precipitation appeared at a temperature lower than the solidification temperature in the Cr6C9 steel. It is presumed that this temperature gap effectively suppressed the formation of the primary carbide, which appeared during solidification. The main alloy design concept of the Cr6C9 steel was to lower the carbide initiation temperature below the solidification temperature as much as possible in order to minimize the actual volume fraction and size of the primary carbide after quenching and tempering.

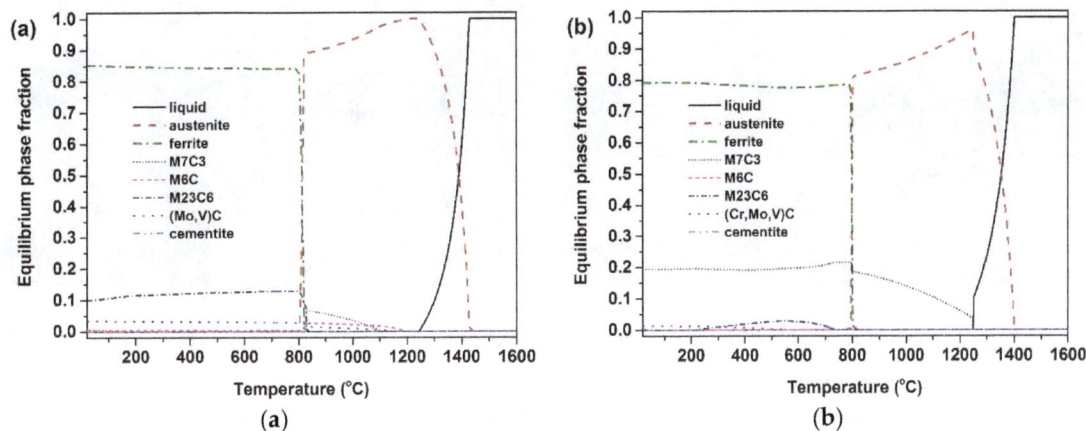

Figure 1. Equilibrium phase fractions calculated by using MatCalc (ver. 5.62) for the two steels: (a) Cr6C9 and (b) STD11.

Figure 2 shows the optical micrographs of the tempered Cr6C9 and STD11 specimens. The carbides in white were observed in the tempered martensite matrix. The volume fraction of carbides in the STD11 specimen was relatively larger than that in the Cr6C9 specimen, which is due to the difference in the volume fraction of the primary carbide. According to the calculated phase fraction in Figure 1, no primary carbide should exist in the Cr6C9 specimen. It is thought that a small amount of the primary carbide was precipitated during solidification in the Cr6C9 steel due to the non-equilibrium reaction and alloy segregation [11]. Figure 3 shows the SEM images of the tempered Cr6C9 and STD11 specimens. Spherical carbide particles with a diameter of less than 0.5 μm were observed in the Cr6C9 specimen whereas carbide particles larger than 1 μm in diameter were observed in the STD11 specimen. These large carbide particles seem to come from the precipitation of primary M_7C_3 carbide in the solidification or austenitization treatment [13,14]. This can be inferred from the equilibrium phase fraction for the STD11 steel as shown in Figure 1b. The smaller size and reduced amount of carbides observed in the Cr6C9 specimen resulted from the reduced amounts of C and Cr [1,11].

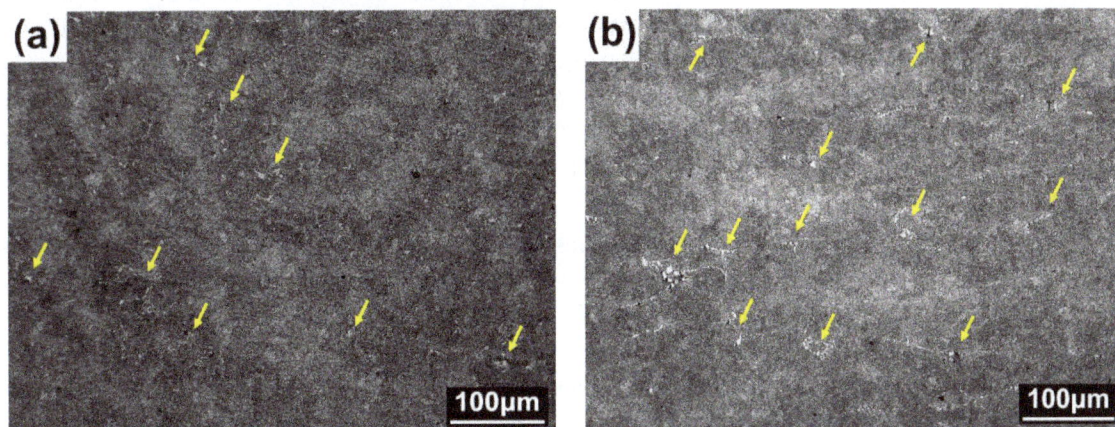

Figure 2. Optical micrographs of the tempered specimens: (a) Cr6C9 and (b) STD11. The arrows indicate the primary carbide in a bright color.

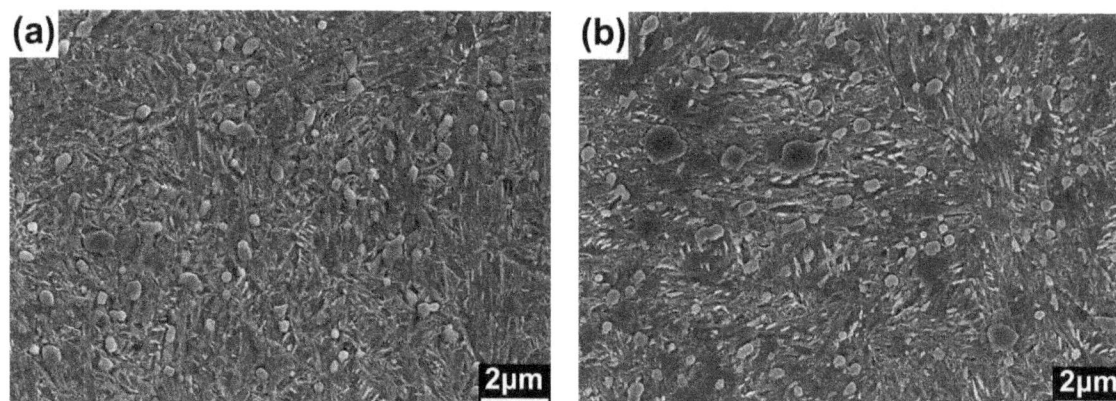

Figure 3. SEM images of the tempered specimens: (**a**) Cr6C9 and (**b**) STD11.

The various carbides with different sizes in the tempered Cr6C9 specimen were investigated by TEM analysis and the results are compared to the corresponding result. The selected area electron diffraction (SAED) pattern and EDS result confirmed that coarse carbide particles larger than 200 nm in diameter had a high Cr level and a trigonal crystal structure, as shown in Figure 4a. It is reported that the M_7C_3 carbide has enriched Cr content with a trigonal crystal structure [3,15]. This carbide is considered as a M_7C_3 precipitate. This is typically precipitated during solidification and is relatively more coarsened compared with other carbides; thus, it is hard to fully dissolve in a high-temperature heat treatment. The chemical composition in the M_7C_3 carbide at 520 °C predicted using MatCalc calculations is 9.8 at. %Fe–53.5 at. %Cr–2.9 at. %Mn–1.1 at. %Mo–2.4 at. %V–0.3 at. %W–30 at. %C. Figure 4b shows a carbide particle with a diameter of about 100 nm, containing high Mo and W contents with a face-centered cubic (FCC) crystal structure confirmed by the SAED pattern and EDS result. This precipitate is considered as the M_6C carbide [3]. The chemical composition in the M_6C carbide at 520 °C predicted using MatCalc calculations is 29.5 at. %Fe–37.7 at. %Mo–13.2 at. %W–3.0 at. %Cr–2.2 at. %V–0.1 at. %Si–14.3 at. %C. The $M_{23}C_6$ carbide also has a FCC crystal structure, but it has a very high Cr content. The chemical composition in the $M_{23}C_6$ carbide at 520 °C predicted using MatCalc calculations is 24.8 at. %Fe–44.3 at. %Cr–9.8 at. %Mo–0.4 at. %Mn–0.7 at. %C. A small carbide particle with a diameter of less than 50 nm is shown in Figure 4c. The SAED pattern confirms the FCC crystal structure and the EDS result indicates a high concentration of Mo and V, but Cr was undetected. This carbide particle is considered as (Mo,V)C, not $M_{23}C_6$, because it is reported that $M_{23}C_6$ contains the enriched Cr [8,16] and the MatCalc calculation also shows enriched Cr in the $M_{23}C_6$. The equilibrium chemical composition in the (Mo,V)C carbide at 520 °C predicted using MatCalc calculations is 36.1 at. %V–15.0 at. %V–2.3 at. %W–46.6 at. %C. It is thought that carbides with a diameter of less than 100 nm, as observed in Figure 4b,c, are the secondary carbides precipitated during tempering at a high temperature of 520 °C. These fine carbide particles consist of high concentrations of Mo, V, and W, which could contribute to the improvement of strength and hardness.

Figure 5 compares the mechanical properties of the tempered Cr6C9 and STD11 specimens. The Cr6C9 specimen indicated a higher hardness value due to the higher contents of Mo, V, and W, even though the C and Cr contents were decreased. The increase of hardness by forming carbide particles with Mo, V, and W is well known as the secondary hardening effect. The impact toughness of the Cr6C9 specimen was two times higher than that of the STD11 specimen. As confirmed by the microstructure images, a higher volume fraction of coarse primary carbides in the STD11 specimen brought about the lower impact toughness value. The Cr6C9 specimen was superior in the tensile properties to the STD11 specimen because of the work-hardening effect related to the fine secondary carbides, which had strong carbide-forming elements such as Mo, V, and W [17]. Also, it is thought that the better elongation of the Cr6C9 specimen results from the suppressed formation of the primary carbide. The measured volume fraction of retained austenite determined by XRD analysis for both the tempered Cr6C9 and STD11 specimens before deformation was less than 2 vol. % since the tempering temperature of 520 °C was high enough to decompose the retained austenite into ferrite and cementite [6,8,9]. Thus, the effect of the retained austenite on the mechanical properties is negligible depending on the steel alloy composition.

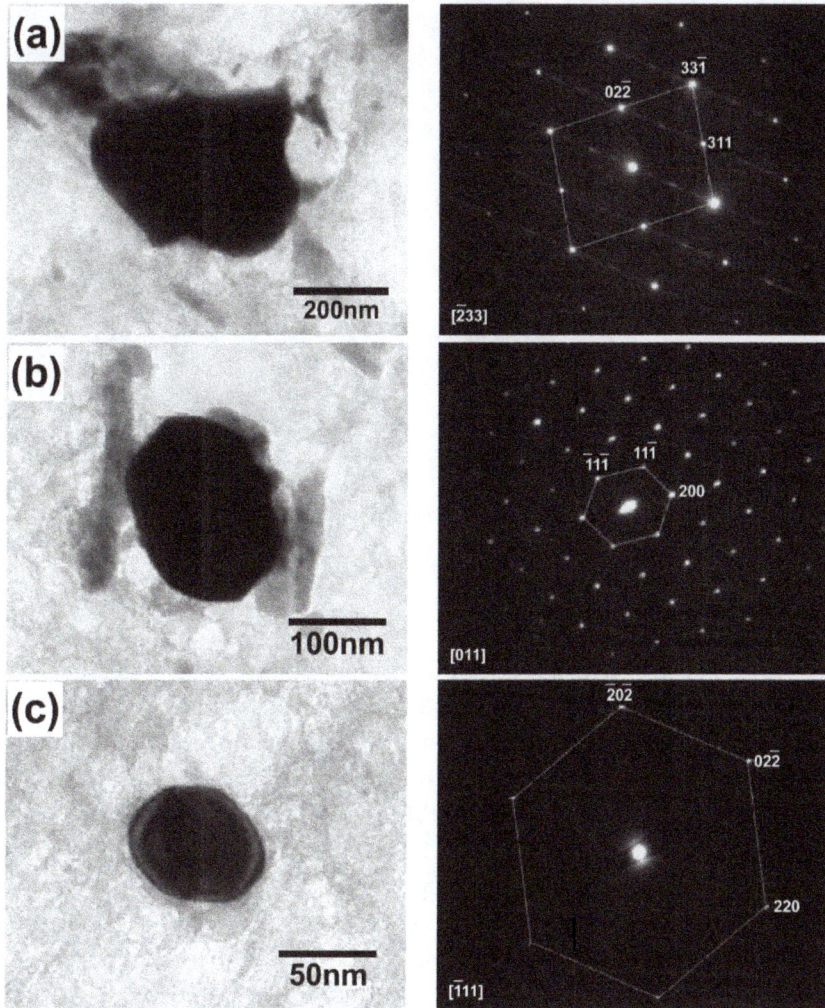

Figure 4. TEM images and the corresponding selected area electron diffraction (SAED) patterns for (a) M_7C_3 precipitate; (b) M_6C precipitate; and (c) (Mo,V)C precipitate.

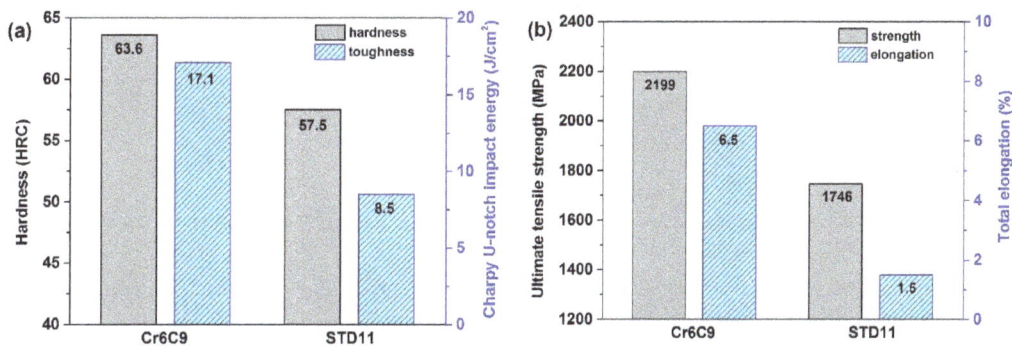

Figure 5. Comparison of the mechanical properties of the Cr6C9 and STD11 specimens: (a) hardness; impact toughness; and (b) tensile properties (ultimate tensile strength and total elongation).

Figure 6 shows the SEM image of the fracture surface of the Charpy impact testing specimen. The fine carbide was observed at the fracture surface in the Cr6C9 specimen whereas the coarse carbide was observed at the fracture surface in the STD11 specimen. It is reported that the crack can easily initiate and propagate near coarse carbide, resulting in fracture, as the stress is applied [18,19]. Hence, the lower impact toughness value was caused by the coarse carbide in the tempered STD11 specimen.

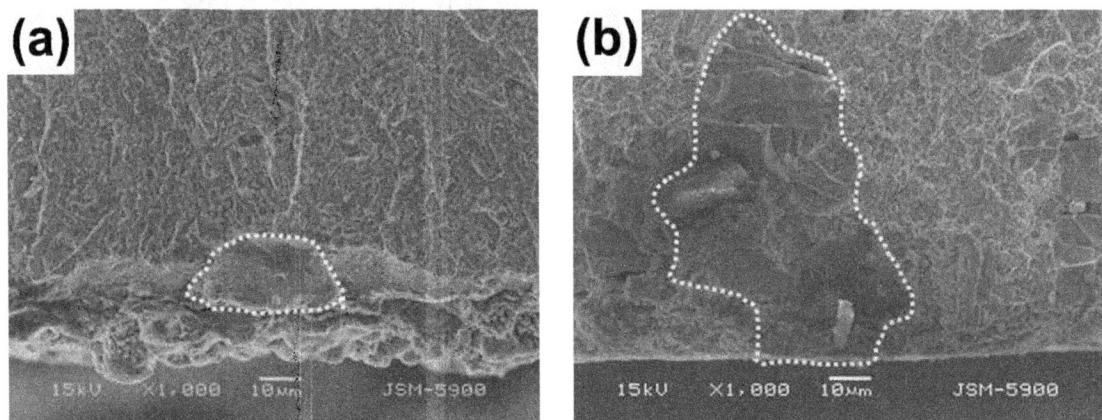

Figure 6. SEM images of fracture surface of the specimens after the Charpy impact test: (**a**) Cr6C9 and (**b**) STD11.

4. Conclusions

A new cold-work tool steel with a reduced Cr content of 6 wt. % has been developed. The alloy composition of the Cr6C9 steel was designed based on thermodynamic calculation in order to avoid the formation of primary carbide as much as possible during solidification. Even though the primary carbide was observed in both the Cr6C9 and STD11 steels, the Cr6C9 steel showed a smaller carbide particle size and a smaller volume fraction of carbides. The tempered Cr6C9 steel showed superior mechanical properties compared to the tempered STD11 steel. The higher values of hardness, impact toughness, tensile strength, and total elongation were the result of the fine secondary carbides' precipitation during tempering due to the higher contents of Mo, V, and W. From the TEM analysis and the thermodynamic calculation, it is considered that M_6C and $(Mo,V)C$ carbides with a diameter below 100 nm contributed to the improved mechanical properties of the tempered Cr6C9 steel.

Acknowledgments: This work (Grant No. 1503002010) was supported by the Business for Academic-industrial Cooperative establishments funded Korea Small and Medium Business Administration in 2015. S.-J.L. appreciates the support by the Ministry of Education through the "Leaders in INdustry-university Cooperation" project.

Author Contributions: Seok-Jae Lee conceived and designed the experiments; Minwook Kim performed the experiments; Singon Kang and Seok-Jae Lee analyzed the data; Singon Kang, Minwook Kim and Seok-Jae Lee contributed reagents/materials/analysis tools; Singon Kang and Seok-Jae Lee wrote the paper.

Conflicts of Interest: The authors declare no conflict of interest.

References

1.   Roberts, G.; Krauss, G.; Kennedy, R. *Tool Steel*, 5th ed.; ASM International: Materials Park, OH, USA, 1998; pp. 7–28.

2.   Totten, G.E. *Steel Heat Treatment: Metallurgy and Technologies*, 2nd ed.; CRC Taylor & Francis: New York, NY, USA, 2006; pp. 651–694.

3.   Krauss, G. *Steels: Processing, Structure, and Performance*; ASM International: Materials Park, OH, USA, 2005; pp. 535–559.

4.   Lee, S.; Kim, S.; Hwang, B.; Lee, B.S.; Lee, C.G. Effect of carbide distribution on the fracture toughness in the transition temperature region of an SA 508 steel. *Acta Mater.* **2002**, *50*, 4755–4762. [CrossRef]

5. Casellas, D.; Caro, J.; Molas, S.; Prado, J.M.; Valls, I. Fracture toughness of carbides in tool steels evaluated by nanoindentation. *Acta Mater.* **2007**, *55*, 4277–4286. [CrossRef]

6. Hong, K.J.; Kang, W.K.; Song, J.H.; Jung, I.S.; Lee, K.A. Study on the anisotropic size change by austenitizing and tempering heat treatment of STD11 tool steel using dilatometry. *Korean J. Met. Mater.* **2008**, *46*, 800–808.

7. Fukaura, K.; Yokoyama, Y.; Yokoi, D.; Tsujii, N.; Ono, K. Fatigue of cold-work tool steels: Effect of heat treatment and carbide morphology on fatigue crack formation, life, and fracture surface observations. *Metall. Mater. Trans. A* **2004**, *35*, 1289–1300. [CrossRef]

8. Kokosza, A.; Pacyna, J. Ewaluation of retained austenite stability in heat treated cold work tool steel. *J. Mater. Proc. Technol.* **2005**, *162–163*, 327–331. [CrossRef]

9. Yaso, M.; Morito, S.; Ohba, T.; Kubota, K. Microstructure of martensite in Fe–C–Cr steel. *Mater. Sci. Eng. A* **2008**, *481–482*, 770–773. [CrossRef]

10. Kang, J.Y.; Heo, Y.U.; Kim, H.Y.; Suh, D.W.; Son, D.M.; Lee, D.H.; Lee, T.H. Effect of copper addition on the characteristics of high-carbon and high-chromium steels. *Mater. Sci. Eng. A* **2014**, *614*, 36–44. [CrossRef]

11. Kim, H.Y.; Kang, J.Y.; Son, D.M.; Lee, D.S.; Lee, T.H.; Jeong, W.C.; Cho, K.M. Microstructure and mechanical properties of cold-wok tool steels: A comparison of 8%Cr steel with STD11. *J. Korean Soc. Heat Treat.* **2014**, *27*, 242–252. [CrossRef]

12. MatCalc Software. Available online: http://www.matcalc.at (accessed on 31 December 2016).

13. Hwang, K.C.; Lee, S.H.; Lee, H.C. Effects of alloying elements on microstructure and fracture properties of cast high speed steel rolls: Part I: Microstructural analysis. *Mater. Sci. Eng. A* **1998**, *254*, 282–295. [CrossRef]

14. Hong, K.J.; Song, J.H.; Chung, I.S. Carbide behavior in STD11 tool steel during heat treatment. *J. Korean Soc. Heat Treat.* **2011**, *24*, 262–270.

15. Shim, J.J.; Kim, Y.H.; Lee, S.Y. A study on the growth characteristics of complex carbides precipitated during spheroidizing in medium chromium steels. *Korean J. Met. Mater.* **1990**, *28*, 409–415.

16. Yamasaki, S. Modelling Precipitation of Carbides in Martensitic Steels. Ph.D. Thesis, University of Cambridge, Cambridge, UK, 2004.

17. Dieter, G.E. *Mechanical Metallurgy*, 3rd ed.; McGraw-Hill: New York, NY, USA, 1990; pp. 95–104.

18. Fukaura, K.; Yokoyama, Y.; Yokoi, D.; Tsuji, N.; Ono, K. Fatigue of cold work tool steels: Effect of heat treatment and carbide morphology on fatigue crack formation, life and fracture surface observations. *Metall. Mater. Trans. A* **2004**, *35A*, 1289–1300. [CrossRef]

19. Kubota, K.; Ohba, T.; Morito, S. Frictional properties of new developed cold work tool steel for high tensile strength steel forming die. *Wear* **2011**, *271*, 2884–2889. [CrossRef]

17

The Effect of Nb/Ti Ratio on Hardness in High-Strength Ni-Based Superalloys

Hiromu Hisazawa [1,*], Yoshihiro Terada [2], Fauzan Adziman [3], David J. Crudden [3],
David M. Collins [3], David E. J. Armstrong [4] and Roger C. Reed [4]

[1] Department of Materials Science and Engineering, Interdisciplinary Graduate School of Science and
     Engineering, Tokyo Institute of Technology, Yokohama 226-8502, Japan
[2] Department of Materials Science and Engineering, School of Materials and Chemical Technology,
     Tokyo Institute of Technology, Yokohama 226-8502, Japan; terada.y.ab@m.titech.ac.jp
[3] Department of Engineering Science, University of Oxford, Parks Road, Oxford OX5 1PF, UK;
     fauzan.adziman@eng.ox.ac.uk (F.A.); david.crudden@materials.ox.ac.uk (D.J.C.);
     david.collins@materials.ox.ac.uk (D.M.C.)
[4] Department of Materials, University of Oxford, Parks Road, Oxford OX5 1PF, UK;
     david.armstrong@materials.ox.ac.uk (D.E.J.A.); roger.reed@eng.ox.ac.uk (R.C.R.)
*   Correspondence: hisazawa.h.aa@m.titech.ac.jp

Academic Editor: Jonathan Cormier

Abstract: The age-hardening behaviour and microstructure development of high strength Ni-based superalloys ABD-D2, D4, and D6 with varying Nb/Ti ratios have been studied. The studied alloys have large volume fractions and multimodal size distributions of the γ' precipitates, making them sensitive to cooling conditions following solution heat treatment. Differential scanning calorimetry was conducted with a thermal cycle that replicated a processing heat treatment. The hardness of these alloys was subsequently evaluated by nanoindentation. The Nb/Ti ratio was not observed to influence the size and distribution of primary and secondary γ' precipitates; however, the difference in those of tertiary γ' and precipitate morphology were observed. The nanoindentation hardness for all alloys reduces once they have been solution-heat-treated. The alloys exhibited specific peak hardness. The alloy with the greatest Nb content was found to have the best increase in hardness among the alloys studied due to its large tertiary γ' precipitate.

Keywords: superalloy; nanoindentation; age-hardening; Nb/Ti ratio

1. Introduction

To satisfy increasing aerospace emissions targets [1], gas turbine engine efficiency needs to be improved. The high temperature mechanical properties and oxidation resistance of the materials selected for use in such application largely limit performance. Currently, nickel-based superalloys are used in the sections of the engine that experience high static stress at elevated temperatures—in particular, turbine discs that rotate at 10,000 rpm or greater; these alloys work very closely to their limit of tolerance of temperature or stress [2]. One of the ways in which the performance of gas turbine engines could be improved is by developing new nickel-based superalloys that could operate under more demanding conditions.

Nickel-based superalloys typically comprise a face-centered-cubic (A1) matrix (γ) solid solution that is strengthened with a dispersion of coherent intermetallic precipitates (γ') with an L1$_2$ crystal structure. During alloy development, the compositions selected are critical, as this will ultimately determine the performance of the material, irrespective of process optimization and its relationship. The effect of different γ' precipitate formers, Al, Ti, Ta, and Nb on the microstructure and mechanical properties in nickel-base superalloys, has been a subject of considerable research [3–6]. For example, the Ti/Al ratio is known to raise the anti-phase boundary (APB) energy of γ' [3] and increase the lattice misfit between the γ and γ' phases [4,5]. Whilst Al has predominantly been used as the γ' former in nickel-based superalloys, Nb has also been used; however, it has been reported to have a limited effect on the γ–γ' lattice misfit [6]. Nb additions will increase the γ' volume fraction and will act as a potent solid solution strengthening element in γ [7]. The effect of combining Ti and Nb is less well known.

Recently, Reed et al. [8] and subsequently Zhu et al. [9] have suggested and employed a systematic alloy composition design procedure. As a part of this design process, they developed several indices for important characteristics, such as creep resistance, density, and cost with a physical understanding between each characteristic and the composition, and new alloys could be selected. However, in terms of this design, Ti and Nb have some similar characteristics to predict the performance of the alloys. Table 1 summarizes a comparison between Ti and Nb in terms of alloy design. Although Ti is a third-row element and Nb is the fourth-row element, they both have the same order of diffusion coefficients in A1–Ni, Vegard coefficients, and metal-d levels. The diffusion coefficient is an important parameter for predicting precipitate coarsening rates and creep resistance. The Vegard coefficient is often used for predicting lattice parameter and lattice misfits; an important parameter that influences coarsening behaviour of γ' precipitates. Metal-d levels are also an important parameter because they are used in simplified estimates of the topologically close-packed (TCP) phase probability, which is developed by elementary band theory [10]. The TCP phase is well associated with the formation of voids, which may potentially act as initiation sites for fractures [11,12].

Table 1. Comparison between the design parameters for alloys containing Ti and Nb.

Parameters	Titanium	Niobium
Atomic number	22	41
Period	4	5
Melting point of pure metal	1941 K	2750 K
Atomic radius	0.160 nm	0.146 nm
Diffusion coefficient in A1–Ni at 1173 K [2]	$5 \times 10^{-16}\,\mathrm{m^2 \cdot s^{-1}}$	$3 \times 10^{-16}\,\mathrm{m^2 \cdot s^{-1}}$
Vegard coefficient for L1$_2$–Ni$_3$Al [2]	2.5×10^{-4} nm/atom %	4.5×10^{-4} nm/atom %
Metal-d levels [10]	2.271 eV	2.117 eV

Microstructures of the alloys for the turbine discs are relatively complex. They can show up to three distinct or more generations of γ' precipitates, with different size ranges, compositions, and precipitation processes each with a different effect on the mechanical properties [13]. Precipitates mostly found at grain boundaries are termed primary γ' and have a diameter of approximately 1 μm. Finer secondary and tertiary γ' precipitates are approximately ~100 nm and ~50 nm in diameter, respectively. Primary γ' precipitates form during alloy manufacturing and processing procedures such as castings and forgings. Secondary and tertiary γ' precipitates form during the cooling from temperatures of solution treatment and through subsequent ageing heat treatments; they form as a result of the competition between the nucleation and growth of the γ' precipitates [13–15]. The cooling conditions from a solution temperature has a great influence on the initial γ' distributions, the subsequent ageing response, and the resulting mechanical properties; careful attention therefore needs to be paid to them in order to understand them.

The plastic deformation of coherent superlattice γ' precipitates require either the shearing of paired dislocations (i.e., leading and trailing dislocations) or the bypassing of a dislocation. It is well established that the shearing of the γ' precipitates is the main contributor to the strength of polycrystalline superalloys, as replicated by a dislocation pair model [16,17]. The models describe that interfacial energy associated with an APB is created by a leading dislocation minimised by the passage of the trailing dislocation. This model considers interactions between dislocation pairs with small (weak pair coupling) and large (strong pair coupling) precipitates with a unimodal size distribution and a low volume fractions of γ' precipitates. Jackson and Reed [18] and subsequently Collins and Stone [19] have employed the weak and strong pair models in the optimization of the microstructures in Udimet 720Li and RR1000, respectively. Their analyses were based on the optimal γ' size where transitions between the weak and strong pair couplings occurred. The maximised critical resolved shear stress (CRSS) is achieved at the optimal size, and it is strengthened by the larger APB energy and volume fractions of the γ' precipitates. The dislocation pair model was extended by Galindo-Nava et al. [20] so as to apply to a multimodal particle distribution; they succeeded in predicting the effect of tertiary γ' precipitates on CRSS, which demonstrates a possibility of improve the mechanical properties of a superalloy. Consequently, the yield behaviour of nickel-based superalloys is strongly influenced by the size distributions and volume fractions of the secondary and tertiary γ' precipitates, whilst the role of the primary γ' is limited to grain boundary pinning, which minimises the growth of γ grains during solution heat treatment. Understanding how to control the secondary and tertiary γ' phases are crucial to maximising the mechanical properties of Ni-based superalloys.

In this study, a range of model alloys are investigated in order to systematically study the effect of Ti and Nb on γ' precipitation and its effect on a simple mechanical property: hardness. In particular, the comparative effects of Ti and Nb on multimodal γ' precipitates size distributions are characterised and compared in this study.

2. Materials and Methods

2.1. Materials

ABD series alloys were used in this study. They are polycrystalline nickel-based superalloys and were manufactured by ATI Powder Metals using a lab-scaled version of a commercial powder metallurgy process. Three different alloys containing varying concentrations of Ti (2.8–4.1 atom %) and Nb (0–1.2 atom %) with these elements are substituted on a 1:1 basis. The nominal composition of each alloy (atom %) is listed in Table 2. A high-temperature synchrotron X-ray diffraction study has been reported in a previous study [21]. The lattice misfits of D2, D4, and D6 are positive at temperatures below 1273 K, i.e., the lattice parameter of the γ' precipitate is larger than that of the γ matrix, whose values for D2, D4, and D6 are approximately 0.07%, 0.09%, and 0.10% at 1123 K, respectively. These alloys have a 40%–45% volume fraction of γ' precipitates at 1123 K. The specimens were received after forging and cut into 1-mm-thick square plates that have dimensions of 2.5×2.5 mm^2 using electro-discharge machining (Brother industries, Ltd., Nagoya, Japan) and a high-speed precision cutting machine (Heiwa Technica, Zama, Japan, and Struers Ltd., Rotherham, UK). The weight was approximately 0.2 g.

Table 2. The nominal alloy compositions in this study (atom %).

Alloy	Ni	Cr	Co	Mo	W	Al	Ti	Ta	Nb	C	B	Zr
D2	Bal.	18.7	18.2	0	0.9	8	4.1	0.6	0	0.127	0.078	0.037
D4	Bal.	18.7	18.2	0	0.9	8	3.6	0.6	0.4	0.127	0.078	0.037
D6	Bal.	18.7	18.2	0	0.9	8	2.8	0.6	1.2	0.127	0.078	0.037

2.2. Heat Treatment

Ni-based superalloys with large volume fractions of γ' precipitates have a strong driving force for precipitation and quite fast precipitation kinetics. The microstructural and mechanical properties of the alloys are very sensitive to solvus temperature and cooling rates following heat treatment [20,22]. Accurate control and homogeneity of the specimen temperature were achieved through the use of a differential scanning calorimeter (DSC) NETSZCH DSC404F1 (NETZSCH Analyzing & Testing, Wolverhampton, UK) for a small specimen in an argon flow. A thermal process typical of a 1 step super-solvus treatment, which is used for current generation turbine disc alloys, and isothermal ageing at 1123 K were conducted on each alloy studied. The rate of heating from room temperature to approximately 30 K above the γ' solvus temperature was controlled to be 10 K·min^{-1}. The subsequent cooling rate was controlled so as to fall to 673 K at a rate of 1 K·min^{-1}. The solution temperatures of the alloys were studied by DSC and approximately confirmed by synchrotron X-ray diffraction [21]. Figure 1 shows the set and measured temperatures together with a typical DSC trace for alloy D2. The other alloys had similar results. The temperature was precisely controlled during cooling to 973 K, at which point significant microstructure change was deemed to be negligible. We also confirmed precipitation of the γ′ phase corresponding to two exothermal peaks, one at 1230 K for the formation of secondary γ′ and the other at 1060 K for the formation of tertiary γ′.

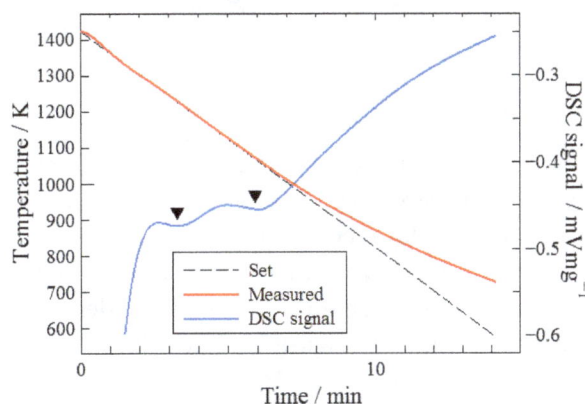

Figure 1. Set and measured temperatures of the D2 alloy together with a typical DSC spectrum; the black downward-pointing triangles indicate exothermal peaks consistent with γ′ precipitation. The other alloys used in this study exhibited similar results.

2.3. Microstructure Observation

To prepare the specimens for microstructure observations, any surface oxide produced by the heat treatment was removed using abrasive media. Microstructure observations were conducted by using a field-emission scanning electron microscope (FE-SEM) (JEOL Ltd., Akishima, Japan) with an accelerating voltage of 15 kV. The specimens analysed by the FE-SEM were prepared using both a traditional metallographic technique and chemical–mechanical polishing by a diluted colloidal silica. Electrolytic etching was performed with phosphoric acid saturated with chromium oxide (VI). The particle size of the γ′ precipitate was also measured using an SEM secondary electron image (SEM-SEI). It is noted that the electrolytic etching dissolves the γ matrix, leaving γ′ precipitates in relief; this results in the γ′ volume fraction being over-estimated and the particle size being underestimated due to overlapping particles. Nevertheless, it is reasonable for us to evaluate the shape of the projected γ′ particles stereologically.

2.4. Nanoindentation

The samples in this study were prepared using DSC; thus, the size of the specimens was limited. It is therefore difficult to reveal the mechanical properties of the samples using conventional testing methods such as a tensile test or even Vickers hardness test. As such, nanoindentation experiments were performed in this study at room temperature. The samples that studied with nanoindentation were also prepared using both a conventional metallographic technique and polishing with diluted colloidal silica. In all of the indentation experiments, a Berkovich tip with a tip angle of 65.3° was used. The load and corresponding displacement were recorded continuously during the indentation. The indentation hardness H_{IN} was derived from the nanoindentation experiments using the following equation:

$$H_{IN} = \frac{F}{A(h)} \approx \frac{F}{26.43h^2} \tag{1}$$

where F is the applied load, $A(h)$ is the projected contact area at that load, and h is the penetration depth. An empirical constant of 26.43 used in Equation (1) is a constant dependent on the indenter shape. The indentation hardness at the maximum depth was selected as being the representative hardness, because at a small depth it might be hardened by either the scale effect or the surface roughness. The maximum h was kept at a constant 2000 nm and the strain rate was kept at 0.1 s^{-1}. The depth of 2000 nm corresponds to an indent width of 15 µm, which is sufficiently smaller than the specimen used but larger than the γ' precipitates. We therefore expect that the hardness measurements are from a volume that is representative of the material, containing the full tri-modal γ' precipitate distribution.

3. Results

3.1. Initial Microstructure

In this study, all of the specimens have a typical multimodal distribution of γ' precipitates. Figure 2 shows micrographs of the as-received (as-forged), and solution-heat-treated specimens. The as-solution-treated D2, D4, and D6 alloys show similar microstructures. The as-received specimens were found to have a heterogeneous grain size, which could be attributed to the heterogeneity of the manufacturing process. The alloys also have large area fractions and densities of primary precipitates at both the grain boundary and the grain interior (Figure 2a). The primary γ' precipitates almost dissolves into the matrix during the solution heat treatment. As there is almost no primary γ' precipitates, the γ grain grew dramatically; uniform grain sizes could be obtained. The solution treatment enabled us to obtain a homogeneous microstructure (Figure 2b). After the solution treatment and cooling, we can observe that the residual primary γ' particles have a diameter of ~1 µm, the secondary γ' particles have a diameter of ~200 nm, and very fine tertiary γ' particles have a diameter of less than 50 nm. (Figure 2c). The primary γ' particles show a lenticular shape and are dispersed along the grain boundaries. The solution heat treatment temperatures were selected to be supersolvus in order to dissolve the primary γ' completely; in fact, the temperature of the real solvus is slightly above the solution heat treatment temperatures. The volume fraction of the primary γ' observed from the FE-SEM is less than 1%. There are two types of secondary γ' particles are observed. Some of the secondary particles are granular, but the others are flower-like or irregular morphologies. The tertiary γ' particles are dispersed around both the γ grain boundary and the primary γ' particles, and they form "pools of tertiary γ". Very few tertiary γ' are observed in the vicinity of the secondary precipitates. The area fraction of the pool or intragranular tertiary γ' is also less than 3% of the total observed area. No distinct differences were found between the alloys in terms of their microstructures.

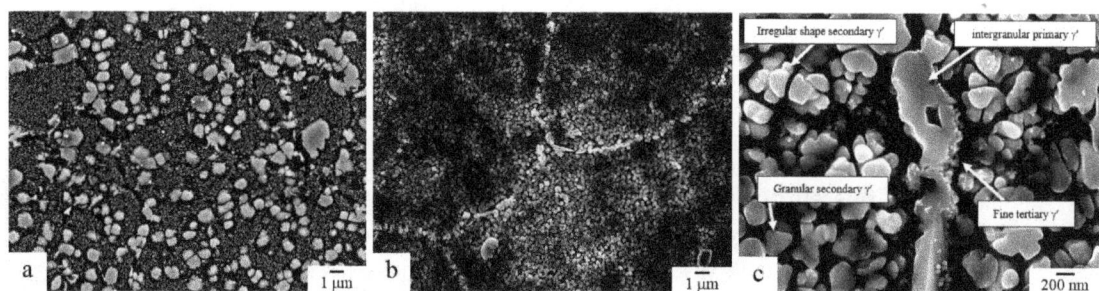

Figure 2. Microstructures of the D2 alloy as-received (**a**), as-solution-heat-treated (**b**), and as-solution-treated viewed at high magnification (**c**).

3.2. Microstructure Evolution with Aging

Figure 3 shows the microstructures of each alloy aged at 1123 K for 8 h. Despite a sufficiently long aging heat treatment, there is no distinct difference between the microstructures with varying Nb/Ti ratios or aging at 1123 K. No coarsening is observed for the primary or secondary γ' precipitates. The shape of the secondary γ' precipitates, however, seems to change slightly; this is discussed in greater detail in Section 4.1. Figure 4 shows the secondary γ' particles in the D4 alloy aged at 1–8 h at 1123 K. After 1 h aging, γ' particles tend to form granular shapes, but by 8 h some of the precipitates form an irregular shape. Irregularly shaped precipitates form "dents" as a result of morphological change (see arrows in Figure 4c). Secondary γ' precipitates slightly coarsen with aging heat treatment, but without significant change. Differences in the microstructures of the samples with Nb/Ti ratio are not observed for the secondary γ' particles.

Figure 3. Microstructures of the D2 (**a**), D4 (**b**), and D6 (**c**) alloys after aging at 1123 K for 8 h.

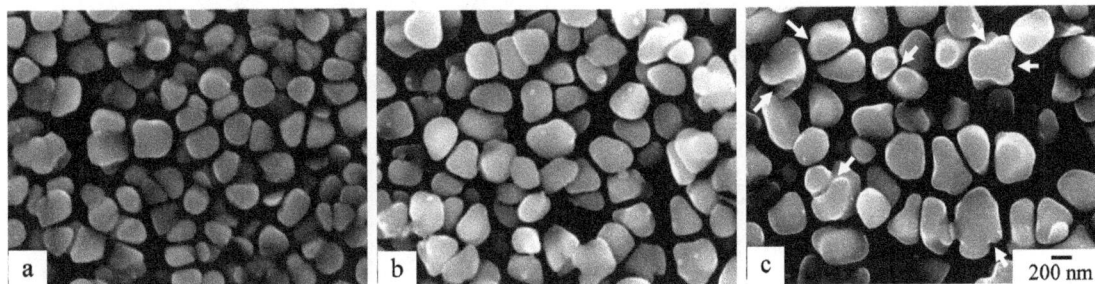

Figure 4. Morphological evolution of the secondary γ' precipitates after aging at 1123 K for 1 h (**a**), 4 h (**b**), and 8 h (**c**) in the D4 alloys. The precipitates tend towards irregular-shape morphology rather than a granular shape as the aging increased. The arrows in (**c**) indicates "dents" of the γ' particles.

In contrast, the tertiary γ' precipitates show a clear difference in their distribution. Figure 5 shows pools of the tertiary γ' particles in specimens aged for 1 h and 8 h. For the 1 h aged specimens, tertiary γ' particles have a diameter of approximately 30 nm, and there are no significant differences observed between the three alloys. The volume fractions of the tertiary γ' precipitates cannot be measured using a microscope, but they are roughly the same for all alloys. However, the size of those precipitates rapidly coarsens with aging. The average diameter of the tertiary γ' particles reach around 80 nm for D2 (Nb/Ti = 0) and 150 nm for D6 (Nb/Ti = 2); these results suggest that Nb might accelerate the growth rate of the tertiary γ' particles, despite having a negligible effect on the initial particle size.

Figure 6 shows a change in the particle size of the primary (a), secondary (b), and tertiary γ' (c) precipitate following heat treatment. The precipitate sizes were evaluated with a radius of an area-equivalent circle; the primary γ' precipitates are >400 nm; the secondary γ' precipitates are 100–400 nm; the tertiary γ' precipitates are <100 nm. As mentioned above, the size of the primary and secondary γ' precipitates are always 800 and 200 nm, respectively, and they grow slightly after 8 h aging. They clearly do not raise to the power 1/3 with respect to time, which is often observed in superalloys [23]. The tertiary γ' precipitates show more rapid growth rates, which are faster in the alloys that have greater Nb/Ti ratios.

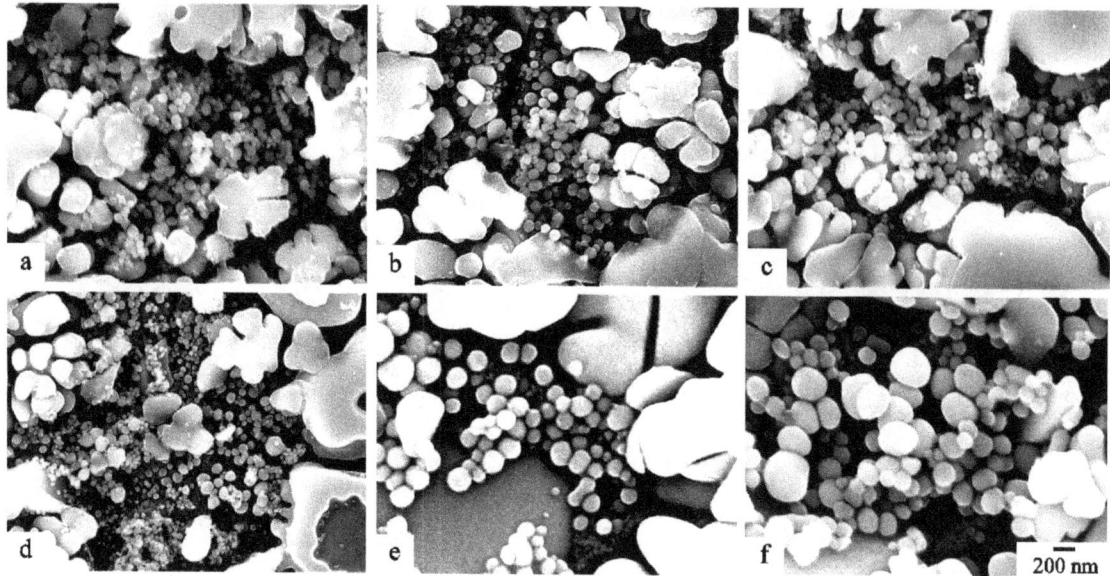

Figure 5. The pool of tertiary γ' precipitates observed in the D2 (**a,d**), D4 (**b,e**), and D6 (**c,f**) alloys aged at 1123 K for 1 h (**a–c**) and 8 h (**d–f**).

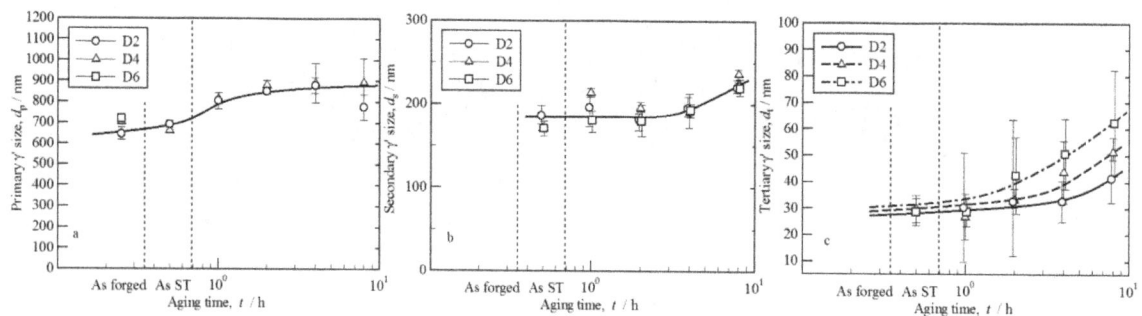

Figure 6. Changes in the particle size of the primary (**a**), secondary (**b**), and tertiary (**c**) γ' particles of ABD alloys following forging, solution treatment, and isothermal aging at 1123 K.

3.3. Nanoindentation Hardness

To evaluate what effects of the microstructures on the mechanical properties of the alloys, nanoindentation hardness measurement was performed. Figure 7 shows the changes in nanoindentation hardness of the D2, D4, and D6 alloys following heat treatment. The as-received specimens with varying Nb/Ti ratios show different hardnesses due to the grain size heterogeneity and coarse primary γ' precipitates described in Figure 2. The difference in hardness is eliminated by solution heat treatment resulting in the same hardness of 3.9 GPa for each alloy. Following the aging treatment, the alloys show typical age-hardening curves, i.e., the hardness increases during precipitation and growth and decreases when the coarsening of the γ' precipitate process dominates. The peak aging time is approximately 4 h for all of the alloys, but the peak hardness increases as the Nb/Ti ratio increases. The D6 alloy, which has an Nb/Ti ratio of 2, exhibits the highest hardness (4.2 GPa). The behaviour of the age-hardening process will be discussed in a Section 4.2.

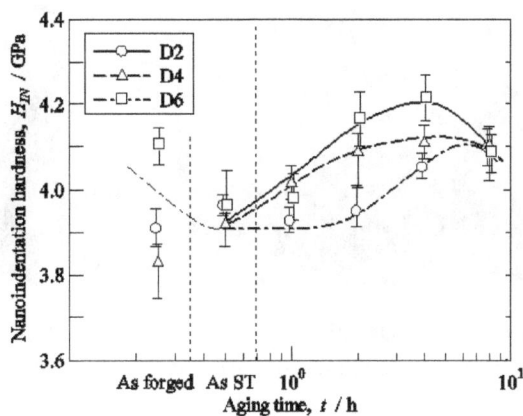

Figure 7. Measured nanoindentation hardness of the ABD alloys following heat treatment: forging, solution treatment, and isothermal aging at 1123 K.

4. Discussion

4.1. The Effect of the Nb/Ti Ratio on the Multimodal Distribution of the γ' Precipitates

Nb and Ti were substituted for the ABD series alloy, and the effect of this on a microstructure of the alloys has been examined. The D2 alloy has the lowest Nb/Ti ratio, the D6 alloy has the highest, and D4 has a ratio that is in the middle of the two. All of the alloys have almost the same mean precipitate sizes and densities of relatively coarse primary and secondary γ' precipitates. A previous study [19] revealed that the volume fractions of γ' precipitate do not differ significantly between these alloys. They also have similar size of tertiary γ' precipitate just after the solution heat treatment and cooling as described in Figure 6. Sarosi et al. [14] have suggested that the multimodal distributions of γ' precipitates form due to competition between the nucleation and growth of γ' precipitates; they predict that a monomodal distribution would be obtained when the growth rate is much faster than the nucleation, and the precipitation would have a wider size distribution when the nucleation is faster. A multimodal distribution can be obtained only when the kinetics of both processes are equivalently fast. There is no significant difference in size distributions of γ' precipitates regardless of Nb/Ti ratio, which suggests that the Nb/Ti ratio does not have a significant influence on the nucleation and growth kinetics during cooling from solution temperature. This is not surprising because the Nb–Ti diffusion rate in the matrix and the thermodynamic stability of the γ' phase are similar.

During aging at 1123 K, the mean size of primary and secondary γ' precipitates do not change. The primary γ' precipitates remained quite large, with little coarsening observed. However, Figure 5 shows the morphological change of the secondary γ' precipitates, which may have occurred due to directional coarsening, which has been reported by Ricks [24] and Qiu et al. [25]. This may also have been due to cyclic and reversal coarsening reported by Chen et al. [26]. It is more likely that the morphological change is due to cyclic coarsening, because the coarsening direction of the precipitates appears to be related to the orientation of the grain where they are situated. The unchanged size of the secondary γ' particles can also be explained by cyclic coarsening. Figure 8 shows additional evidence of the cyclic coarsening phenomenon. During the microstructure observation, granular-shape precipitates (Figure 8a) were observed over almost the entire area of the specimen, but irregular-shape precipitates (Figure 8b) were observed in some locations within the γ grain. A few residual intragranular γ' particles show an irregular shape (Figure 8c), which may have been a result of the growth and splitting of the γ' precipitates. The "humps" of the intragranular primary γ' particles are similar in size to that of the secondary γ' precipitates, and we estimated the maximum precipitate size under the constraint of a lattice misfit. The coarsening direction depends on the elastic interaction between particles. In other words, the morphological change is a result of the directional coarsening in conjunction with the bypassing of the elastic field generated by neighboring particles and of their own splitting [27].

Figure 8. Microstructures of the D4 alloy aged at 1123 K for 1 h showing microstructure heterogeneity: Granular-shaped precipitates (**a**), irregular-shape precipitates (**b**), and residual intragranular precipitates (**c**).

Only the tertiary γ' precipitates are influenced by the Nb/Ti ratio, and the alloys with higher Nb/Ti ratios show faster growth rates of the tertiary γ' precipitates. The differences between the growth rates are not simply attributed to the diffusion rate; Nb has a similar or slightly smaller diffusion coefficient than that of Ti. The other potential reason for the difference is related to the partitioning of the Nb and Ti between the matrix and the precipitate. For instance, the Nb would be enriched near the pools of tertiary γ' particles and would enhance the growth of the tertiary γ' if the Nb is in the partitioned matrix around the solution temperature. A compositional study is therefore needed in order to reveal the reason why the tertiary precipitates grow faster in the D6 alloy. However, it is very difficult to analyse the composition around the tertiary γ' particles because of a requirement of high spatial resolution and accuracy.

4.2. Age Hardening Behaviour of the ABD Alloys

The Nb/Ti ratio does not influence the nanoindentation hardness of the solution-treated alloys; however, it impacts the subsequent age hardening behaviour. The alloys with higher Nb/Ti ratios show higher peak aging hardness. It is known that precipitate shearing is the main deformation mechanism in alloys for turbine discs [17], and it is expected that Nb would have a significant effect on the APB energy, lattice misfit, microstructure, volume fraction of the γ' precipitates, and several other factors influencing strength. However, a previous study of X-ray synchrotron diffraction [21] shows that the lattice misfit almost never changes and that the volume fraction slightly increases as the Nb/Ti ratio increases. Because the hardness increased with volume fraction, this reported tendency contrasts with our experimental result. It is likely that the solid solution strengthening of the matrix and the increase in APB energy by Nb in the γ' precipitate is negligible because the hardness of the solution-treated alloys are similar. The alloys that have different Nb/Ti ratios show similar microstructures, except for the size of tertiary γ' precipitate, as mentioned in Section 4.1.

We can therefore see that the growth of tertiary γ' precipitate is the most potent factor for improving age-hardening ability. Jackson and Reed [18] have shown that the tertiary γ' precipitates play an important role in the mechanical property of Udimet 720Li, and Galindo-Nava et al. [20] demonstrated that small amounts of fine tertiary γ' precipitates are able to significantly strengthen the CRSS in Udimet 720Li. The area fraction of the pool of tertiary γ' precipitates is less than 3%; however, intragranular tertiary γ' is not observed; it is possible that they were washed out by the electrolytic etching. The fact that only a small amount of tertiary γ' is required to improve the hardness of the alloys encourages us to investigate the process of a multimodal distribution of γ' precipitates and its dependence on the composition, such as the Nb/Ti ratio.

5. Conclusions

In this study, the effect of Nb/Ti ratio on the microstructure formation and hardness during a typical solution treatment and aging is investigated in high strength superalloys. We were able to precisely control the cooling rates from the solution heat treatment, which enabled us to evaluate the role of Nb and Ti in the formation of multimodal distributions of γ' precipitates. Our conclusions are summarised as follows:

- Near-supersolvus solution-treated and cooled alloys show multimodal distributions of γ' precipitates containing relatively coarse primary and secondary γ' precipitates at the grain boundary and interior, respectively, and fine tertiary γ' precipitates form a pool of them around the grain boundary or primary γ' precipitates.
- All of the alloys studied show typical age-hardening behaviour at 1123 K, but higher peak hardness is obtained in the alloys that have higher Nb/Ti ratios.
- The microstructures of solution-treated alloys are similar and imply that they are independent of the Nb/Ti ratio; however, faster growth rates are observed for the tertiary γ' precipitate in the alloy with higher Nb/Ti ratios. Larger tertiary γ' precipitate are believed to be the reason why the higher hardness of the alloys studied.
- The Nb/Ti ratio does not influence the morphologies of primary and secondary precipitates, but it does influence tertiary γ' precipitates; this might be due to Nb and Ti partitioning at high temperatures.
- Several pieces of the evidences are observed for cyclic coarsening or splitting in all of the alloys. The size of the secondary γ' precipitates remained constant, which might be due to cyclic coarsening of the precipitates.

Acknowledgments: The authors wish to thank Paraskevas Kontisa, Enrique Alabort, and André Németh for experimental support and good discussion. This work was supported by a Grant-in-Aid for JSPS Research Fellow Number JP2611840, Japan and in part by Program for Leading Graduate Schools "Academy for Co-creative Education of Environment and Energy Science", MEXT, Japan.

Author Contributions: D.J. Crudden, D.M. Collins, and R.C. Reed designed the experiments and supplied the materials; H. Hisazawa performed the experiments, analysed the data, and wrote the paper; D.E.J. Armstrong contributed nanoindentation experiments; F. Adziman and Y. Terada support to writing this paper.

Conflicts of Interest: The authors declare no conflict of interest. The founding sponsors had no role in the design of the study; in the collection, analyses, or interpretation of data; in the writing of the manuscript; or in the decision to publish the results.

References

1. IATA—Carbon-Natural Growth by 2020. Available online: http://www.iata.org/pressroom/pr/Pages/2009-06-08-03.aspx (accessed on 13 December 2016).
2. Reed, R.C. *The Superalloys: Fundamentals and Applications*; Cambridge University Press: Cambridge, UK, 2006.
3. Miller, R.F.; Ansell, G.S. Low Temperature Mechanical Behavior of Ni–15Cr–Al–Ti–Mo Alloys. *Metall. Trans. A* **1977**, *8*, 1979–1991. [CrossRef]
4. Maniar, G.N.; Bridge, J.E.; James, H.M.; Heydt, G.B. Correlation of Gamma-Gamma Prime Mismatch and Strengthening in Ni/Fe-Ni Base Alloys Containing Aluminum and Titanium as Hardeners. *Metall. Trans. A* **1970**, *1*, 31–42.
5. Xu, Y.; Zhang, L.; Li, J.; Xiao, X.; Cao, X.; Jia, G.; Shen, Z. Relationship between Ti/Al Ratio and Stress-Rupture Properties in Nickel-Based Superalloy. *Mater. Sci. Eng. A* **2012**, *544*, 48–53. [CrossRef]
6. Guo, E.C.; Ma, F.J. The Strengthening Effect of Niobium on Ni–Cr–Ti Type Wrought Superalloy. In Proceedings of the Fourth International Symposium on Superalloys, Champion, PA, USA, 21–25 September 1980; pp. 431–438.
7. Mishima, Y.; Ochiai, S.; Hamao, N.; Yodogawa, M.; Suzuki, T. Solid Solution Hardening of Nickel—Role of Transition Metal and B-subgroup Solutes—. *Trans. Jpn. Inst. Met.* **1986**, *27*, 656–664. [CrossRef]
8. Reed, R.C.; Tao, T.; Warnken, N. Alloys-By-Design: Application to Nickel-Based Single Crystal Superalloys. *Acta Mater.* **2009**, *57*, 5898–5913. [CrossRef]
9. Zhu, Z.; Höglund, L.; Larsson, H.; Reed, R.C. Isolation of Optimal Compositions of Single Crystal Superalloys by Mapping of a Material's Genome. *Acta Mater.* **2015**, *90*, 330–343. [CrossRef]

10. Morinaga, M.; Yukawa, N.; Adachi, H.; Ezaki, H. New PHACOMP and Its Applications to Alloy Design. In Proceedings of the Fifth International Symposium on Superalloys, Champion, PA, USA, 7–11 October 1984; pp. 523–532.

11. Shi, Z.; Liu, S.; Yue, X.; Wang, X.; Li, J. Effect of Nb content on Microstructure Stability and Stress Rupture Properties of Single Crystal Superalloy Containing Re and Ru. *J. Cent. South Univ.* **2016**, *23*, 1293–1300. [CrossRef]

12. Rae, C.M.F.; Reed, R.C. The precipitation of Topologically Close-Packed Phases in Rhenium-Containing Superalloys. *Acta Mater.* **2001**, *49*, 4113–4125. [CrossRef]

13. Singh, A.R.P.; Nag, S.; Chattopadhyay, S.; Ren, Y.; Tiley, J.; Viswanathan, G.B.; Fraser, H.L.; Banerjee, R. Mechanisms Related to Different Generations of γ' Precipitation during Continuous Cooling of a Nickel Base Superalloy. *Acta Mater.* **2013**, *61*, 280–293. [CrossRef]

14. Sarosi, P.M.; Wang, B.; Simmons, J.P.; Wang, Y.; Mills, M.J. Formation of Multimodal Size Distributions of γ' in a Nickel-base Superalloy during Interrupted Continuous Cooling. *Scr. Mater.* **2007**, *57*, 767–770. [CrossRef]

15. Radis, R.; Schaffer, M.; Albu, M.; Kothleitner, G.; Pölt, P.; Kozeschnik, E. Multimodal Size Distributions of γ' Precipitates during Continuous Cooling of UDIMET 720Li. *Acta Mater.* **2009**, *57*, 5739–5747. [CrossRef]

16. Raynor, D.; Silcock, J.M. Strengthening Mechanisms in γ' Precipitating Alloys. *Metal Sci. J.* **1970**, *4*, 121–130. [CrossRef]

17. Reppich, B. Some New Aspects Concerning Particle Hardening Mechanisms in γ' Precipitating Ni-Base Alloys—I. Theoretical Concept. *Acta Metall.* **1982**, *30*, 87–94. [CrossRef]

18. Jackson, M.; Reed, R.C. Heat Treatment of UDIMET 720Li: The Effect of Microstructure on Properties. *Mater. Sci. Eng. A* **1999**, *259*, 85–97. [CrossRef]

19. Collins, D.; Stone, H. A Modelling Approach to Yield Strength Optimisation in a Nickel-Base Superalloy. *Int. J. Plast.* **2014**, *54*, 96–112. [CrossRef]

20. Galindo-Nava, E.I.; Connor, L.D.; Rae, C.M.F. On the Prediction of the Yield Stress of Unimodal and Multimodal γ' Nickel-Base Superalloys. *Acta Mater.* **2015**, *98*, 377–390. [CrossRef]

21. Collins, D.M.; Crudden, D.J.; Alabort, E.; Connolley, T.; Reed, R.C. Time-Resolved Synchrotron Diffractometry of Phase Transformations in High Strength Nickel-Based Superalloys. *Acta Mater.* **2015**, *94*, 244–256. [CrossRef]

22. Mitchell, R.J.; Preuss, M.; Hardy, M.C.; Tin, S. Influence of Composition and Cooling Rate on Constrained and Unconstrained Lattice Parameters in Advanced Polycrystalline Nickel-Base Superalloys. *Mater. Sci. Eng. A* **2006**, *423*, 282–291. [CrossRef]

23. Baldan, A. Review Progress in Ostwald Ripening Theories and Their Applications to the γ'-Precipitates in Nickel-Based Superalloys. *J. Mater. Sci.* **2002**, *37*, 2379–2405. [CrossRef]

24. Ricks, R.A.; Porter, A.J.; Ecob, R.C. The Growth of γ' Precipitates in Nickel-base Superalloys. *Acta Metall.* **1983**, *31*, 43–53. [CrossRef]

25. Qiu, C.L.; Andrews, P. On the Formation of Irregular-Shaped Gamma Prime and Serrated Grain Boundaries in a Nickel-Based Superalloy during Continuous Cooling. *Mater. Charact.* **2013**, *76*, 28–34. [CrossRef]

26. Chen, Y.; Prasath Babu, R.; Slater, T.J.A.; Bai, M.; Mitchell, R.; Ciuca, O.; Preuss, M.; Haigh, S.J. An Investigation of Diffusion-Mediated Cyclic Coarsening and Reversal Coarsening in an Advanced Ni-Based Superalloy. *Acta Mater.* **2016**, *110*, 295–305. [CrossRef]

27. Zhao, X.; Duddu, R.; Bordas, S.P.A.; Qu, J. Effects of Elastic Strain Energy and Interfacial Stress on the Equilibrium Morphology of Misfit Particles in Heterogeneous Solids. *J. Mech. Phys. Solids* **2013**, *61*, 1433–1445. [CrossRef]

18

Effect of Sintering Time on the Densification, Microstructure, Weight Loss and Tensile Properties of a Powder Metallurgical Fe-Mn-Si Alloy

Zhigang Xu [1,2], Michael A. Hodgson [1], Keke Chang [3], Gang Chen [4], Xiaowen Yuan [5] and Peng Cao [1,*]

[1] Department of Chemical and Materials Engineering, University of Auckland, Private Bag 92019, Auckland 1142, New Zealand; zxu886@aucklanduni.ac.nz (Z.X.); ma.hodgson@auckland.ac.nz (M.A.H.)

[2] School of Automotive Engineering, Wuhan University of Technology, Wuhan 430070, China

[3] Materials Chemistry, RWTH Aachen University, D-52056 Aachen, Germany; chang@mch.rwth-aachen.de

[4] State Key Laboratory of Porous Metal Materials, Northwest Institute for Non-ferrous Metal Research, Xi'an 710016, China; mychgcsu@163.com

[5] School of Engineering and Advanced Technology, Massey University, Private Bag 102904, Auckland 0745, New Zealand; xw.yuan@massey.ac.nz

* Correspondence: p.cao@auckland.ac.nz

Academic Editor: Hugo F. Lopez

Abstract: This work investigated the isothermal holding time dependence of the densification, microstructure, weight loss, and tensile properties of Fe-Mn-Si powder compacts. Elemental Fe, Mn, and Si powder mixtures with a nominal composition of Fe-28Mn-3Si (in weight percent) were ball milled for 5 h and subsequently pressed under a uniaxial pressure of 400 MPa. The compacted Fe-Mn-Si powder mixtures were sintered at 1200 °C for 0, 1, 2, and 3 h, respectively. In general, the density, weight loss, and tensile properties increased with the increase of the isothermal holding time. A significant increase in density, weight loss, and tensile properties occurred in the compacts being isothermally held for 1 h, as compared to those with no isothermal holding. However, further extension of the isothermal holding time (2 and 3 h) only played a limited role in promoting the sintered density and tensile properties. The weight loss of the sintered compacts was mainly caused by the sublimation of Mn in the Mn depletion region on the surface layer of the sintered Fe-Mn-Si compacts. The length of the Mn depletion region increased with the isothermal holding time. A single α-Fe phase was detected on the surface of all of the sintered compacts, and the locations beyond the Mn depletion region were comprised of a dual dominant γ-austenite and minor ε-martensite.

Keywords: Fe-Mn-Si alloy; isothermal holding time; powder sintering; density; weight loss; tensile properties

1. Introduction

Fe-Mn-Si alloys have been intensively investigated due to the so-called shape memory effect (SME) caused by the reversible phase transformation between face-cantered cubic (fcc) γ-austenite and hexagonal close-packed (hcp) ε-martensite [1–3]. In the family of metallic shape memory alloys (SMAs), Fe-Mn-Si SMAs exhibit relatively low costs of both raw materials and processing in comparison with their Ni-Ti alloys and Cu-based counterparts [4,5]. This makes Fe-Mn-Si shape memory alloys promising candidates for various civil engineering applications such as pipe joints and rail couplings [6–10]. Recently, temporary biomedical devices such as cardiovascular stents and bone fixation plates have been discussed as potential applications of Fe-Mn-Si alloys due to their reasonable biodegradability, good biocompatibility, and mechanical properties [11–15]. So far, Fe-Mn-Si SMA alloys are traditionally fabricated by melting and casting as it is favourable to obtain fully dense bulk materials with homogenized composition.

Powder metallurgy (PM) is a cost-effective metal forming technology that provides various benefits for industrial production in comparison to melting and casting. The products manufactured by PM techniques exhibit a near net shape that requires few or no further machining steps [16–19]. Moreover, it presents the ability to synthesise products with controlled porosity and microstructure [20]. Mechanical milling (MM) is an efficient technique to refine powder particles, which is beneficial for improving the densification and mechanical properties of the PM alloys in the subsequent sintering process [21–24].

In the past, only a few published works have discussed the powder preparation and the sintering behaviors of Fe-Mn-Si alloys using elemental Fe, Mn, and Si powders [25–27]. We sintered mechanically milled Fe-28 wt. %-xSi powder mixtures with different Si contents (x) at 1200 °C in a high vacuum furnace, and compared their mechanical and corrosion properties to wrought alloys [11].

It is noted that the sublimation of Mn must be considered when sintering Mn-containing Fe-based alloys in high vacuum, especially those with high Mn concentration ($\geq$20 wt. %). The high vapor pressure of Mn at high sintering temperatures may lead to the serious sublimation/evaporation of Mn, and consequently change the composition of the sintered Fe-Mn-based alloys [28]. This is harmful to the resulting microstructure and mechanical properties of the sintered compacts. Thus, it is of importance to study the sublimation behaviour of Fe-Mn-Si compacts as a function of the isothermal holding time during vacuum sintering. Currently, the available references on the sublimation/evaporation behaviour of Mn-containing Fe-based alloys mainly focuses on alloys with low Mn content ($\leq$5 wt. %) [28–30]. We recently reported on the sublimation behaviour of Fe-28Mn-3Si alloys [31] at different sintering temperatures ranging from 1000 °C to 1200 °C. For the first time, we calculated the sublimation rate of the sintered samples as a function of sintering temperature and discussed the factors that affect the sublimation of the Fe-Mn samples. However, the effect of the isothermal holding time at the sintering temperature on the microstructure and mechanical properties has not yet been reported. It is therefore necessary to reveal how the sublimation rate changes over the isothermal holding time. In this study, we also aim to explore the effect of the isothermal holding on the microstructure and tensile properties of the sintered Fe-28Mn-3Si alloys.

2. Experiment

2.1. Powder Preparation

Three elemental powders were selected as starting materials in this study: Fe (99.7 wt. % purity), Mn (99.7 wt. % purity), and Si (99.9 wt. % purity). The mean particle sizes of the Fe, Mn, and Si powders are 38.6, 38.7, and 43.4 μm, respectively. The details on the morphologies and purities of these particles are summarised in our previously published work [11], where the Fe powder exhibits an irregular shape, while both the Mn and Si powders are of angular shape.

Powder mixtures with a nominal composition of 69% Fe, 28% Mn, and 3% Si (all in weight percent) were prepared by 10 h mixing and subsequent 5 h mechanical milling (MM) in a planetary ball mill (Pulverisette 6, Fritsch, Idar-Oberstein, Germany) under Ar protection to avoid the oxidation of the powders. More details on the MM parameters are available elsewhere [11].

2.2. Press and Sinter

The MM powder mixtures were then compacted in a rectangular die under a pressure of 400 MPa at room temperature. The dimensions of the green compacts were 40 mm × 16 mm × 4.2 mm, and the green density of the ball milled Fe-Mn-Si compacts was ~65%, as determined in reference [15]. The green compacts were subsequently sintered in a high vacuum furnace with a vacuum level of 5×10^{-3} Pa. The heating ramp was 10 °C/min below 800 °C and 5 °C/min above 800 °C. All samples were sintered at 1200 °C with a wide range of isothermal holding times from 0 to 3 h, as shown in

Figure 1. The cooling was carried out in the furnace under high vacuum (5×10^{-3} Pa) with an average cooling rate of 60 °C/min if the temperature was above 600 °C.

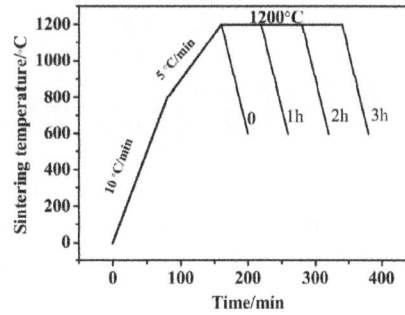

Figure 1. Heating profile of the Fe-28Mn-3Si powder mixture sintered at 1200 °C for different isothermal holding times.

2.3. Characterisation and Analysis

The density and open porosity of all the sintered alloys were evaluated by the Archimedes' principle according to the standard ASTM B962-14 [32]. The measurement used distilled water as the immersing medium in air at ~25 °C. The theoretical density of a pore-free Fe-28Mn-3Si alloy is 7.50 g/cm^3 [11]. The relative density, R (percentage of theoretical density), was introduced to reflect the densification of the sintered compacts, and calculated by $R = \rho_s / \rho_t$ where ρ_t is the theoretical density of the fully dense Fe-28Mn-3Si alloy (i.e., 7.50 g/cm^3), and ρ_s is the measured absolute density of sintered porous Fe-28Mn-3Si alloys. The weight of both the green and sintered compacts was measured using a precision electronic balance (KINO, Norcross, GA, USA). The weight loss is expressed as $(m_0 - m_1)/m_0$, where m_0 is the weight of the green compacts before sintering, while m_1 is the weight of the corresponding specimens after sintering. The microstructure and morphologies of the sintered PM alloys were examined using a scanning electron microscope (SEM, Quanta 200F, FEI, Hillsboro, OR, USA) attached with an X-ray energy dispersive spectrometer (EDS). An X-ray diffractometer (XRD D2 Phaser, Bruker, Karlsruhe, Germany) with Cu Kα radiation (λ = 1.54 Å) was used to analyse the phase compositions of the PM compacts at room temperature. A universal testing machine (Instron 3367, Norwood, MA, USA) equipped with an extensometer was used to measure the tensile properties of the sintered PM parts. The tensile bars were flat dog-bone shaped specimens cut from the sintered blocks. The gauge length of the tensile bars was 8 mm and the size of the cross-section was 13 mm × 3 mm. The tensile testing was measured at a cross-head speed of 0.2 mm/min, equivalent to an initial strain rate of 4.2×10^{-3} s^{-1}.

2.4. Principle to Calculate the Vapor Pressure

The equilibrium partial pressure (P_i^e) of an element in the ternary system can be expressed as:

$$P_i^e = \alpha_i \cdot P_i \tag{1}$$

where α_i and P_i are the activity and equilibrium vapor pressure of a pure element i, respectively. The values of P_{Fe}, P_{Mn}, and P_{Si} are 6.3×10^{-2}, 78, and 8.7×10^{-2} Pa at 1200 °C, which is determined according to reference [33].

The activity of element i is given as:

$$a_i = \exp\left(\frac{\mu_i - \mu_i^0}{RT}\right) \tag{2}$$

where μ_i is the chemical potential of i in a certain state, μ_i^0 is the chemical potential of i in the standard state, R is the gas constant, and T is the temperature in K. Therefore, the activity is 1 for a pure element in the standard state.

The chemical potential of element i is calculated from the Gibbs energy:

$$\mu_i = \left(\frac{\partial_G}{\partial_{ni}} \right)_{T,P,n_j(j \neq i)} \tag{3}$$

where the temperature (T), pressure (P), and composition of the other elements are constants.

At equilibrium, μ_i is the same value for all of the phases. For instance, the following relationship is fulfilled for a three-phase equilibrium ($\alpha + \beta + \gamma$ phases) in a ternary A-B-C system:

$$\mu_i^\alpha = \mu_i^\beta = \mu_i^\gamma \tag{4}$$

The Gibbs energy of a solution phase (φ) in a ternary A-B-C system can be described by the Redlich–Kister polynomial [34]:

$$^0G_m^\varphi = x_A{}^0G_A^\varphi + x_B{}^0G_B^\varphi + x_C{}^0G_C^\varphi + RT(x_A \ln x_A + x_B \ln x_B + x_C \ln x_C) + x_A x_B L_{A,B}^\varphi + x_B x_C L_{B,C}^\varphi$$
$$+ x_A x_C L_{A,C}^\varphi + {}^0G_{A,B,C}^\varphi + {}^\Delta G_{mag}^\varphi \tag{5}$$

where x_i is the molar fraction of element i, $^0G_i^\varphi$ is Gibbs energy at the standard state, the terms $L_{i,j}^\varphi$ (i, j = A, B, C) are the interaction parameters from the binary systems, $^0G_{A,B,C}^\varphi$ is the excess Gibbs energy, and $^\Delta G_{mag}^\varphi$ is the magnetic contribution to the Gibbs energy.

The Gibbs energies of all phases in the Fe-Mn-Si system were given in Reference [35] using the CALPHAD method [36], used to calculate the activities of Fe, Mn, and Si of the ternary system in the present study.

3. Results

3.1. Weight Loss and Chemical Composition

Figure 2 shows the weight loss rate of the alloys sintered for different isothermal holding times. In general, the weight loss rate increases with the increase of the isothermal holding time. In detail, the weight loss of the sintered MM alloys with no isothermal holding is very mild, only ~2 wt. %. When the isothermal holding time increases to 1 h, the weight loss increases significantly to ~7.5 wt. %, which is a ~4-fold increase compared to its counterpart with no isothermal holding. However, the weight loss rate of the alloys sintered for 2 h and 3 h increases by only ~2 wt. % and 3 wt. %, as compared with that sintered for 1 h. This indicates that the weight loss mainly happens during the first hour of isothermal holding, and the weight loss rate decreases with the increase of the isothermal time.

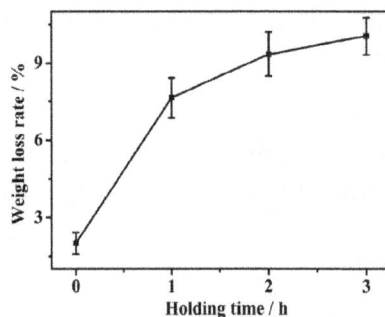

Figure 2. Weight loss rate of the sintered MM Fe-Mn-Si samples as a function of the isothermal holding time.

EDS line scans were performed on the cross-sections of the sintered alloys to examine the distribution of Fe, Mn, and Si in the sintered Fe-Mn-Si alloys. As shown in Figure 3, the EDS line scan direction is perpendicular to sample surface-resin interface, and all scans were recorded from the sample surface to the middle of the cross-sections. Figures 4 and 5 illustrate two typical EDS line scan results of the alloys sintered for no isothermal holding and for 1 h isothermal holding. It can be seen that a Mn depletion layer exists on the surface of the sintered Fe-Mn-Si samples. The Mn content increases parabolically with the increase of the scan distance, starting from the resin-sample interface of the sintered samples until it is stabilized at ~27.5 wt. %. By contrast, the Si content is constant at ~3.1 wt. % for the scale of the entire distance. We define the scan distance (x) between the resin-sample interface ($x = 0$) of the sintered samples and the location where the Mn content reaches a stable value (~27.5 wt. %) as the length of the Mn depletion region (LD). Table 1 presents the average surface chemical compositions and LD of the sintered alloys for different isothermal holding times at 1200 °C. Interestingly, the LD increases with the increase of the isothermal holding time, as shown in Table 1. It is noted that the LD regions were removed from all the samples for the following microstructure observations (Section 3.4) and tensile testing (Section 3.5).

Table 1. The average chemical composition and Mn depletion region (LD) of the MM Fe-Mn-Si alloys sintered at 1200 °C for different isothermal holding times.

Holding Time/h	Chemical Composition on the Surface $x = 0$ (wt. %)				LD (µm)	Chemical Composition (wt. %) at Positions $\geq$LD			
	Mn	Si	O	Fe		Mn	Si	O	Fe
0	1.12 ± 0.12	3.21 ± 0.03	0.45 ± 0.04	Bal.	12 ± 5	27.65 ± 0.38	3.07 ± 0.04	0.45 ± 0.03	Bal.
1	1.14 ± 0.08	3.18 ± 0.06	0.43 ± 0.07	Bal.	405 ± 29	27.58 ± 0.41	3.09 ± 0.06	0.39 ± 0.06	Bal.
2	1.19 ± 0.09	3.09 ± 0.05	0.46 ± 0.07	Bal.	445 ± 31	27.67 ± 0.38	3.11 ± 0.04	0.47 ± 0.08	Bal.
3	1.13 ± 0.07	3.11 ± 0.06	0.47 ± 0.05	Bal.	500 ± 36	27.49 ± 0.45	3.05 ± 0.09	0.42 ± 0.09	Bal.

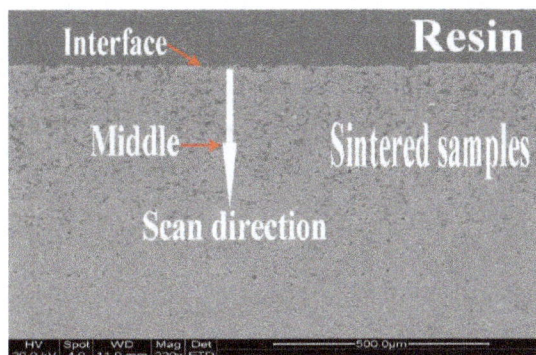

Figure 3. Morphologies of the cross-section of the MM Fe-28Mn-3Si alloys sintered at 1200 °C for 1 h.

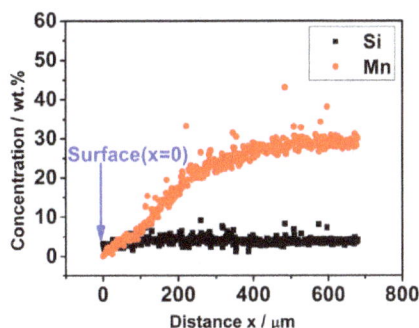

Figure 4. Mn and Si concentration of the cross-section of the MM Fe-Mn-Si alloys sintered at 1200 °C for 1 h isothermal holding.

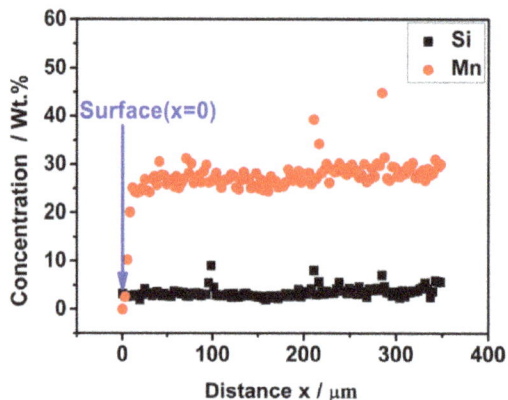

Figure 5. Mn and Si concentration on the cross-section of the MM Fe-Mn-Si alloys sintered at 1200 °C with no isothermal holding.

3.2. Density and Porosity

Figure 6 shows the evolution of the relative densities and open porosities of the Fe-28Mn-3Si alloys sintered at 1200 °C as a function of the isothermal holding time. A drastic increase of the relative density occurs in the alloy sintered for 1 h. In detail, the relative density of the alloys sintered for 1 h is approximately 80%, an increase of 11% as compared to those with no isothermal holding. However, the excessive holding time only plays a limited role in improving the sintered density. For example, the density in the alloy sintered for 3 h is ~85%, which increases by only ~4% and ~1% with respect to their counterparts sintered for 1 and 2 h, respectively.

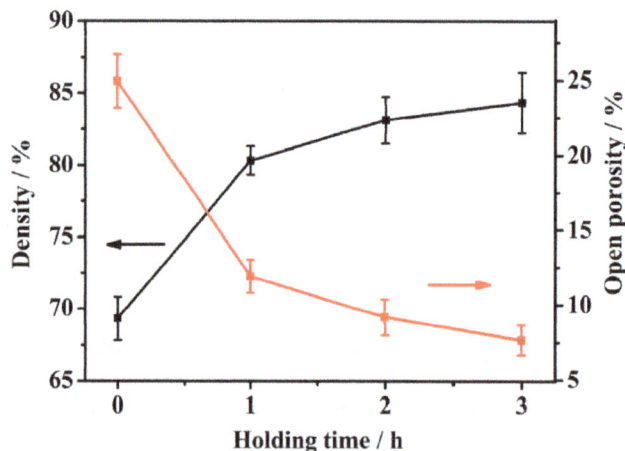

Figure 6. Relative densities and open porosities of the Fe-Mn-Si compacts sintered at 1200 °C as a function of the isothermal holding time.

In general, the open porosities of all the sintered alloys decrease with the increase of the isothermal holding time. The open porosity in alloys sintered with no isothermal holding is approximately 25%. When the isothermal holding time extends to 1 h, the open porosities sharply reduce to 12%. However, further increasing the isothermal holding time to 2 h or 3 h only has a limited effect on eliminating the open porosities.

3.3. Phase Identification

Figures 7 and 8 present the XRD results on the surface and the middle parts (outside the LD region) of the sintered ternary Fe-Mn-Si alloys, respectively. The surface region of all the sintered alloys presents a single α-Fe phase, regardless of the duration of the isothermal holding time (Figure 7). Figure 8 reveals that no peaks belonging to Fe, Mn, or Si are observed in the locations beyond LD. This indicates that both Mn and Si have been dissolved in the Fe-Matrix at the ramp stage. It is also noted that all the sintered alloys consist of a duplex major γ-austenite and minor ε-martensitic phase, as shown in Figure 8.

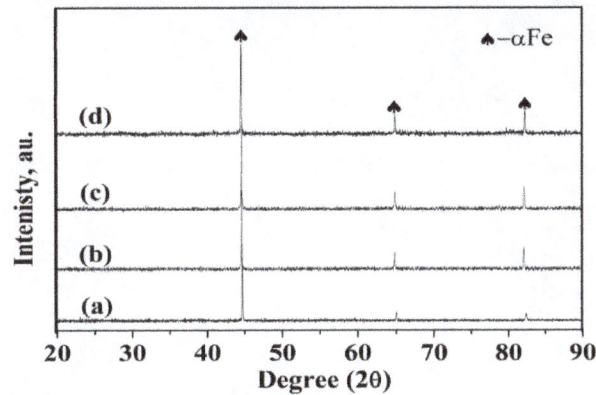

Figure 7. X-Ray Diffraction (XRD) results on the surface of the ternary Fe-28Mn-3Si alloys sintered at 1200 °C for different isothermal holding times. (**a**) MM alloys without isothermal holding; (**b**) MM alloys with 1 h isothermal holding; (**c**) MM alloys with 2 h isothermal holding; (**d**) MM alloys with 3 h isothermal holding.

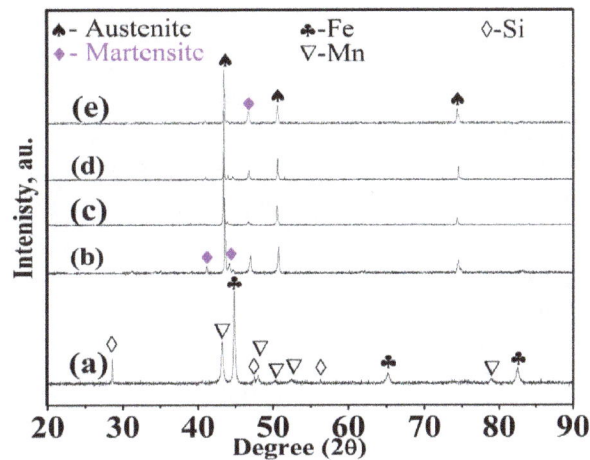

Figure 8. XRD results on the middle parts (outside the LD region) of the ternary Fe-28Mn-3Si alloys sintered at 1200 °C for different isothermal holding times. (**a**) Green compact before sintering; (**b**) MM alloys without isothermal holding; (**c**) MM alloys with 1 h isothermal holding; (**d**) MM alloys with 2 h isothermal holding; (**e**) MM alloys with 3 h isothermal holding.

3.4. Microstructure

Figure 9 shows the SEM graphs of the Fe-28Mn-3Si alloys sintered at 1200 °C as a function of the holding time. Figure 9a illustrates that a large number of interconnected irregular pores are distributed in the alloys with no isothermal holding, indicating that sintering at this stage is incomplete. When the holding time increases to 1 h, the pore size as well as the overall porosities, especially the open porosities, reduce to a large extent. However, most of the pores still exhibit an irregular shape. As long as the isothermal holding time increases to 2 h, the pore size further decreases, and some pores become spherical and isolated. The morphology of the alloys sintered for 3 h is similar to that sintered for 2 h.

However, more isolated and spherical pores are observed in the alloys sintered for 3 h, as shown in Figure 9d.

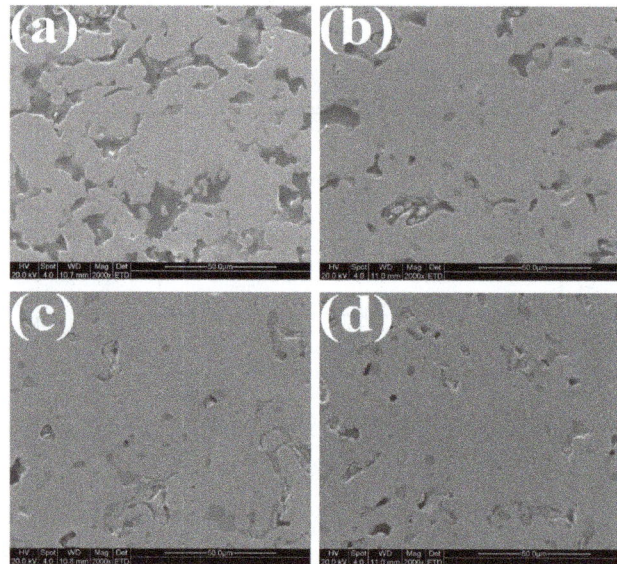

Figure 9. Morphology of the MM Fe-Mn-Si compacts sintered at 1200 °C for various isothermal holding times: (**a**) no isothermal holding; (**b**) 1 h; (**c**) 2 h; (**d**) 3 h.

3.5. Tensile Properties

Figure 10 and Table 2 illustrate the stress-strain curves and the tensile properties of the sintered alloys with different holding times. As shown in Table 2, the isothermal holding time plays an important role in upgrading the tensile properties of the sintered Fe-Mn-Si compacts. In detail, the ultimate tensile strength (UTS) and fracture strain of the alloys sintered for 1 h are ~258 MPa and 6.4%, which are ~2 times and ~3 times higher, respectively, than their counterparts with no isothermal holding. However, a further extended sintering time only has a limited effect on improving the tensile properties. Taking the UTS for example, the UTS of the alloys sintered for 3 h is 330 MPa, which is slightly higher than that of the sintered sample for 2 h (310 MPa).

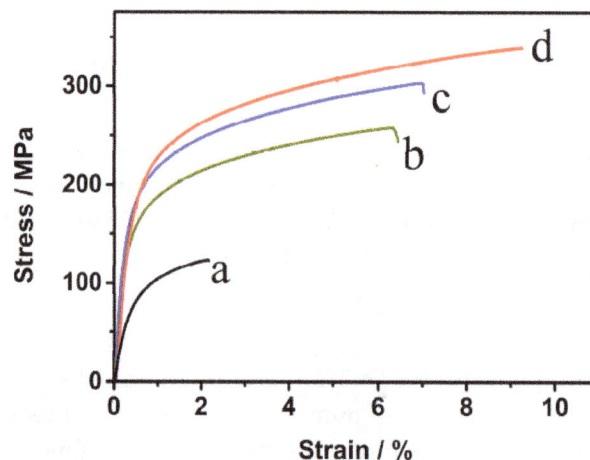

Figure 10. Tensile stress-strain curves of the MM Fe-28Mn-3Si compacts sintered at 1200 °C for various isothermal holding times: (**a**) no isothermal holding; (**b**) 1 h; (**c**) 2 h; and (**d**) 3 h.

Table 2. Average static tensile properties of Fe-Mn-Si alloys sintered at 1200 °C for different isothermal holding times.

Sintering Temperature/°C	Holding Time/h	Ultimate Tensile Strength/MPa	Fracture Strain/%	Young's Modulus/GPa
	0	125 ± 9	2 ± 1	53 ± 4
1200	1	258 ± 9	6 ± 1	69 ± 3
	2	310 ± 15	8 ± 1	71 ± 3
	3	330 ± 23	9 ± 1	74 ± 3

4. Discussion

4.1. Weight Loss Mechanism

Figure 2 shows that the weight loss rate of the samples sintered with no isothermal holding is ~2%, which is only ~1/4 of the sintered sample sintered for 1 h. This reveals that the weight loss mainly occurs during the isothermal holding stage at 1200 °C. In other words, the weight loss in the entire sintering process can be approximately equal to that in the isothermal holding stage. In this case, we only discuss the weight loss in the isothermal holding stage.

Table 1 reveals that Fe, Mn, and Si are homogenously distributed, and the concentration of Fe, Mn, and Si is stabilized at ~69 wt. %, 27 wt. %, and 3.1 wt. % if the detection point is located at a distance from the surface $\geq$LD. This can be further supported by the EDS line scan results in Figures 4 and 5. According to the XRD result (Figure 8) and the SEM micrographs (Figure 9), this composition is a typical solid γ-austenite phase. At the distance from the surface <LD, the Si content stabilizes at ~3.1%, while the Mn content increases with increasing distance until it stabilizes at ~27 wt. %. The isothermal section of the ternary Fe-Mn-Si phase diagram [37] at 1200 °C reveals that these compositions are solid α-Fe and γ-austenite depending on the composition. Based on the above discussion, the sublimation of Fe, Mn, and Si contributes to the weight loss of the ternary Fe-28Mn-3Si compacts at 1200 °C.

The sublimation rate of a certain component in the bulk materials can be determined by the Langmuir theory, which is given as [38–40]:

$$N_i = -K_L \cdot \varepsilon \cdot P_i^e \sqrt{M_i / T} \tag{6}$$

where N_i is the sublimation rate (g·cm^{-2}·s^{-1}) of component i, K_L is the Langmuir constant, ε is the condensation coefficient depending on the materials (for metals $\varepsilon = 1$), P_i^e is the partial pressure of component i in the Fe-Mn-Si system, M_i is the molecular weight of component i, and T is the absolute temperature in K.

In our case, all the samples were sintered at the same temperature (1200 °C). Therefore, according to Equation (6), the sublimation rate of component i in the ternary Fe-Mn-Si system is determined by the partial pressure P_i^e and M_i.

The partial pressure of Fe, Mn, and Si as a function of the Mn concentration (1 wt. %–28 wt. %) in the ternary Fe-Mn-Si alloys was calculated and is presented in Figure 11.

To evaluate the difference in the sublimation rate between these three components in the LD region of the ternary Fe-Mn-Si compacts, the sublimation rate ratio of the two components i and j can be calculated with Equation (7). As the K_L, ε, and T are constant in all of the Fe-Mn-Si compacts during sintering, the sublimation rate ratio (B) is then given as,

$$B = \frac{N_i}{N_j} = \frac{P_i^e \sqrt{M_i}}{P_j^e \sqrt{M_j}} \tag{7}$$

where P_{Fe}^e, P_{Mn}^e, and P_{Si}^e at 1200 °C are shown in Figure 11, while M_{Fe}, M_{Mn}, and M_{Si} are 56 g·mol^{-1}, 55 g·mol^{-1}, and 28 g·mol^{-1}, respectively.

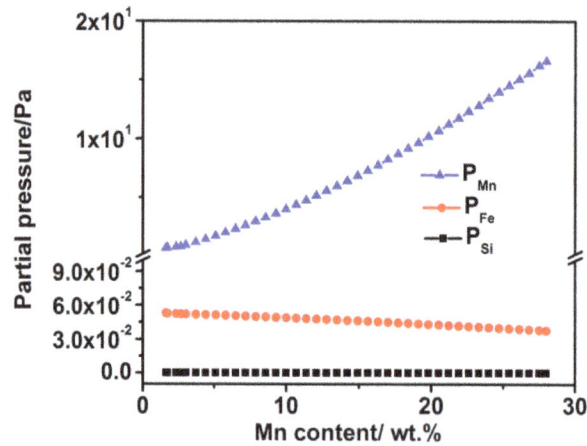

Figure 11. The partial pressure of Fe, Mn, and Si in Fe-Mn-Si sintered at 1200 °C as a function of Mn concentration.

Figure 12 shows that the sublimation rate of Mn is at least 3.41×10^5-fold higher than that of Si at all compositions in the ternary Fe-28Mn-3Si system, and the N_{Mn}/N_{Fe} ratio ranges from 15 to 420 depending on the composition at a certain location in the Fe-28Mn-3Si compacts during sintering. A high ratio of N_{Mn}/N_{Si} and N_{Mn}/N_{Fe} reveals that the sublimation of the sintered Fe-Mn-Si compacts is mainly caused by the sublimation of Mn. Therefore, the Mn depletion regions (LD) exist in all the sintered compacts, as shown in Table 1 and Figure 4.

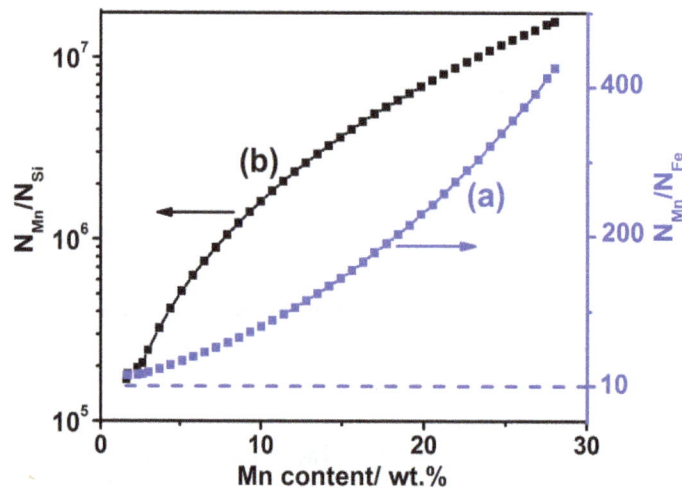

Figure 12. Sublimation rate ratio between the two different components in Fe-Mn-Si alloys at 1200 °C. (a) N_{Mn}/N_{Fe}; (b) N_{Mn}/N_{Si}.

It is noted that the weight loss of the sintered compacts isothermally held at the second and third hour is ~1.7% and ~1.1%, respectively, which is only ~1/3 and ~1/5 of that of the holding at the first hour, as illustrated in Figure 2. This demonstrates that the sublimation rate of Mn decreases with the increase of the isothermal holding time. The change to the sublimation rate of the compacts sintered for different isothermal holding times may be caused by their different open porosities during sintering. This is because open pores create large amounts of walls in the interior of the compacts, which contribute to the sublimation of these components. Moreover, open pores are the flowing channels that move the Mn vapor from the interior areas to the outside of the porous samples.

The composition of the starting powder mixtures and sintering conditions (i.e., sintering temperature and heating rate) for all the samples are identical during isothermal holding, and the only difference between the compacts during sintering is the open porosities. It is therefore believed that a different volume of the open porosities leads to the decrease in the sublimation rate during isothermal holding.

The porous Fe-Mn-Si alloys with higher open pores present a higher sublimation rate due to the larger areas of open pore walls and more flowing channels for Mn vapor transfer. In our case, the volume of the open pores in the compacts sintered at the beginning of isothermal holding is ~25%, and it significantly decreases to ~12%, 9%, and 8% after 1 h, 2 h, and 3 h of isothermal holding, respectively. The remarkably higher open pore volume during holding at the first hour gives rise to a significantly higher weight loss compared to that sintered at the second and third hour. By contrast, the number of open pores sintered for 2 h is only ~3% higher than its 3 h counterparts. Thus, the weight loss of the Fe-Mn-Si compacts sintering at the second and the third hour is very limited compared with that sintered at the first hour.

4.2. Densification

It is recognised that the elimination of surface energy is the driving force for powder densification [20]. It is noted that the green density of the ball milled Fe-Mn-Si compacts was ~65%, as determined in reference [15]. The density of the sintered alloys with no holding was ~5% higher than their green compacts. Alloying during heating is further demonstrated in the XRD result; a single γ-austenite phase was formed and the diffraction peaks indexed to α-Fe, Mn, and Si all disappear in the samples sintered without holding. Although MM does not cause alloying during milling [15], the much refined particles, larger particle surface areas, and stored strain energy during MM might assist in densification in the temperature rising stage.

Interestingly, a rapid densification of the Fe-Mn-Si compacts occurs at the first hour of isothermal holding. Figure 6 shows that the density of the compacts after 1 h isothermal holding increased by 11% compared to that with no isothermal holding. As discussed above, alloying was completed before the start of isothermal holding; therefore, the rapid densification of compacts at the first hour of isothermal holding may be driven by the diminution in the surface energy due to the reduction in the volume and the surface areas of the pores.

However, a further increase in isothermal holding time (>1 h) has a limited promotion in densification. This may be attributed to the following factors. Firstly, the remarkable decrease in open porosities contributes to the slow densification of the compacts when the isothermal holding time >1 h. This is confirmed in Figure 6. The significant decrease in the open porosities of the compacts sintered for longer than 1 h results in significant reduction in the surface area, and hence reduces the driving force. In addition, the Mn sublimation in the pores, especially in isolated pores, plays an important role in preventing the densification of the sintered alloys. Densification is a process of pore elimination. Published work reveals that the pore elimination depends on a balance between the surface energy in the curved surface of the pores and the gas pressure trapped in the pores [20]. For example, the pore shrinkage stops if the gas pressure trapped in the pores is larger than the surface energy on the surface of the pores. In our case, as shown in Figure 11, the partial pressure of the component Mn in the ternary Fe-Mn-Si alloys is as high as 16.6 Pa. In other words, the pressure in the pores is kept at ~16.6 Pa during isothermal holding in high vacuum conditions. This inhibits the pore shrinkage, especially for the isolated pores, and consequently prevents the densification of the sintered compacts.

4.3. Tensile Properties and Fracture

As show in Table 2, the tensile properties of all the sintered PM alloys increase gradually with the increase of the isothermal holding time. The variation in tensile properties is attributed to the different porosities of the sintered alloys, which can be illustrated by the Gibson-Ashby model [41]:

$$\frac{\sigma}{\sigma_0} = C_1 \left(\frac{\rho}{\rho_0} \right)^{n_1}$$

(8)

$$\frac{E}{E_0} = C_2 \left(\frac{\rho}{\rho_0} \right)^{n_2} \tag{9}$$

where σ and σ_0 are the tensile stress of the sintered and pore-free alloys, respectively, and E and E_0 are the modulus of elasticity of the sintered and pore-free alloys, respectively. $(1 - \rho/\rho_0)$ is the porosity and ρ/ρ_0 is the relative density, and C_1, C_2, n_1, and n_2 are material constants depending on the pore structure. The Gibson-Ashby relationship demonstrates that both the tensile strength and modulus of elasticity of the PM alloys increase with the decrease of porosities. Figure 6 shows that the porosities of all the sintered alloys decrease with the increase of the isothermal holding time. It is then expected that the tensile properties of the sintered alloys increase as the isothermal holding time increases. Data fitting reveals a linear relationship between log (ρ/ρ_0) and logσ, with $R^2 = 1.00$ (see Figure 13 for details). This linear relationship is also observed for the plot of log (ρ/ρ_0) vs. logE with $R^2 = 0.988$ (see Figure 14 for details). The extrapolation of these double logarithmic plots suggests a Young's modulus and tensile strength for a fully dense Fe-28Mn-3Si alloy as ~96 GPa and 740 MPa, respectively. This extrapolated Young's modulus is lower than the reported value of $E_0 = 175$ GPa for a (Twining-induced plasticity) TWIP steel [42], and the discrepancy might be because the Gibson-Ashby model is more suitable for porous materials with a porosity level $\geq 70\%$ [41].

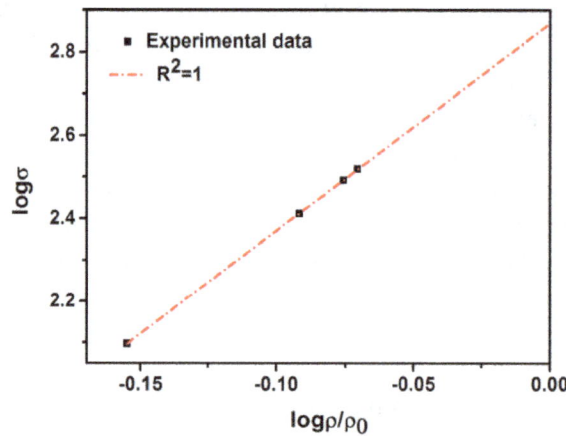

Figure 13. Change of the tensile strength (σ) with relative density (ρ/ρ_0) for the sintered alloys with different isothermal holding times.

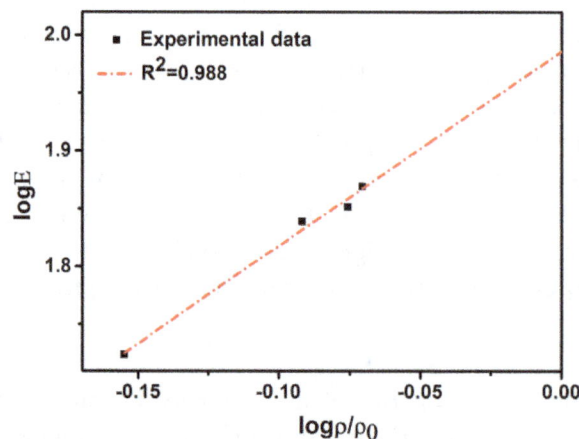

Figure 14. Change of the Young's modulus (E) with relative density (ρ/ρ_0) for the sintered alloys with different isothermal holding times.

Figure 15 presents the SEM images of the sintered Fe-Mn-Si alloys with various isothermal times. Shallow dimples are observed on the fracture surface of all of the sintered alloys. This indicates that all the sintered alloys exhibit ductile fracture. As shown in Figure 15, the number of dimples was similar in all the sintered alloys if the isothermal holding time is >1 h. This again reveals that the extension of the isothermal holding time has a limited effect on increasing the ductility of the Fe-Mn-Si compacts.

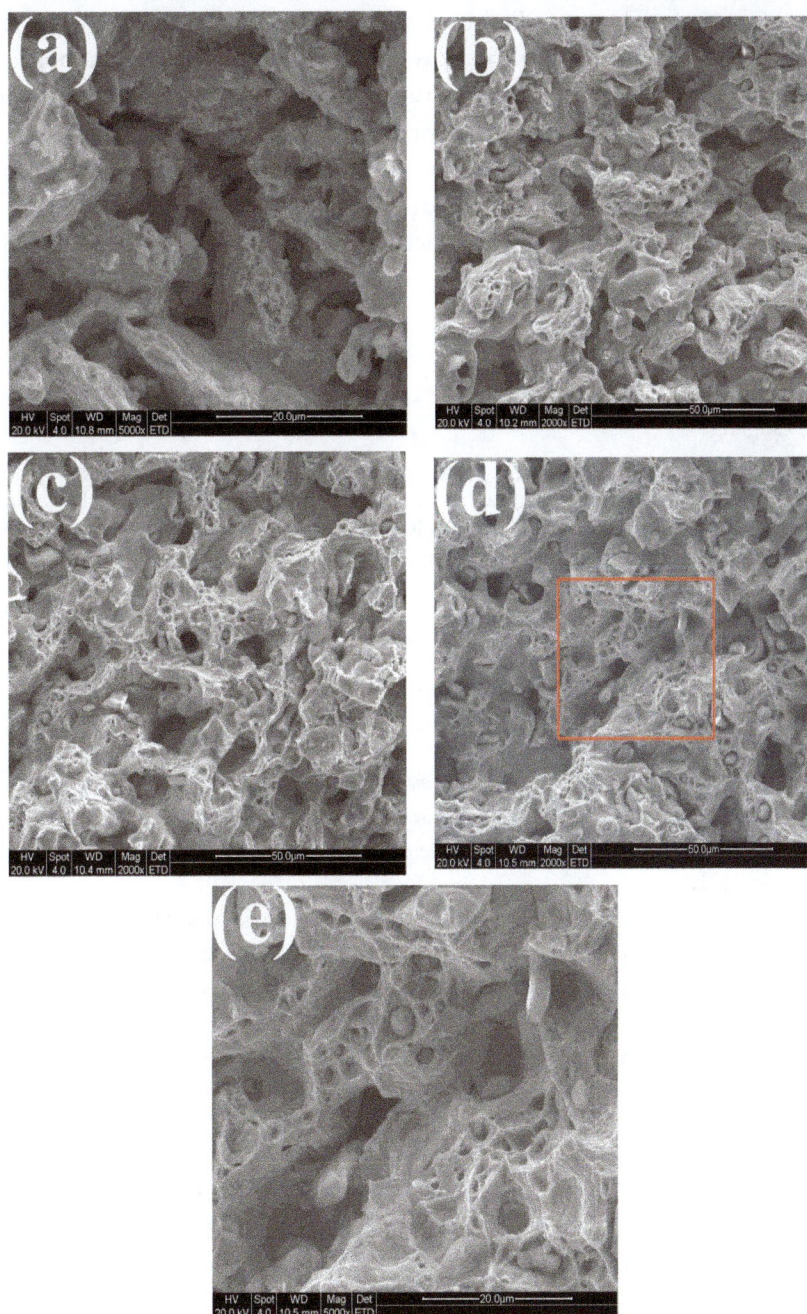

Figure 15. Fractography of the MM Fe-Mn-Si compacts sintered at 1200 °C for different isothermal holding times: (**a**) no isothermal holding; (**b**) 1 h; (**c**) 2 h; (**d**) 3 h; (**e**) enlarged area in (**d**).

5. Conclusions

This work presents the effect of the isothermal holding time on the densification, microstructure, weight loss behavior, and tensile properties of the sintered Fe-Mn-Si compact at 1200 °C. The following key conclusions can be summarised.

(1)   The weight loss of the sintered Fe-Mn-Si compacts with no isothermal holding is only ~2%. The weight loss of the sintered Fe-Mn-Si compacts increases significantly to ~7.6% after the first hour of isothermal holding. A further increment in weight loss is very limited when the isothermal holding time is >1 h. An Mn depletion region exists on the surface layer of all of the sintered compacts. The length of the Mn depletion region (LD) increases with the increase of the isothermal holding time. The weight loss is mainly caused by the sublimation of Mn.

(2)   The density of the sintered ternary Fe-Mn-Si alloys increases drastically during the first hour of isothermal holding, while densification slows down when the isothermal holding time is >1 h. The drop in open porosities mainly occurs during the first hour of isothermal holding.

(3)   The surface of the sintered Fe-Mn-Si is comprised of a single α-Fe phase. The sintered compacts in locations ≥LD consist of a major γ-austenite and minor ε-martensite.

(4)   The tensile properties of the sintered compacts increase with the increase of the sintering time. The tensile strength, elongation, and elasticity drastically increase from 125 MPa, 2.1%, and 52 GPa for the samples with no isothermal holding to 258 MPa, 6.4%, and 69 GPa for the samples with isothermal holding for 1 h.

Acknowledgments: The China Scholarship Council (CSC) is gratefully acknowledged for providing a doctoral scholarship to Xu, Zhigang.

Author Contributions: All authors were involved in designing the experiments. Zhigang Xu performed the sample preparation, data analysis, and manuscript writing and editing. Michael Hodgson contributed to manuscript proofreading. Dr. Keke Chang contributed to the thermodynamic calculation. Peng Cao contributed to data analysis and manuscript revision.

Conflicts of Interest: The authors declare no conflicts of interest.

References

1.   Ölander, A. An electrochemical investigation of solid cadmium-gold alloys. *J. Am. Chem. Soc.* **1932**, *54*, 3819–3833. [CrossRef]

2.   Jani, J.M.; Leary, M.; Subic, A.; Gibson, M.A. A review of shape memory alloy research, applications and opportunities. *Mater. Des.* **2014**, *56*, 1078–1113. [CrossRef]

3.   Rao, A.; Srinivasa, A.; Reddy, J. *Design of Shape Memory Alloy (SMA) Actuators*, 1st ed.; Springer: New York, NY, USA, 2015; pp. 1–30.

4.   Lee, W.; Weber, B.; Leinenbach, C. Recovery stress formation in a restrained Fe-Mn-Si-based shape memory alloy used for prestressing or mechanical joining. *Constr. Build. Mater.* **2015**, *95*, 600–610. [CrossRef]

5.   Dasgupta, R. A look into Cu-based shape memory alloys: Present scenario and future prospects. *J. Mater. Res.* **2014**, *29*, 1681–1698. [CrossRef]

6.   Dong, Z.; Kajiwara, S.; Kikuchi, T.; Sawaguchi, T. Effect of pre-deformation at room temperature on shape memory properties of stainless type Fe-15Mn-5Si-9Cr-5Ni-(0.5–1.5) NbC alloys. *Acta Mater.* **2005**, *53*, 4009–4018. [CrossRef]

7.   Tanaka, Y.; Himuro, Y.; Kainuma, R.; Sutou, Y.; Omori, T.; Ishida, K. Ferrous polycrystalline shape-memory alloy showing huge superelasticity. *Science* **2010**, *327*, 1488–1490. [CrossRef] [PubMed]

8.   Sato, A.; Kubo, H.; Maruyama, T. Mechanical properties of Fe-Mn-Si based SMA and the application. *Mater. Trans.* **2006**, *47*, 571–579. [CrossRef]

9.   Cladera, A.; Weber, B.; Leinenbach, C.; Czaderski, C.; Shahverdi, M.; Motavalli, M. Iron-based shape memory alloys for civil engineering structures: An overview. *Constr. Build. Mater.* **2014**, *63*, 281–293. [CrossRef]

10.   Sawaguchi, T.; Nikulin, I.; Ogawa, K.; Sekido, K.; Takamori, S.; Maruyama, T.; Chiba, Y.; Kushibe, A.; Inoue, Y.; Tsuzaki, K. Designing Fe-Mn-Si alloys with improved low-cycle fatigue lives. *Scr. Mater.* **2015**, *99*, 49–52. [CrossRef]

11.   Xu, Z.; Hodgson, M.A.; Cao, P. A comparative study of powder metallurgical (PM) and wrought Fe-Mn-Si alloys. *Mater. Sci. Eng. A* **2015**, *630*, 116–124. [CrossRef]

12.   Xu, Z.; Hodgson, M.A.; Cao, P. Microstructure and degradation behavior of forged Fe-Mn-Si alloys. *Int. J. Mod. Phys. B* **2015**, *29*, 1–6. [CrossRef]

13. Liu, B.; Zheng, Y.; Ruan, L. In vitro investigation of Fe30Mn6Si shape memory alloy as potential biodegradable metallic material. *Mater. Lett.* **2011**, *65*, 540–543. [CrossRef]

14. Xu, Z.; Hodgson, M.A.; Cao, P. Effect of Immersion in Simulated Body Fluid on the Mechanical Properties and Biocompatibility of Sintered Fe-Mn-Based Alloys. *Metals* **2016**, *6*, 309. [CrossRef]

15. Xu, Z.; Hodgson, M.A.; Cao, P. Effects of Mechanical Milling and Sintering Temperature on the Densification, Microstructure and Tensile Properties of the Fe-Mn-Si Powder Compacts. *J. Mater. Sci. Technol.* **2016**, *32*, 1161–1170. [CrossRef]

16. Cintas, J.; Cuevas, F.; Montes, J.; Herrera, E. High-strength PM aluminium by milling in ammonia gas and sintering. *Scr. Mater.* **2005**, *53*, 1165–1170. [CrossRef]

17. McNeese, M.D.; Lagoudas, D.C.; Pollock, T.C. Processing of TiNi from elemental powders by hot isostatic pressing. *Mater. Sci. Eng. A* **2000**, *280*, 334–348. [CrossRef]

18. Angelo, P.; Subramanian, R. *Powder Metallurgy: Science, Technology and Applications*; PHI Learning Pvt. Ltd.: New Delhi, India, 2008.

19. Rahimian, M.; Parvin, N.; Ehsani, N. The effect of production parameters on microstructure and wear resistance of powder metallurgy Al-Al$_2$O$_3$ composite. *Mater. Des.* **2011**, *32*, 1031–1038. [CrossRef]

20. German, R.M. *Powder Metallurgy and Particulate Materals Processing: The Processes, Materials, Products, Properties and Applications*, 1st ed.; Metal Powder Industries Federation: Princeton, NJ, USA, 2005; pp. 221–260.

21. Suryanarayana, C. Mechanical alloying and milling. *Prog. Mater Sci.* **2001**, *46*, 1–184. [CrossRef]

22. Zhu, M.; Dai, L.; Gu, N.; Cao, B.; Ouyang, L. Synergism of mechanical milling and dielectric barrier discharge plasma on the fabrication of nano-powders of pure metals and tungsten carbide. *J. Alloys Compd.* **2009**, *478*, 624–629. [CrossRef]

23. Gheisari, K.; Javadpour, S.; Oh, J.; Ghaffari, M. The effect of milling speed on the structural properties of mechanically alloyed Fe-45%Ni powders. *J. Alloys Compd.* **2009**, *472*, 416–420. [CrossRef]

24. Garroni, S.; Enzo, S.; Delogu, F. Mesostructural refinement in the early stages of mechanical alloying. *Scr. Mater.* **2014**, *83*, 49–52. [CrossRef]

25. Zhang, Z.; Sandström, R.; Frisk, K.; Salwén, A. Characterization of intermetallic Fe-Mn-Si powders produced by casting and mechanical ball milling. *Powder Technol.* **2003**, *137*, 139–147. [CrossRef]

26. Liu, T.; Liu, H.; Zhao, Z.; Ma, R.; Hu, T.; Xie, Y. Mechanical alloying of Fe-Mn and Fe-Mn-Si. *Mater. Sci. Eng. A* **1999**, *271*, 8–13. [CrossRef]

27. Saito, T.; Kapusta, C.; Takasaki, A. Synthesis and characterization of Fe–Mn–Si shape memory alloy by mechanical alloying and subsequent sintering. *Mater. Sci. Eng. A* **2014**, *592*, 88–94. [CrossRef]

28. Šalak, A.; Selecka, M.; Bureš, R. Manganese in ferrous powder metallurgy. *Powder Metall. Prog.* **2001**, *1*, 41–58.

29. Hryha, E.; Dudrova, E.; Nyborg, L. Critical aspects of alloying of sintered steels with manganese. *Metall. Mater. Trans. A* **2010**, *41*, 2880–2897. [CrossRef]

30. Hryha, E.; Gierl, C.; Nyborg, L.; Danninger, H.; Dudrova, E. Surface composition of the steel powders pre-alloyed with manganese. *Appl. Surf. Sci.* **2010**, *256*, 3946–3961. [CrossRef]

31. Xu, Z.; Hodgson, M.A.; Cao, P. Weight loss behavior of a vacuum sintered powder metallurgical Fe-Mn-Si alloy. *J. Mater. Res.* **2017**, *32*, 644–655. [CrossRef]

32. American Society for Testing and Materials (ASTM) International. *ASTM B962–14, Standard Test Methods for Density of Compacted or Sintered Powder Metallurgy (PM) Products Using Archimedes' Principle*; ASTM International: West Conshohocken, PA, USA, 2014; p. 7.

33. Lide, D.R. *CRC Handbook of Physics and Chemistry*, 85th ed.; CRC Press: Boca Raton, FL, USA, 2004.

34. Redlich, O.; Kister, A. Algebraic representation of thermodynamic properties and the classification of solutions. *Ind. Eng. Chem.* **1948**, *40*, 345–348. [CrossRef]

35. Forsberg, A.; Ågren, J. Thermodynamic evaluation of the Fe-Mn-Si system and the γ/ε martensitic transformation. *J. Phase Equilib.* **1993**, *14*, 354–363. [CrossRef]

36. Saunders, N.; Miodownik, A.P. *CALPHAD (Calculation of Phase Diagrams): A Comprehensive Guide*; Pergamon: Oxford, UK, 1998; Volume 1, pp. 1–100.

37. Raghavan, V.; Raynor, G.V.; Rivlin, V.G. *Phase Diagrams of Ternary Iron Alloys*; Indian Institute of Metals: New Delhi, India, 1987; pp. 363–377.

38. Jones, H.A.; Mackay, G. The rates of evaporation and the vapor pressures of tungsten, molybdenum, platinum, nickel, iron, copper and silver. *Phys. Rev.* **1927**, *30*, 201. [CrossRef]

39. Langmuir, I. Vapor pressures, evaporation, condensation and adsorption. *J. Am. Chem. Soc.* **1932**, *54*, 2798–2832. [CrossRef]

40. Chen, G.; Cao, P.; He, Y.; Shen, P.; Gao, H. Effect of aluminium evaporation loss on pore characteristics of porous FeAl alloys produced by vacuum sintering. *J. Mater. Sci.* **2012**, *47*, 1244–1250. [CrossRef]

41. Gibson, L.J.; Ashby, M.F. *Cellular Solids: Structure and Properties*, 2nd ed.; Cambridge University Press: Cambridge, UK, 1999; pp. 100–200.

42. Kim, J.; Lee, S.-J.; de Cooman, B.C. Effect of Al on the stacking fault energy of F-18Mn-0.6C twinning-induced plasticity. *Scr. Mater.* **2011**, *65*, 363–366. [CrossRef]

Permissions

The contributors of this book come from diverse backgrounds, making this book a truly international effort. This book will bring forth new frontiers with its revolutionizing research information and detailed analysis of the nascent developments around the world.

We would like to thank all the contributing authors for lending their expertise to make the book truly unique. They have played a crucial role in the development of this book. Without their invaluable contributions this book wouldn't have been possible. They have made vital efforts to compile up to date information on the varied aspects of this subject to make this book a valuable addition to the collection of many professionals and students.

This book was conceptualized with the vision of imparting up-to-date information and advanced data in this field. To ensure the same, a matchless editorial board was set up. Every individual on the board went through rigorous rounds of assessment to prove their worth. After which they invested a large part of their time researching and compiling the most relevant data for our readers.

The editorial board has been involved in producing this book since its inception. They have spent rigorous hours researching and exploring the diverse topics which have resulted in the successful publishing of this book. They have passed on their knowledge of decades through this book. To expedite this challenging task, the publisher supported the team at every step. A small team of assistant editors was also appointed to further simplify the editing procedure and attain best results for the readers.

Apart from the editorial board, the designing team has also invested a significant amount of their time in understanding the subject and creating the most relevant covers. They scrutinized every image to scout for the most suitable representation of the subject and create an appropriate cover for the book.

The publishing team has been an ardent support to the editorial, designing and production team. Their endless efforts to recruit the best for this project, has resulted in the accomplishment of this book. They are a veteran in the field of academics and their pool of knowledge is as vast as their experience in printing. Their expertise and guidance has proved useful at every step. Their uncompromising quality standards have made this book an exceptional effort. Their encouragement from time to time has been an inspiration for everyone.

The publisher and the editorial board hope that this book will prove to be a valuable piece of knowledge for researchers, students, practitioners and scholars across the globe.

List of Contributors

Shengke Zou, Shuyuan Ma, Changmeng Liu, Cheng Chen, Jiping Lu and Jing Guo
School of Mechanical Engineering, Beijing Institute of Technology, Beijing 100081, China

Limin Ma
Beijing Aeronautical Science & Technology Research Institute of COMAC, Beijing 102211, China

Omer Eyercioglu
Department of Mechanical Engineering, Gaziantep University, Gaziantep 27310, Turkey

Ahmed Samir Anwar
Department of Mechanical Engineering, Salahaddin University, Erbil 44001, Iraq

Kursat Gov
Department of Aeronautics and Astronautics Engineering, Gaziantep University, Gaziantep 27310, Turkey

Necip Fazil Yilmaz
Department of Mechanical Engineering, Gaziantep University, Gaziantep 27310, Turkey

Ming Qian
School of Construction Engineering, Jilin University, Changchun 130026, China

Shaoming Ma, Youhong Sun and Baochang Liu
School of Construction Engineering, Jilin University, Changchun 130026, China
Key Laboratory of Drilling and Exploitation Technology in Complex Conditions, Ministry of Land and Resources, China No. 938 Ximinzhu Street, Changchun 130026, China

Yinlong Ma and Chi Zhang
School of Construction Engineering, Jilin University, Changchun 130026, China
Key Laboratory of Automobile Materials of Ministry of Education & School of Materials Science and Engineering, Jilin University, No. 5988 Renmin Street, Changchun 130025, China

Xiaoshu Lü
School of Construction Engineering, Jilin University, Changchun 130026, China
Department of Civil and Structural Engineering, School of Engineering, Aalto University, Helsinki 02015, Finland

Huiyuan Wang
Key Laboratory of Automobile Materials of Ministry of Education & School of Materials Science and Engineering, Jilin University, No. 5988 Renmin Street, Changchun 130025, China

Zhonghong Zhang
School of Materials Science and Engineering, Anhui University of Technology, Ma'anshan 243002, China

Hui Zhang
School of Materials Science and Engineering, Anhui University of Technology, Ma'anshan 243002, China
The Advanced Manufacturing Technology Research Centre, Department of Industrial and Systems Engineering, Hong Kong Polytechnic University, Hung Hom, Hong Kong, China

T. M. Yue
The Advanced Manufacturing Technology Research Centre, Department of Industrial and Systems Engineering, Hong Kong Polytechnic University, Hung Hom, Hong Kong, China

Qing Yuan, Guang Xu , Mingxing Zhou, Bei He and Haijiang Hu
The State Key Laboratory of Refractories and Metallurgy, Hubei Collaborative Innovation Center for Advanced Steels, Wuhan University of Science and Technology, 947 Heping Avenue, Qingshan District, Wuhan 430081,China

Vladimir Sarychev, Sergey Nevskii and Victor Gromov
Physics Department, Siberian State Industrial University, 42 Kirova str., Novokuznetsk 654007, Russia

Sergey Konovalov
Physics Department, Siberian State Industrial University, 42 Kirova str., Novokuznetsk 654007, Russia
School of Mechanical and Electrical Engineering, Wenzhou University, Wenzhou 325035, China

Department of Metals Technology and Aviation Materials, Samara National Research University, Moskovskoye Shosse 34, Samara 443086, Russia

Xizhang Chen
School of Mechanical and Electrical Engineering, Wenzhou University, Wenzhou 325035, China

Milan Trtica
School of Mechanical and Electrical Engineering, Wenzhou University, Wenzhou 325035, China
VINCA Institute of Nuclear Sciences, University of Belgrade, P.O. Box 522, Belgrade 11001, Serbia

Yonglin Kang
School of Materials Science and Engineering, University of Science and Technology Beijing, Beijing 100083, China

Zaiwang Liu
School of Materials Science and Engineering, University of Science and Technology Beijing, Beijing 100083, China
Shougang Research Institute of Technology, Beijing 100043, China

Zhimin Zhang and Xiaojing Shao
Shougang Research Institute of Technology, Beijing 100043, China

L'uboslav Straka and Slavomíra Hašová
Department of Manufacturing Processes Operation, The Technical University of Košice, Štúrova 31, 08001 Prešov, Slovakia

Ivan Čorný Ján
Department of Science and Research, The Technical University of Košice, Bayerova 1, 08001 Prešov, Slovakia

Pitel'
Department of Mathematics, Informatics and Cybernetics, The Technical University of Košice, Bayerova 1, 08001 Prešov, Slovakia

Ma. Isabel Reyes-Valderrama, Eleazar Salinas-Rodríguez, J. Fabian Montiel-Hernández, Isauro Rivera-Landero, Eduardo Cerecedo-Sáenz, Juan Hernández-Ávila and Alberto Arenas-Flores
Área Académica de Ciencias de la Tierra y Materiales, Universidad Autónoma del Estado de Hidalgo, Carretera Pachuca—Tulancingo km. 4.5, C.P. 42184, Mineral de la Reforma, Hidalgo, México

Enes Akca
Department of Mechanical Engineering, Faculty of Engineering and Natural Sciences, International University of Sarajevo, Hrasnička cesta 15, 71210 Sarajevo, Bosnia and Herzegovina

Ali Gursel
Department of Mechanical Engineering, Faculty of Engineering and Natural Sciences, International University of Sarajevo, Hrasnička cesta 15, 71210 Sarajevo, Bosnia and Herzegovina

Department of Mechanical Engineering, Faculty of Engineering, Duzce University, 81620 Duzce, Turkey

František Lukáč, Monika Vilémová, Barbara Nevrlá, Jakub Klečka and Tomáš Chráska
Institute of Plasma Physics, Czech Academy of Science, Za Slovankou 3, 18200 Prague, Czech Republic

Orsolya Molnárová
Mathematics and Physics Faculty, Charles University, Ke Karlovu 3, 12116 Prague, Czech Republic

Jeong-Min Kim and Tae-Hyung Ha
Department of Advanced Materials Engineering, Hanbat National University, 125 Dongseo-daero, Yuseong-gu, Daejeon 34158, Korea

Il-Hyun Kim and Hyun-Gil Kim
Light Water Reactor Fuel Technology Division, Korea Atomic Energy Research Institute, 989-111 Daedek-daero, Yuseong-gu, Daejeon 34057, Korea

Zhiqiang Liu, Feifei Ji, Mingqiang Wang and Tianyu Zhu
School of Mechanical Engineering, Jiangsu University of Science and Technology, Zhenjiang 212003, China

Zhong-Min Wang, Song Tao, Jia-Jun Li, Jian-Qiu Deng, Huaiying Zhou and Qingrong Yao
School of Material Science and Engineering, Guilin University of Electronic Technology, Guilin 541004, China

Junyu Tian, Guang Xu, Mingxing Zhou, Haijiang Hu and Xiangliang Wan
The State Key Laboratory of Refractories and Metallurgy, Hubei Collaborative Innovation Center for Advanced Steels, Wuhan University of Science and Technology, 947 Heping Avenue, Qingshan District, Wuhan 430081, China

Singon Kang
Advanced Steel Processing and Products Research Center, Department of Metallurgical and Materials Engineering, Colorado School of Mines, Golden, CO 80401, USA

Minwook Kim and Seok-Jae Lee
Division of Advanced Materials Engineering, Research Center for Advanced Materials Development, Chonbuk National University, Jeonju 561-756, Korea

Hiromu Hisazawa
Department of Materials Science and Engineering, Interdisciplinary Graduate School of Science and Engineering, Tokyo Institute of Technology, Yokohama 226-8502, Japan

Yoshihiro Terada
Department of Materials Science and Engineering, School of Materials and Chemical Technology, Tokyo Institute of Technology, Yokohama 226-8502, Japan

Fauzan Adziman, David J. Crudden and David M. Collins
Department of Engineering Science, University of Oxford, Parks Road, Oxford OX5 1PF, UK

David E. J. Armstrong and Roger C. Reed
Department of Materials, University of Oxford, Parks Road, Oxford OX5 1PF, UK

Michael A. Hodgson and Peng Cao
Department of Chemical and Materials Engineering, University of Auckland, Private Bag 92019, Auckland 1142, New Zealand

Zhigang Xu
Department of Chemical and Materials Engineering, University of Auckland, Private Bag 92019, Auckland 1142, New Zealand
School of Automotive Engineering, Wuhan University of Technology, Wuhan 430070, China

Keke Chang
Materials Chemistry, RWTH Aachen University, D-52056 Aachen, Germany

Gang Chen
State Key Laboratory of Porous Metal Materials, Northwest Institute for Non-ferrous Metal Research, Xi'an 710016, China

Xiaowen Yuan
School of Engineering and Advanced Technology, Massey University, Private Bag 102904, Auckland 0745, New Zealand

Index